BIOELECTRODYNAMICS AND BIOCOMMUNICATION

BIOELECTRODYNAMICS AND BIOCOMMUNICATION

Editors

Mae-Wan Ho
Open University, U.K.

Fritz-Albert Popp
International Institute of Biophysics, Kaiserslautern, Germany

Ulrich Warnke
University of Saarland, Saarbrücken, Germany

World Scientific
Singapore • New Jersey • London • Hong Kong

Published by

World Scientific Publishing Co. Pte. Ltd.

P O Box 128, Farrer Road, Singapore 9128

USA office: Suite 1B, 1060 Main Street, River Edge, NJ 07661

UK office: 73 Lynton Mead, Totteridge, London N20 8DH

BIOELECTRODYNAMICS AND BIOCOMMUNICATION

ISBN: 981-02-1665-3

Printed in Singapore by Utopia Press.

Preface

The English translation of the Soviet biologist, A.S. Presman's book, *Electromagnetic Fields and Life* (Plenum Press, New York, 1970), first brought to the attention of the scientific community world-wide, a whole class of fascinating phenomena suggesting that organisms are sensitive to extremely weak electromagnetic fields, the energies of which are below thermal threshold. At around the same time, solid-state physicist, Herbert Fröhlich, in Britain, put forward a theory of energy storage as 'coherent excitations' to account for the long range dynamical order in living systems which is difficult to reconcile with classical, equilibrium physics.

Since then, many new findings concerning the sensitivity of organisms to weak electromagnetic fields have come to light, concommitantly with significant advances in our understanding of the biophysical basis for the phenomena of electromagnetic sensitivity in terms of cooperativity and coherence in systems far from thermodynamic equilibrium. The idea is gaining acceptance that electromagnetic fields in the entire range of frequencies are intimately involved in biological functioning, and that bioelectromagnetic, or, more generally, bioelectrodynamical phenomena, may provide a key to understanding the organization of living systems. The subject is especially important at a time of increasing concern over the possible health hazards from high tension power lines and other electrical installations in the environment, while electromedicine in all its forms is becoming increasingly popular without adequate scientific monitoring.

These considerations have convinced us that there is a need for a comprehensive treatment of the subject from both an empirical and a conceptual perspective, more so, because it is a rapidly expanding research area, and the relevant publications are scattered widely over specialist journals belonging to many disicplines. We have, therefore, gathered a representative collective of the available research findings in one volume, together with a clear description of the historical and scientific background for the subject. Our immediate objective is to promote greater understanding and communication among workers in different parts of the electromagnetic spectrum, and so give impetus to further research. The volume is also suitable for introducing the subject to the uninitiated undergraduates, graduates, and scientists in other disciplines.

This book can be read in many ways. For those already familiar with the subject to some extent, they can start anywhere, although they may find that the first Chapter gives a good thorough perspective for the origins of the subject that is not found elsewhere. Equally, they may find the last Chapter a useful preview of the contents. For the totally uninitiated, we recommend that they should proceed from the beginning, but to read the last Chapter before proceeding with Chapter 3 onwards. Finally, for the adventurous who do not wish to know the ending before beginning, they should definitely leave the last Chapter at the end, where it belongs!

The Chapters vary somewhat in technical detail. For most people, the most demanding are probably Chapters 16 and 17, on account of the density of mathematical equations. Those who have no difficulty with mathematical symbols will find these Chapters very inspiring. The rest should skip the equations, and they will find that a lot can still be gleaned from the clearly written text.

Many people have contributed to this volume. We would like to thank all the authors for their excellent efforts in producing clearly written manuscripts, and for their forebearance in dealing with our queries. In addition, we are very grateful to Peter T. Saunders and Wei-Ping Mei for editorial advice and assistance, and to Judith Deny for expert word-processing and layout for the book.

The publication of this collection is a joyous occasion tempered with sadness, as one of our authors, the distinguished Ukrainian scientist, A.S. Davydov, has passed away in February, about one month after sending us his manuscript. This collection may contain the last paper he has written in a long and brilliant career. He is responsible for the idea of solitons which has applications in many fields of bioenergetics and dynamics. Two of the Editors met him, for the first, and last time, at a conference in Kaiserslautern towards the end of 1992, and were much inspired by a series of stimulating discussions with him. He is sorely missed by all of us, and by the rest of the scientific community world-wide.

Mae-Wan Ho, Fritz-Albert Popp
and Ulrich Warnke
May, 1994.

Authors' Addresses

Mr. M. Bischof Gutenbergstr. 6, CH-4562 Biberist, Switzerland.

Prof. J.J. Chang Institute of Biophysics, Chinese Academy of Science, Beijing 100101, P.R. China.

Prof. A.S. Davydov† Institute Theor. Phys., 252143 Kiev, Ukraine.

Dr. D. Edmonds University of Oxford, Dept. of Physics, Clarendon Laboratory, Parks Road, Oxford OX1 3PU, U.K.

Mr. A. French Bioelectrodynamics Laboratory, Open University, Walton Hall, Milton Keynes, MK7 6AA, U.K.

Dr. C. Gross Dept. of Biochemistry, The Hong Kong University of Science and Technology, Clear Water Bay, Kowloon, Hong Kong.

Prof. Q. Gu International Institute of Biophysics, Technology Center, Opelstr. 10, D-67661 Kaiserslautern, F.R. Germany.

Mr. J. Haffegee Bioelectrodynamics Laboratory, Open University, Walton Hall, Milton Keynes, MK7 6AA, U.K.

Dr. M.W. Ho Bioelectrodynamics Laboratory, Open University, Walton Hall, Milton Keynes, MK7 6AA, U.K.

Dr. A.A. Ioannides Dept. of Physics, Open University, Walton Hall, Milton Keynes, MK6 6AA, U.K.

Dr. R.P. Liburdy Building 74, Biology and Medicine Division, Lawrence Berkeley Laboratory, University of California, Berkeley, California 94720, USA.

Dr. W.P. Mei International Institute of Biophysics, Technology Center, Opelstr. 10, D-67661 Kaiserslautern, F.R. Germany.

Dr. R. Pethig IMBE, University College of North Wales, Dean St., Bangor, Gwynedd LL57, 1UT, U.K.

Prof. F.A. Popp International Institute of Biophysics, Technology Center, Opelstr. 10, D-67661 Kaiserslautern, F.R. Germany.

Prof. P.T. Saunders Dept. Mathematics, University of London, King's College, Strand, London, WC2R 2LS, U.K.

Dr. C.W. Smith Department of Electrical Engineering, University of Salford, Manchester, SM5 4WT, U.K.

Dr. T.Y. Tsong Dept. of Biological Sciences, 140 Gortner Laboratory, 1479 Gortner Ave., St. Paul, Minnesota 55108, U.S.A.

Dr. U. Warnke Fachbereich Biologie (15.4) der Universität des Saarlands, D-6600 Saarbrücken, F.R. Germany.

Prof. T.M. Wu Dept. of Physics, State University of New York, Binghamton, P.O. Box 6000, USA.

Prof. C.L. Zhang Dept. of Biology, Hangzhou University, P.O. Box 15, Hangzhou 310028, P.R. China.

Contents

Chapter 1

The History of Bioelectromagnetism

Marco Bischof

1.1 Introduction

The history of bioelectromagnetism is very much also a history of the detection and measurement technology. We follow the suggestion of the eminent historian of physiology, Karl E.Rothschuh[3], to define the three main periods of development as: the pre-Galvanic, Galvanic and instrumental eras. Besides Brazier's volumes on the history of neurophysiology, and many original papers and books, the major sources I have used are *Electricity and Medicine* by Rowbottom and Susskind[4], and Gillispie's *Dictionary of Scientific Biography*[5]. The *Subject Catalogue of the History of Medicine and Related Sciences*[6] of the Wellcome Institute has been useful for locating many of the papers.

1.2 The Pre-Galvanic Period

1.2.1 Electric Fish

The first bioelectric phenomenon to come to human awareness was the electric discharge of certain fishes: the Nile catfish (*Malopterurus electricus*), the electric ray (torpedo), the common skate (*Mormyrus*) and the electric eel (*Gymnotus*, now *Electrophorus electricus*), which are capable of delivering a very painful and paralyzing shock. The output of the electric eel is between 0.5-0.75 amperes, averaging 350 volts, but can be as high as 650 volts.

Although the Nile catfish is already depicted in Egyptian murals of the 3rd millenium B.C., the first written account of its numbing effects were found in the *Corpus Hippocraticum* almost four thousand years later. They are described by Plato, Aristotle, Plutarch, Pliny and many other antique authors, but the curative value of the electric discharges was not introduced into medicine until about A.D. 46 when the eminent Roman pharmacist and physician Scribonius Largus recommended it as a cure for headache and gout. This is the first recorded use of electrotherapy.

Many Arab physicians in the period before the Renaissance recommended the same therapy for the induction of sleep, the alleviation of migraine, melancholy, headache, vertigo and epilepsy. From the seventeenth century, we have a report on its use from Ethiopia.

Instances of Europeans using electric fish for medical purposes are to be found in the literature up to about 1850. When in 1745, the Leyden jar was invented, the similarity between the shock it delivered and the discharge of the electric fish was pointed out, which led to a wave of experiments with electric fish. Henry Cavendish in 1776 made experiments that convinced him that the variations in the intensity of the shock released by electric fish were understandable in terms of electric fields. In attempting to clarify the concept of electrical resistance, he made an artificial torpedo which simulated the action of the living animal when excited by a battery of Leyden jars.

1.2.2 Electromagnetism Becomes a Science

Before 1745, the observations on the effects of the discharge of electric fish seemed unconnected with observations on electric effects of inorganic matter. However, *triboelectricity*, generated by rubbing or striking dielectric materials such as amber, agate, rock crystal and horn, has probably been observed and used since at least the Stone Age. The medical use of the special properties of magnetic materials is documented even much earlier than that of electric fish[7]. Already in the *Papyrus Ebers* (3600 B.C.) there is a recommendation for the use of lodestone, and some museums display Egyptian scarab-amulets made from magnetite and magnetic iron. Hippocrates prescribes a preparation with magnetite powder against infertility and gastric problems.

However, the scientific era of bioelectromagnetics did not start until 1600, when William Gilbert defined and named electricity. This inaugurated the pre-Galvanic period of bioelectromagnetics[1,3,4,8]. Later, the Englishman Stephen Gray first differentiated good and bad conductors of electricity, and Charles Du Fay in France recognized, in 1733, the opposite effects of vitreous (+) and resinous (-) electricity, for which F.W.Lichtenberg proposed the terms ‘positive’ and ‘negative’ electricity. An important advance was made in 1745 with the invention of the Leyden jar by E.G. von Kleist and Pieter van Musschenbroek, which made it possible to store large amounts of electricity, and to demonstrate the effects on living organisms by its discharges.

Electrical experimentation in the Baroque salons

The discovery was a great stimulus for biological experiments. They soon became the favourite entertainment in the scientific salons of the aristocracy of the Enlightenment. One of the leading experimenters of this time was the Abbé Jean-Antoine Nollet. As much showman as physicist, Nollet had 180 guardsmen at Versailles to join hands in a row, while the last two men touched a Leyden jar. This made them all jump up simultaneously, much to the amusement of the king. In his more serious experiments, Nollet found that the pulse and respiration of electrically stimulated humans increased perceptibly and that the evaporation rate of plants under a conducting wire was higher than that of the controls. In a long series of experiments with flowers, seeds, birds and cats he claimed to have established the growth-enhancing effect of electrification and thus became the founder of what would later be called 'electroculture'.

1.2.3 Vital fluids and Etheric Spirits: Early Hypotheses

In pre-scientific times, not only the discharges of electric fish, but also inorganic electrical and magnetic phenomena were taken to be expressions of some mysterious divine presence or vital force that was supposed to be present not only in living beings but also in cult objects, at certain places and in special moments, and was believed to endow shamans, healers, chiefs and kings with their special powers.

With the discovery of the Leyden jar, the concept of 'animal spirits' or a 'vital fluid', inherited from antiquity, was immediately identified with electricity. Newton linked the animal spirits with the ether, while for Descartes they were a subtle fluid generated in the brain, from where they flow in hollow nerve-tubes to the muscles to cause motion. Swiss polymath, Albrecht von Haller adopted the concept of 'irritability' - the property of living matter to react to stimulation or 'irritation' - from Francis Glisson, and made it widely known. Haller showed experimentally that contraction is initiated by a stimulus and established the use of muscle-nerve preparations in physiology. His Italian pupils Leopoldo Caldani and Felice Fontana introduced electrical stimulation into physiology.

1.2.4 Pre-Galvanic Electro- and Magnetotherapy

In the 18th century, the use of magnets for healing became widespread. In 1723 Claisault had first produced artificial magnets, thus making obsolete the unwieldly use of magnetic stones. Systematic observations were now undertaken, and detailed protocols drawn up for investigations and for recording the results. As for electrotherapy, Nollet was the first, in 1746, to try to reestablish movement or feeling in paralysed body parts by electric shock at the Charité hospital, without success. But two years later, Swiss physicist and mathematician Jean Jallabert reported on the first successful treatment of paralysis by electricity.

1.3 The Galvanic Period

1.3.1 The Galvani-Volta Controversy: Does 'Animal Electricity' Exist ?

Luigi Galvani graduated in medicine at Bologna in 1759, where Caldani's and Fontana's views in support of Haller were strongly discussed. One of the first papers Galvani delivered was on Hallerian irritability (1772). We know that by 1780 he was experimenting with a frictional electric machine and Leyden jars.

Galvani's experiments

It is not generally known that Galvani actually performed three series of experiments, of which the famous one with the metal hook was least suitable to prove the existence of animal electricity[8,9]. One day when he was about to electrostimulate the usual frog muscle-nerve preparation in which legs and feet were attached to a stump of the vertebral column by the sciatic nerves, he accidentally discovered that the muscles contracted strongly when the nerve was touched by a scalpel, and only when sparks were drawn simultaneously from the electrostatic machine across the room.

The most cited of his experiments concerns the second series[10]. Having noticed that frog preparations hung by copper hooks from the iron railing of a balcony contracted not only through thunderstorms but also in good weather, he began to determine the conditions under which the contractions occurred. When at some point he pressed the hook fastened to the vertebral column against the railing, he also got contractions. Subsequently, he tried pressing the hook against an iron plate, and using various other combinations of metals, which all gave contractions, but of different intensities, depending on the metals used. Poor conductors did not cause any reaction.

Until 1786, Galvani considered these phenomena to be expressions of 'metal electricity', but later changed his mind and believed he was dealing with the compensation of animal electricity by means of the metallic connection. When he published the results in 1791 in his famous *De viribus electricitatis*, a veritable storm was aroused among physicists, physiologists and physicians, so that everywhere, experiments with frogs and dissimilar metals were conducted. Many hoped the solution of the problem of the nerve fluids was now at hand, and much more besides.

But not everone agreed with his interpretation. His main opponent was Alessandro Volta, professor of physics at Pavia, who had successfully replicated Galvani's experiments before demonstrating, in a masterful series of further experiments which culminated in the invention of the voltaic pile, that the contraction caused by metals was due to stimulation of the muscle or nerve with a minute electrical current generated by the contact of dissimilar metals. Galvani's answer was a third series of experiments made without any metals, which definitely proved the existence of an animal electricity. The same frog preparation was held by one foot and and the other was made to touch the vertebral column. In that condition, or when the vertebral column was made to fall on the thigh, the muscles contracted vigourously. The third experiment Volta tried to explain as the result of heterogenous tissues, which he maintained where producing the currents.

This argument was disproved in a series of brilliant experiments, published in 1797, by the young Alexander von Humboldt, who was not satisfied with Volta's interpretation. Humboldt repeated Volta's and Galvani's experiments and extended them. He concluded that Galvani had uncovered two different phenomena, both genuine: bimetallic electricity and intrinsic animal electricity, which were not mutually exclusive.

1.3.2 The Therapeutic Use of Galvanism

Ironically, bimetallic electricity, whose existence Galvani denied to the end of his life, soon became known as 'Galvanic electricity' or 'Galvanism'. 'Galvanotherapy' is the name used up to the present day for the therapeutic use of the new DC currents drawn from voltaic piles. This form of electrotherapy was not least pioneered by Galvani's nephew, Giovanni Aldini, professor of experimental physics at Bologna. In his *Improvements in Galvanism* (1803), Aldini praised the utility of galvanic currents in the ressuscitation of the drowned and asphyxiated, and maintained he had successfully treated asthmatics with electricity.

Galvani's discovery of bioelectricity was soon swept away by the invention of the voltaic pile in 1800, which increasingly began to shift attention to the chemical aspects of bioelectricity.

1.4 The Instrumental Era

1.4.1 Electromagnetic Theory Leads to New Instruments

A new era of bioelectromagnetics began when in 1820 Hans Christian Oersted accidentally discovered the long-suspected connection between electricity and magnetism. The observation of the deflection of a magnetic needle by the current in a nearby platinum wire parallel to the needle marks the beginning of a period of rapid innovation. It was found that the deflection of the needle could be used to indicate current strength, and in 1820, a first instrument for this purpose was developed by J.S.C.Schweigger. A few weeks later, André Marie Ampére established the concepts we now call current and voltage, and proposed the use of a *galvanometer* based on them. Schweigger and J.C.Poggendorff then developed this instrument for detecting and measuring electrical currents

Ampére showed that magnetism could be caused by electricity and invented the solenoid. He also developed the first mathematical theory of electrodynamics. Michael Faraday discovered in 1821 that if a current-carrying wire could cause a magnetic needle to turn, a magnet could conversely cause rotation of a wire - the principle underlying all electric motors and generators. Henley, Cavendish, Ritter and Ohm elucidated the laws of electrical resistance. The electromagnet was developed by Sturgeon and Watkins, from ideas of Arago and Ampére, and finally Faraday in 1831 discovered the basis for all transformers and related devices, (mutual) induction, and Henry stated the principle of self-induction.

These discoveries, all made within a dozen years of Oersted's observation, laid the basis for a new type of electrical generator. By rotating a magnet in the vicinity of stationary coils, or *vice versa*, a current was induced in the coils that alternated in direction according to the magnet's position. Soon, this magneto-electrical machine was used in electrotherapy, only to be rivaled a few years later by the induction-coil, invented by Faraday. The technology for the new era was complete when induction coils incorporating various forms of interrupters appeared on the market, and, finally, in 1851, Heinrich Daniel Rühmkorff designed the powerful induction coil bearing his name. It was to become the standard instrument for a long time to come.

1.4.2 Duchenne and Faradic Electrotherapy

Owing to the introduction of induced currents from magnetoelectric and electromagnetic machines in the late 1830's and early 1840's, electrotherapy, which had fallen somewhat into disfavour by the end of the 18th century, gradually returned to greater prominence. Guillaume Benjamin Duchenne, called "Duchenne de Boulogne", began to work scientifically on electrotherapy, and to develop electricity into a powerful diagnostic tool in neurology. He was the first to observe that the physiological effects of induced current were different from that of galvanic current, and proposed in 1851 that the application of the former should be called 'faradization'. The new ability to direct and control the action of electricity opened up an unexplored area, which he systematically undertook to map. Based on his new technique of electrodiagnostics, he also described a number of new neurological pathologies that, to this day, still carry his name.

The little known beginnings of electroacupuncture

The first decades of the 19th century also saw the beginnings of the application of electrical currents through acupuncture needles[4,11], a method which, in 1958, would be introduced into China, and is nowadays, the standard instrumentation of acupuncture analgesia.

Acupuncture first arrived in the West with the publication of Willem Ten Rhyne's *Dissertatio de arthritide* (1683) and Engelbert Kaempfer's *Amoenitatum exoticarum politico-physico-medicarum* (1712). Both were surgeons of the Dutch East-Indian Company and described their own observations from Japan. However, until the early 19th century, acupuncture received only scant attention.

Renewal of interest came in 1809 when Louis V.J.Berlioz, father of the composer, performed experiments that triggered a veritable acupuncture fad comparable to that of the early 1970s. The method was not adopted in the form practised in the Far East, instead, the needles were used in conjunction with electricity to treat refractory bone fractures, tumors, aortic aneurism, pain, and a host of other disorders, sometimes with considerable success. In 1835 electroacupuncture was even adopted by the famous physiologist Franois Magendie, professor of medicine at the Collége de France and teacher of Claude Bernard. By mid-nineteenth century, electroacupuncture had spread to Italy and Gemany.

1.4.3 The Rise of Instrumental Electrophysiology

Since about 1820, interest in the phenomenon of animal electricity had generally receded; only Aldini, Nobili and particularly Matteucci continued to make measurements and to theorize about it[4]. Nonetheless, they improved the instrumentation and conclusively demonstrated the existence of bioelectric currents, thus providing a confirmation of 'animal electricity'.

1.4.4 The Classical Period: Du Bois-Reymond and the Berlin School

Matteucci's 1840 *Essai sur les phénomène électrique des animaux* was given by Alexander von Humboldt to Johannes von Mller, the foremost physiologist of his time and professor at the Berlin Medical University, who believed in some 'organic creative force' directing life functions according to the same chemical and physical laws that govern nature throughout. He passed Matteucci's book to his pupil Emil Du Bois-Reymond[4,9] with the commission of reproducing the Italian's experiments.

What set Du Bois-Reymond apart from all the earlier researchers was his meticulousness and his high standards of experimental technique. He developed most of the methods and apparatus needed to eliminate errors in measurement in order to obtain clearcut results in electrophysiology. For example, he introduced the technique of electrode compensation, and was the first to develop an unpolarizable electrode, which remained in use until about 1942. Another of his technical innovations was the *rheotome* ('flow slicer') invented in 1849, which alternately delivered a stimulus to the tissue and measured the resulting current through it. Du Bois-Reymond also discovered many of the basic laws of bioelectricity.

In 1842 he used his new galvanometer to reproduce Matteucci's injury current on muscle. He found that the place of injury is always negative relative to the uninjured areas, as is the interior of the muscle fibers with respect to the surface. The resulting current he called the "resting current". From 1843 onwards, he studied the contracted muscle during tetanus (prolonged contraction) and found a decrease of the current, which he called "negative variation". He showed that tetanus consisted of a series of rapid contractions accompanied by a rapid variation of muscle currents. In 1845 he established a law which was important for electrotherapy: the excitation of contractions does not depend on current density but on the rate of change of the current per unit time; the more rapid the change, the greater the excitation.

Even more significant were Du Bois-Reymond's findings on nerve physiology. In 1843 he discovered the resting current in nerves, which was much weaker than that of muscles, and found that tetanized nerves also show the 'negative variation'. He further discovered the polarization at the points of entry and exit during the passage of direct current through a nerve. In 1849, he demonstrated that the passage of the nerve impulse could be detected electrically. His main publication is *Investigations into Animal Electricity* (Part I, 1848; Part II, 1849). In the years 1860-1884, he added Part III.

In 1850 another member of the Berlin school and Du Bois' friend, Hermann von Helmholtz, succeeded in precisely measuring the velocity of the nerve impulse. Using the ballistic galvanometer, he got a value of 27.25 m/sec - about the same as measured today. This showed that the passage of the nerve impulse was much slower than the flow of electric current along a wire, so that the conviction arose that it was an entirely different phenomenon. The mode of propagation was soon to be elucidated by some pupils of Du Bois.

One of them, Julius Bernstein, was author of the membrane theory of nerve excitation, first proposed in 1871. It was to become the paradigm of scientific medicine, and has contributed substantially to the demise of bioelectricity that was soon to follow. According to Bernstein's hypothesis, what propagates along the excited nerve fiber is a depolarization of the membrane; the electric excitation wave is explained by changes in the ionic composition within and outside the membrane. The success of the hypothesis, still accepted today, led to the conviction that this kind of electric activity is the only kind that exists in the organism. By this time, microscopists had found a gap between nerve and muscle, the 'synaptic gap', and the transmission of the nerve impulse across this gap remained the last domain of bioelectricity. But in

1921, Austrian physiologist Otto Loewi (together with H.H.Dale) postulated that the transmission of the nerve impulse across the synaptic gap is chemical, setting neurophysiology definitely on a chemical path.

The Berlin school of physiology, consisting mainly of Du Bois-Reymond, Hermann von Helmholtz, Carl Ludwig, Carl Reichert, Ernst Brücke and Du Bois' pupils, among which were also Eduard Pflüger, Hermann Munk and E. Boruttau, led German physiology to uncontested leadership at the end of the 19th and the beginning of the 20th century. The influence of the Berlin group was to extend far beyond physiology. In the first decades of the 20th century their methods and concepts were adopted worldwide. In the United States, 'scientific medicine' was established in the 1880's by some pupils of Helmholtz, W.H.Welch, H.P.Bowditch and J.J.Abel. Their influence spread from Johns Hopkins University and the Rockefeller Institute, such that by 1910, the Flexner Report was to recommend a general reform of medicine and science on the basis of experimental science, which was accomplished by 1930. A deadly blow was dealt to practices such as colour therapy, homeopathy and electrotherapy. Whereas in the 1880s, around 10 000 medical doctors were reported to use some form of electrotherapy in the USA, by the 1930s, anyone referring to bioelectricity jeopardized his career.

1.4.5 Tesla, D'Arsonval and Early Twentieth Century Electrotherapy

The 1880s witnessed a number of fundamental discoveries in physics that had a strong impact on bioelectromagnetics. At the same time, many technological innovations laid the foundations for the electrification of Western civilisation. Edison built the first electric power station at Pearl Street in 1882. His perfection of Swan's incandescent lamp was the breakthrough for electric lighting. However, the long-distance transmission of electric power would never have been possible with DC current, as only high-voltage AC current could prevent the otherwise huge power losses. It was Nikola Tesla who, in 1888, invented the multiphase system based on a rotating magnetic field, that made possible the industrial use of AC current. In the same year, Heinrich Hertz, a pupil of Helmholtz, demonstrated the existence of the electromagnetic waves predicted by James Clerk Maxwell in 1864.

The promises of the electrical industry to provide electric lighting in homes, factories, offices and streets and to revolutionize transportation with electric tramways, and many other features of electrification were presented at an exhibition on the occasion of the first International Congress of Electricians held in Paris in 1881. Among the organizers of the congress was Jacques Arséne d'Arsonval[4,13], physiologist and leading developer of electrical technology. He was instrumental in passing a resolution at the congress requiring electrophysiological experimenters to make exact measurements of current stength, potential differences, etc., with instruments calibrated in cgs units.

D'Arsonval had played an important part in the development of microphones and the telephone. In 1880, he had modified the moving-coil galvanometer into a highly sensitive instrument that was to lead to the Einthoven string galvanometer and to the electrocardiograph. From 1873 to 1878, d'Arsonval was assistant to Claude Bernard, one of the founders of experimental physiology and originator of the concept of the *milieu intérieur*. His main contribution was in the electrical stimulation of muscles and nerve. For this purpose he developed a special instrument. He found that the physiological effect of any current was the same whatever the electrical source, as long as the waveform was the same. With sinusoidal currents he got the most striking results, and therefore recommended them for use in electrotherapy. He showed that higher frequencies above 5000 Hz did not cause any more neuromuscular excitation. In 1890 d'Arsonval started to use Hertzian oscillations, which, as he soon realized, could be greatly extended in frequency above the range that was possible by mechanical means.

High-frequency currents became possible through Tesla's work. Tesla's method for obtaining them was similar, but much more efficient and powerful than d'Arsonval's and achieved a higher range of frequencies. He produced them not by means of a Hertzian oscillator, but used his own transformer with primary and secondary coils. He demonstrated that currents at frequencies of 400 KHz to more than 10 GHz were harmless to humans. Tesla tended to ascribe the healing effect of high-frequency currents to the action of heat, while d'Arsonval, although recognizing the existence of heating, was convinced that the healing was due to a specifically electrical effect and not a consequence of heating.

In 1892, d'Arsonval modified the Tesla circuit and produced an arrangement that came to be widely used for medical purposes. In order to apply high-frequency currents to humans, he built a large solenoid resembling a cage, in which a high-frequency oscillating magnetic field induced strong currents in the body of the patient. Together with Paul Marie Oudin, he also constructed, in 1897, another form of the device producing a powerful brush discharge. Clinical trials with large numbers of patients were carried out. Positive results were obtained for cases of arthritis, rheumatism, and gout.

Walther Nernst[14] refuted Tesla's idea that high-frequency energy did not enter the body and confirmed d'Arsonval's observation that the stimulating effect of high-frequency currents decreased with increasing frequency. He postulated that the stimulating effect of alternating current is proportional to current intensity and inversely proportional to the square root of the frequency of the current[15,16]

Diathermy

In 1899, the Austrian chemist Zeynek, a pupil of Nernst, determined quantitatively the heating rate of tissue as a function of frequency and current density and first proposed the use of high-frequency currents for the therapeutic heating of tissue deep inside the body[17]. Together with Bernd, he built the first high-frequency apparatus for deep-tissue heating[18]. Independently, the German physician Karl Franz Nagelschmidt introduced in 1909 the concept of diathermy: the development of heat right through the body by molecular agitation caused by high-frequency currents[19]. He also developed the prototype of his diathermy apparatus, which used condenser discharges to produce high-frequency oscillations. His 1913 textbook on diathermy inaugurated a new era of high-frequency electrotherapy[20]. The concept of diathermy has also greatly contributed to the conviction that the biological effects of electromagnetic fields were restricted to thermal effects, a notion which has dominated the field until recently and severely hindered bioelectromagnetic research.

Biological effects of short waves

The first work on the biological effects of short waves was done by the Franco-Russian engineer and electrotherapist, Georges Lakhovsky. Using his own 'Radio-cellulo-oscillator', built in 1923 and based on 2 triode tubes to generate waves of 2-10 m, he showed at the Salpétriére clinic that 2 m-waves (150 MHz) could be used to destroy experimental plant tumours. His experiments[21], inspired by d'Arsonval, were based on the assumption that cells are electrical oscillators and that the chromosomes work as short-wave oscillating coils emitting a weak electromagnetic field. According to Lakhovsky, the disturbance of the natural oscillations of the cell is the cause of illness. To re-establish the oscillatory equilibrium of the organism Lakhovsky invented, first, his 1923 device, then, in 1931, the 'Multiple-Wave Oscillator' supposed to cover the resonance frequencies of all the body's cells, from 10 cm to 400 m (750 KHz-3 GHz). The latter was used in clinical trials conducted in various Paris, Rome and New York hospitals, in a number of cases, including cancer, allegedly with success.

Pioneering work on the effects of short waves on animals was done by Joseph W.Schereschewsky[22] in the USA and Erwin Schliephake in Germany, who together with physicist Abraham Esau, undertook the first therapeutic trials with ultra-short wave, using condenser fields. Based on his experimental work with bacteria, seeds etc., Schliephake and some of his pupils supported the view that specific, non-thermal, short-wave effects existed[23], while others disputed this. One of the main problems was the lack of reliable methods of dose measurement.

The introduction of electron-tube oscillators initiated a new form of the treatment, called alternatively fever therapy, electropyrexia, or hyperthermia[4]. The hyperthermia method was first proposed by Neymann and Osborn in 1929[24] and mainly developed and used in the USA and France, whereas the Schliephake school devoted itself more to local short-wave and ultra-shortwave applications.

The development of diathermy concluded with microwave diathermy (2450- MHz), originated soon after the invention of radar during World War II[25], while non-thermal high-frequency therapy went into its next stage with the work of Abraham J. Ginsberg[26]. In 1940 he presented the novel 'Diapulse' device to the New York

Academy of Sciences, which was produced industrially from 1959 onwards by Remington Rand. It used a pulsating field of 20 MHz and a repetition cycle of 60 Hz, and was applied with apparent success to inflammatory conditions, bone fractures, cardiovascular, gynecological and geriatric problems and in sports medicine and dermatology. It was supposed to stimulate immune and regenerative mechanisms of the organism.

Investigations of electrodermal activity

The discovery of the electrical activity of the skin is generally ascribed to Romain Vigouroux, Charles Féré and Ivan Romanovich Tarkhanov[27]. Vigouroux was a medical electrotherapist and electrodiagnostician, pupil of Duchenne de Boulogne. He was a collaborator of the famous clinical neurologist Jean-Martin Charcot at the Salpétriére hospital in Paris. Charcot had been studying hysterical symptoms produced by hypnotical suggestion. Charcot also used metallotherapy and magnetotherapy in the treatment of hysterics. Vigouroux attributed the effect of metallotherapy and electrotherapy to changes in the electrical tension of the body parts treated, caused by changes in blood flow and possibly also in nerve conduction. Vigouroux was commissioned by Charcot to study the electrophysiological changes in a patient during hysterical anaesthesia under the influence of magnetotherapy. This was how he came to do the first measurement of what is called today, 'basal skin resistance' (BSR). Up to this moment, electrical skin resistance had merely been considered as a disturbing factor to be eliminated in applying current for diagnostic or therapeutic purposes. Vigouroux found that hysterical insensitivity was accompanied by high BSR values and these shifted abruptly together with the hysterical symptom. He interpreted the shifts as changes in conductivity due to changes in the vascular tonus which in turn caused shifts in the electrolyte content of tissues.

Féré, neurologist and student of Charcot, was also was among those trying to defend Charcot's electrical theory of hypnosis and hysteria. He began to use galvanic (DC) currents to measure skin resistance on grounded subjects, known in electrodermal research today as the 'exosomatic method'. This was the first study of the 'galvanic skin response' (GSR). Féré noted that closing the eyes diminished the current flow, which he concluded was a sign that the absence of excitation increased the SR.

He did not continue his studies, but Prince Ivan Romanovich Tarkhanov[2], professor of physiology at St.Petersburg, in 1889 and 1890 published studies on skin potentials, in which he synthesized the work of Vigouroux, Féré and D'Arsonval with that of Du Bois-Reymond and Hermann, whose measurement techniques he applied to Féré's experimental problem . Using a very sensitive galavanometer, he measured changes in skin potential without the influence of external current under the same set of sensory and emotional stimuli as those applied by Féré, but adding suggestions of memories of the stimuli. This method today is known as the 'endosomatic method'[28].

International attention to BSR and GSR came only as a result of work done more than a decade later in Switzerland. In 1904 Erich Konrad Müller, electrotherapist and little known Swiss pioneer of several branches of bioelectromagnetics, independently found changes in the resistance to a galvanic current passed through the body, which to him seemed to be correlated not only with meteorological and geomagnetic factors, but also with psychological processes[29]. He noted that alternating currents produced different results from direct currents.

Müller directed the attention of the Zürich neurologist Otto Veraguth to the phenomenon, who added photographic recording and also used the method to monitor physiological reactions in Carl Gustav Jung's word association tests.

Veraguth believed he had discovered a new reflex, which he dubbed "psychogalvanic reflex" (PGR) and considered it to be of great value for the objective study of psychological, psychiatric and neurological problems. His monograph on the subject was published in 1909[30]; Jung welcomed the method as a means to study objectively the psychological 'complex' and its 'feeling tone'. Jung improved on the recording equipment and, after having done some work himself[31], got some of his American friends, Peterson, Morton Prince and Ricksher, interested. The many papers published in the next two decades (1910-1930) established the field as a major research domain. The registration of electrodermal activity is routinely used today for investigating physiological correlates of psychological processes[32]. In the form of the polygraph ('lie detector') it is also widely used, at least within the US, in the forensic field and for staff selection. Its application to medical diagnostics is mainly based on acupuncture and has been developed later (see 4.6.8.).

1.4.6 The Era of Modern Instrumentation

Instrumentation in bioelectromagnetics remained quite unreliable and inaccurate well into the 1920s when vacuum-tube amplifiers and oscilloscopes came into more general use. The diode tube for rectifying high-frequency alternating currents, based on a discovery by Edison in 1882, had been invented by J.A.Fleming. The oscillating vacuum-tube triode by Lee de Forest (1906) was the first device to produce a completely continuous waveform. However, the early vacuum tubes were still noisy and their characteristics drifted with time. Some of these problems were not overcome until the introduction of solid-state components in the 1960s.

Electroencephalography

After Du Bois-Reymond had demonstrated the electrical nature of the nerve impulse in 1848, speculations arose as to whether sensory impulses in the brain were also electrical. In 1875, Richard Caton of Liverpool, showed that in experimental animals, variations in electrical potential were produced in specific parts of the cortex by corresponding movements. Caton also noted that his galvanometer registered unexpected fluctuations even when no muscular activity occurred. Independently of Caton, "spontaneous fluctuations in the electrical activity of the brain" were also discovered in 1877 by Vasili Y. Danilevsky at Kharkov and in 1891 by Adolf Beck at Cracow, who also observed their cessation at the incidence of an external stimulus, such as a flash of light or a clap. While these recordings were done on animal brains, the recording of human brainwaves from the surface of the skull was not achieved until 1924, when Hans Berger, a German psychiatrist, was lucky enough to have available a superior instrument, a double-coil galvanometer made by Siemens. Berger was interested to find a physiological basis for psychic phenomena, especially telepathy. This led him to attempt for years to evoke changes in the brain potential of human subjects by sensory stimulation, before succeeding in 1924, when he was able to discriminate between alpha- and beta-waves. He developed a method to determine the extent of damage in cases of brain injury, and published all these discoveries only in 1929, in a paper whose title coined the term 'electroencephalogram'. He also applied the method to study epilepsy and the effects of mental effort. In 1935 Edgar Douglas Adrian confirmed Berger's findings and discovered delta waves. Theta waves were discovered in 1943 by the American neurologist William Grey Walter. He and his coworkers also discovered the visual evoked potentials produced by rhythmical stimuli such as strobe lights shone on open or closed eyes[33]. They observed that at certain frequencies the stimulus not only evoked changes in the visual center but spread over the entire brain and entrained it to oscillate synchronously.

Electrocardiography

In 1878, Theodore Wilhelm Engelmann at Utrecht, using Bernstein's differential rheotome, obtained the first primitive electrocardiogram, a sawtooth waveform that clearly showed a positive rise and fall followed by a negative excursion. Ten years later, Waller produced the first ECG of the human heart. But the differential rheotome was too slow and sluggish. The situation changed in 1901 with the invention of the string galvanometer by Willem Einthoven, who for this and many other contributions to the field has earned the title of 'father of electrocardiography'. His string galvanometer was a unwieldy machine weighing over a quarter ton and needed five people to operate it. But he produced ECGs of remarkable quality. The string galvanometer principle continued to be the basis of ECG instrument design until the the 1940s, at least in Britain, although Siemens and Halske introduced a commercial instrument based on vacuum tubes and oscillography as early as 1921.

Electromyography

The first electromyogram was obtained in 1929 by Edgar Douglas Adrian and Detlev Wulf Bronk[34] and in the same year, independently, by Derek Ernest Denny-Brown. The great advance was the invention of the coaxial-needle electrode by Bronk, making possible the recording of the activity of a single motor unit. The first extensive clinical EMG study was undertaken by Hans Piper whose report, *Elektrophysiologie menschlicher Muskeln*[35] was the first book on the subject.

Electrophysiology with modern instruments

By 1920, biomedical researchers were increasingly using electronic amplifiers and oscillators, followed in the 1930s by instruments and devices for monitoring and enhancing certain physiological functions. In 1921, Gasser together with Joseph Erlanger invented the cathode-ray oscilloscope, the first inertia-less recording device with modern amplification. They used it for their investigations on the transmission velocity of nerve impulses in different types of nerve fibres, for which they got the Nobel prize for physiology in 1944.

Biological effects of light and phototherapy

Heliotherapy

The therapeutic use of sunlight[36] and colours[37,38] is as old as medicine itself. Egyptian physicians used rooms with coloured walls to heal various afflictions. Heliotherapy is documented by several antique authors. Hippocrates mentions the favourable influence of sunlight warmth on wounds and open fractures in his *Aphorisms*. The doctors of Roman imperial times recommended to "subject patients to the sun" and laid down precise indications for heliotherapy

In the Middle Ages, the knowledge of the therapeutic value of the sunlight got lost with the general decline of hygiene, except in Islamic culture where Avicenna, around 1000 A.D., mentions that exposure to sunlight and fresh air protects from illness. From Galenos to the end of the 18th century, the subject of light was banned from the *Materia medica*, although we can assume it never ceased to be used in folk medicine.

With Rousseau's "back to nature" and the new emphasis placed on hygiene at the time of the French revolution, a number of medical papers on heliotherapy were published from the 1770s to the 1790s. In 1815, the first comprehensive works on heliotherapy were published by Loebel and Cauvin. Cauvin recommended light therapy for depression and asthenia, rickets and scurvy, ulcers, swellings, rheumatic conditions and paralysis. In 1816 Doebereiner, professor of chemistry at Jena, laid down the basic tenets of modern light therapy. In his papers, he analyses the effects of light according to its spectral components and differentiates the thermal effects of light from chromotherapy.

Sunbaths became very popular in the second half of the 19th century in Germany within the framework of the naturopathy movement initiated by Vincenz Priessnitz and taken up by many German physicians between 1830 and 1840.

UV phototherapy

One of the few branches of phototherapy to find general recognition, UV therapy (then called "actinic therapy") was founded as a consequence of the discoveries by Pasteur and Koch on the role of bacteria in illness. In 1877, Arthur Henry Downes and Thomas Porter Blunt discovered the bactericidal properties of sunlight, which they ascribed especially to the short-wave end of the spectrum. Marshall Ward in 1892 showed that the UV band has the strongest antibacterial effects.

Ward's results became the basis of the Nobel-prize winning work of Niels R. Finsen who is regarded to be the founder of scientific sunlight and artificial-light therapy[4]. In 1895, he published a study on the stimulating properties of different frequencies of light, but his most famous paper was on the treatment of *lupus vulgaris* (skin tuberculosis) by concentrated UV light. In his Medical Light Institute established in Copenhagen in 1896, he successfully treated hundreds of cases. He also used red light against scarring after chicken-pox and smallpox, and green light for tuberculosis. Most of his work was done with light from carbon-arc lamps with quartz lenses he developed himself. In 1903 he got the Nobel prize for medicine for his pioneering work.

During the 1910s, the value of both local and general UV irradiation came to be increasingly recognized. By the 1920s the use of UV irradiation for the treatment of rickets, lupus and other forms of tuberculosis, for certain skin conditions, its germicidal properties and its value in the healing of wounds became widely recognized.

Visible light Chromotherapy

Although there has been sporadic research on the effects of different frequencies of visible light, chromotherapy has largely remained outside the mainstream. This may be partly due to the fact that the scientific community has always been reluctant to deal with therapeutic methods believed to be tainted with vitalist or occult notions, even when they were based on sound and rich empirical data. This association has often been made for colour therapy[38,39].

Inspired by experiments on the stimulation of plant growth by coloured light in the 1870s, practitioners of light therapy - who had until then used only direct sunlight - began to look at colour and its effects[39.40]. In General Augustus J.Pleasanton's book *Blue and Sun-Lights*, a series of observations and theories were advanced that both enraged and inspired the botanists and horticulturists of his day and excited succeeding light therapists. He claimed that the quality, yield and size of grapes in his greenhouse had been significantly increased by panes of blue glass. Blue light, according to him, also cured diseases in animals and humans, and increased fertility and sped up maturation in animals. Pleasanton considered organisms to be living energy systems consistently kept in balance by sunlight.

By far more sophisticated was the light therapy system expounded by another physician, Edwin D.Babbitt from New York, in his classic work *Principles of Light and Color*[41]. Unlike his predecessors, Babbitt not only used red and blue, but incorporated many different frequencies obtained from both natural and artificial light sources which he applied by means of several different devices with coloured filters and lenses. He developed elaborate and far-reaching theories based on the effects of colours.

The foundation of present-day chromotherapy was laid in the first two decades of the 20th century in the USA, by Dinshah P.Ghadiali and Harry Riley Spitler. Ghadiali[42] used twelve colours produced by sunlight or a light source such as a slide projector shone through a set of glass filters. He irradiated the coloured light on certain body zones, at certain times of the day. To determine the appropriate zones and type of treatment he also used his 'Itisometer' developed in 1923, a kind of bolometer to measure the temperature (IR) radiation of the body surface. He theorized that the absorption of light was essential for the organism as is nutrition; the balance of the various 'colour energies' in the organism he considered fundamental for health.

Many of Ghadiali's indications were much sounder than some of his theorizing and have been confirmed by later research; nevertheless in the wake of the Flexner report (1910), colour therapy was attacked heavily by the AMA and Ghadiali persecuted by the FDA as a charlatan. In 1925, he was arrested during a lecture tour and sentenced to 5 years imprisonment.

While Ghadiali's highly influential work was not scientific by modern standards, the work of physician Harry Riley Spitler makes him the founder of modern scientific chromotherapy. In 1909, after having obtained four doctorates and studied the works of Babbitt and Ghadiali, he began to research and apply the therapeutic use of coloured light especially by visual perception through the eyes. His method of treatment was more comprehensive than those of his predecessors, and involved the use of more than 30 different combinations of colour filters which were specific to every patient according to his or her bodily and psychological constitution. In 1933, when the clinical effectiveness of his method was proven, he founded the College of Syntonic Optometry as a research and teaching facility for his work; it still exists today. His principal work, *The Syntonic Principle*, was published in 1941.

With the dissemination of antibiotics in the late 1940s, chromotherapy, even less established than UV light therapy, lost its reputation and was increasingly considered unscientific, if not outright quackery.

The influence of coloured light on muscle tonus

This interesting by-field of photobiological research was inaugurated in 1900 by Charles Féré, who found that in hysterics, red light increases muscular power, breathing, and circulation[43]. Grip strength was strongest under red light and decreased with decreasing wavelength. Stefanescu-Goanga (1911) found that red, yellow, orange and purple produced various physiological tension effects, whereas green, blue, indigo, and violet produced relaxation[44]. Metzger (1925) showed that the tonus-affecting effect is different for different frequency bands[45]. The frontally raised arms of people irradiated with red and yellow light involuntarily moved towards the light source, while green and blue light caused movement away from the source. The colours also seemed to influence subjective time. Neurologist Kurt Goldstein[46] and Goldstein and Rosenthal[47] confirmed these observations and found effects also for UV and IR light. Ehrenwald showed that the reaction also takes place when the eyes are completely covered, and in blind people, which indicates it is not mediated by the eyes, but by the skin[48]. Goldstein concluded that the effects of coloured light on the organism are a field-like reaction of the whole organism, characterized by either a reaching out and extending towards the stimulus or a shrinking back from it.

Impressive confirmation for these early findings came in 1958 from Robert Gerard's doctoral dissertation at the University of California at Los Angeles[49]. He demonstrated that the autonomic nervous system and visual cortex were significantly less aroused when stimulated by blue or white light than by red light. Each colour also elicited significantly different feelings in the subjects. Stimulation with blue light was associated with increased relaxation, less anxiety and less hostility, while stimulation with red definitely increased tension and excitement. Since then, Aaronson[50] and Plack and Schick[51] reported similar effects of specific colours on activation and arousal. Finally, Wolfahrt and Sam[52] in 1982 not only confirmed the effects of selected colours on behaviour and physiology, but also corroborated Ehrenwald's finding that blind subjects were as affected as sighted ones.

Light emission by organisms

Living organisms are not only affected by light, but they also emit light. The bioluminescence of bacteria, moulds and fungi, fire-flies, certain jellyfish and deep-sea fish is well known for centuries[53]; The luciferin-luciferase system responsible for light generation was isolated by Dubois in 1913[54].

However, there is an altogether different kind of biological luminescence which is much less obvious and therefore was only discovered much later by the Russian biologist Alexander G.Gurvich[55] in the course of investigations on the hypothesis of a 'biological field' regulating the growth and development of biological form. He observed the stimulation of mitoses at a spot of an onion root at which the tip of a second onion root was pointed. Gurvich was able to exclude chemical stimulation by shielding the roots from each other with glass; he found that the effect only ocurred if quartz glass, but not ordinary glass, was used. Gurvich concluded that the agent stimulating cell division was UV radiation. According to its effect, he called it, "mitogenetic radiation" (MR). Subsequently, the same effects were discovered in many other plant and animal tissues. Mitosis could also be triggered by very weak artificial UV light, Furthermore, mitogenetic radiation impinging upon biological material could also trigger a wave of propagating 'secondary emissions'. Changes in the intensity of MR seemed to be related to the activity of the cells; the MR of muscles, for example., was found to increase during contraction, and tumour tissue emitted more strongly than normal tissue. Gurvich's method of detecting MR was based on a very sensitive biological assay system of yeast cell cultures.

In the 1930s, mitogenetic research had spread to other European countries with mixed results. The physical demonstration of the existence of MR was not possible before World War II; the 'Gurvich radiation' was therefore suspect. There were other reasons for its unpopularity: the disastrous consequences of Stalin's science policy for biology (resulting in the Lysenko affair) and the fact that, at the time, biological science became dominated by molecular biology, which made Gurvich's approach obsolete. In Eastern Europe, however, the research continued during the war, although nothing of it was published in any Western language publication till the early 1960s.

By the 1940s, photoelectronic multipliers, some tens to hundreds of times more sensitive than previous instruments, became available and finally made the detection of very weak light physically possible. In the Soviet Union this led to an intensification of mitogenetic research in the 1950s. Only a few of the mitogenetic researchers - the 'biophysical school' - were willing to retain the whole Gurvichian heritage, including the theory of the biological field. They maintain that the organism is able to act as an emitter and receiver of a broad range of electromagnetic frequencies and use these coherently for communication, regulation and internal structuring. In the same tradition are biologists, Dombrovskii, and his pupil Victor M.Inyushin at Kazakh State University, Alma-Ata[56], and medical researcher Vlail P.Kaznacheev of the Academy of Medical Sciences at Novosibirsk[57]. Inyushin, who is also one of the pioneers of therapy by means of very weak laser irradiation, postulated, in 1967, that the main source of MR was the cell nucleus and that it was coherent - a notion that gave the field a new dimension.

The first Western scientists using the photomultiplier for investigating ultraweak light emission were the Italians, L.Colli and U.Facchini, from Milano[58]. They showed that a number of plant seedlings, among them wheat, lentils, beans, and maize, emit light. However, it was not UV light, but visible light from the green to the red with an intensity from ten to several hundred photons per second/m^2. Further Western research into the subject was not done until 1967, when Australians Metcalf and Quickenden[59] started their investigations, and in 1976 with the beginning of the work of Popp and coworkers[60]. To-day, biophoton research is a thriving and growing area[61].

Bioelectrical potentials

Inspite of the increasingly adverse scientific climate in the early decades of the 20th century, a few researchers chose to pursue unpopular lines of inquiry. One of these was concerned with steady (D.C.) bioelectrical potentials in plants, animals and humans[62,63]. In contrast to action potentials which are well-understood, they are relatively steady, undergoing only slow temporal changes, with magnitudes of up to 150 millivolts.

In 1903, Mathews demonstrated that hydras were electrically polarized; the head was positive, the tail negative[64]. He suggested a possible role of these potential differences in the process of regeneration, as the rate of regeneration depends on the position of the cut relative to the poles, with the point of most rapid regeneration corresponding to the point of maximum negativity. Mathews concluded it should be possible to control the rate of regeneration by using externally applied current of equal strength. Similar potential differences were subsequently found in developing eggs[65], and it was observed that electrical currents, passed through aquarium water, seemed to speed up the regeneration of larval salamanders[66]. In 1920, Sven Ingvar of Yale University reported on experiments with very weak galvanic current (2-4 x 10^{-9} ampere) on tissue cultures[67]. Whereas in controls, the cell and fiber outgrowths occurred in all directions, under electrical influence, cell movements took place entirely along the lines of force in the field, and cell processes growing towards the anode showed morphological differences from those growing towards the cathode. Ingvar concluded that electrical forces may play a role in formative processes during morphogenesis.

In the early 1920s, Elmer J.Lund of the University of Texas[68] demonstrated that the polarity of regeneration in species related to hydra could be controlled, or even reversed, by small direct currents passed through the animal's body. A head could be caused to form where a tail should have regrown, and *vice versa.* His finding was confirmed by others, and Lund went on to study the bioelectric potentials of eggs, embryos and trees, using a Compton electrometer, which draws no current fom the system measured.

Harold Saxton Burr of Yale University School of Medicine took up biopotential research in the 1930s[69], stimulated by Lund's work. From 1936 onwards he used a sensitive vacuum-tube microvoltmeter he had developed with C.T.Lane and L.F.Nims. Burr and his coworkers found steady electric potential gradients on the surface of many different organisms which were characteristic for each species. That of corn seeds was correlated with the viability of the variety and its prospective growth rate, spadix weight and height of the full-grown stalk. Long-term measurements on trees made every hour for the years 1943 to 1966 showed that trees possess bioelectric fields that varied in response not only to physiological activity, light intensities and moisture, but also to changes in atmospheric electricity and geomagnetic field strength related to thunderstorms, sunspot cycles and phases of the moon. Burr and his coworkers correlated changes of bioelectric potential to growth, regeneration, tumor formation, the effects of drugs, hypnotic states and sleep.

However, the work of people like Lund and Burr was almost completely ignored by fellow scientists. The reason was not only that the scientific climate of the time was not very sympathetic towards this kind of research. As in the case of mitogenetic radiation, the devices they used were not sensitive enough, and produced too much noise to furnish reliable results. Furthermore, the elaborate conclusions they made from their measurements were not backed up by the biological knowledge of the period.

It was orthopedic surgeon at the Veteran's Administration Hospital in Syracuse, N.Y., Robert O.Becker, that was to bring the field to the attention of a larger community of scientists[70,71]. Becker showed in the early 1960s that living animals indeed displayed on their body surface measurable D.C.potentials, but more complex than the simple polar pattern found by Lund and Burr. The pattern was spatially related to the anatomical arrangement of the nervous system and the potentials could be measured directly on the peripheral nerves, where they showed polarity differences depending on their motor or sensory functions. Becker also demonstrated that the D.C. currents implied by the potential measurements exist, flowing along the neural elements. They are, as Becker showed, the 'injury currents' found by Matteucci and play a central role in the process of regeneration. Soon he had confirmed the existence of a potential over the stumps of freshly amputated animal limbs. The healing of the wound was consistently accompanied by characteristic changes in electric potential. He was able to relate regenerative healing in a causal fashion to the current of injury. Animals lacking regenerative ability, such as the frog, differ from those who possess it (e.g., the salamander) in inadequate local current of injury due to inferior peripheral innervation. With increased cerebralization, peripheral innervation is diminished in animals higher up the evolutionary ladder, explaining their lack of regenerative ability. Experiments at restoring either the innervation or substituting stimulation by externally applied D.C. currents have demonstrated the feasibility of reactivating regeneration even in higher animals and humans.

In the 1960s, Becker and Andrew Basset showed that living bone piezoelectrically generates electric potentials[72], and demonstrated that osteogenesis could be stimulated electrically, which had already been reported in 1849 by Stanley. The introduction of currents induced by pulsed electromagnetics fields (PEMFs) made it possible to develop electrostimulation with weak D.C. currents into a successful therapeutic method for the healing of non-uniting bone fractures.

Local anomalies of electrical skin conductivity and acupuncture points

In the 1950s, the discovery that the acupuncture points of traditional Chinese medicine possess electrical characteristics different from their surroundings, led to a new branch of electrodermal research. Several attempts to find electrical differences between acupuncture points and the surrounding skin were undertaken. In contrast to the galvanic skin response studies, these investigations were made with small electrodes of less than 0.1 cm diameter.

The first of these attempts was probably that of German physician Richard Croon who, from 1947 onwards, undertook to localize and measure "special areas of small diameter of the human skin with high electrical conductivity"[73]. In 1953, another German physician, Reinhold Voll, used a small DC current to locate acupuncture points and developed a diagnostic method based on the changes of the electrical conductivity upon stimulation, which he interpreted as an indication of the regulative capacity of the organisms[74].

The Japanese, Yoshio Nakatano, independently discovered 'good conductivity points' with a 12 volt DC current in 1956[75]. He linked these points with imaginary 'good conductivity lines' (*Ryodoraku*) and developed a diagnostic method with this name. Similar findings soon followed in Russia, France, and later, Austria. In China, Zhang *et al*[76] and Zeng *et al*[77] found that the location of points of high conductivity coincided amazingly well with the acupuncture points.

Transcutaneous and percutaneous neuroelectric stimulation

While its main development belongs in the 20th century, there are numerous episodes of electrical stimulation of the nervous system for pain relief in the pre-1900 history of electromagnetics[78,79].

The first organized study of electricity as an anesthetic agent was performed by French physician and physicist Stéphane Leduc in 1902-03[80]. Leduc had his associates try the method on himself and reported on a dream-like state with inability to move and speak and insensitivity to painful stimuli. In the 1920s and 1930s the method was much used for stunning animals in slaughterhouses, and in 1938 was developed into convulsive electroshock treatment for mental illness by Roman neuropsychiatrist Ugo Cerletti.

Modern interest in cutaneous electrostimulation was inaugurated in 1967 by Patrick Wall and his associates[81–83]. Meanwhile, it was recognized that the induction of sleep was not necessary for therapeutic effectiveness, and in 1969, Wageneder proposed to use the term "Cerebral (or Transcerebral) Electrotherapy" (TCET or CET)[84]. In the early 1970s the method was introduced into the US[85]. Independently, percutaneous electroanalgesia was introduced by clinicians in the People's Republic of China as part of acupuncture analgesia[86].

In 1976, neurosurgeon Irving S.Cooper found that the electrostimulation of the cerebellum in his patients treated for various neurological disorders had a number of interesting side effects, which also appeared when the bodily symptoms did not improve. These effects included the reduction of fear, stress and tension, an improvement of the thinking faculty and of the fluidity of speech, lower depression tendency and more optimistic outlook, lesser tendency for fits of rage and aggressive behaviour, and generally a heightened emotional control.

By the late 1970s, the hypothesis that the effectiveness of electrostimulation may be based on its effects on endorphine metabolism was backed by increasing evidence[87], and the first reports came in on its successful application to the treatment of addiction to alcohol[88], barbiturates and opiates[89]. Even dependency-related brain dysfunction has been treated successfully. One pioneer in this field was Scottish surgeon Margaret Patterson[90], who discovered by chance, in 1972, that electroacupuncture for pain relief also made withdrawal symptoms disappear. Her 'Neuro-Electric Therapy' (NET), has become known for its successes in the treatment of rock musicians such as Eric Clapton and Keith Richards.

1.5 The Era of Biophysics: The Rise of Modern Bioelectromagnetics

Much of the preceding account covers the time after World War II, which actually marks the threshold to yet another era of bioelectromagnetics. Just before the war, Otto Loewi's Nobel prize, in 1936, for the chemical transmission of the nerve impulse and Gerhard Domagk's Nobel prize in 1939 for the discovery of the sulfonamides, had definitely established the biochemical view of life. In 1940, German electrophysiologist Hans Schaefer noted in his textbook that "in many questions of general importance electrophysiology does not seem to take the key position accorded to it uncontestedly twenty years ago. The weight of many electrical hypotheses of life processes decreases due to new knowledge on chemical and other processes"[91]. Also, with the beginning of the war, the controversy over the deleterious side effects of electromagnetic energy was drawn to a close. Non-ionizing radiation was now thought to have no biological effects.

After the war, many new electromagnetic technologies, originally developed for military purposes, became available to research laboratories. Among these were the microwave technologies that followed from the development of radar, the photoelectric multipliers used in bioluminescence research and the electronic transistor.

When the discipline of biophysics was founded in the 1920s and 1930s, it had been mainly concerned with ionizing and non-ionizing radiation. One of the founders of biophysics, Boris Rajewsky of the University of Frankfurt am Main, held that the main task of biophysical research was to apply physical thought and physical methods to biological phenomena and the elucidation of physical relationships in life processes. He investigated the effects of light and of electromagnetic fields on the intact organism and also made a contribution to the research on ultraweak bioluminescence. He was convinced that living tissue reacts with "uncanny precision" to all physical influences. In the 1950s, the new discipline underwent a change of emphasis typical for the scientific climate of the time. When it seemed clear that non-ionizing radiation did not have biological effects, it began to concern itself with the application of new physical methods such as x-ray diffraction (first used by J.D.Bernal in 1934) and electron miscroscopy (developed in 1931 by H.H.Knoll and E.Ruska and improved in 1939 by V.Zworykin) to study the molecular structure of proteins and was losing sight of the living organism. The term "molecular biology" was coined in 1951 by W.T.Astbury and independently in 1952 by Paul Weiss[92].

However, the 1940s and 1950s also saw the beginnings of a number of new developments, which are now again shifting the emphasis towards biophysics, and may lead to a new 'biophysical era of bioelectromagnetics', or, 'bioelectromagnetic era of biophysics'. In 1941, Albert Szent-Györgyi, 1937 Nobel prize winner for his discoveries on biological oxidation and of vitamin C, suggested that energy in living tissues may be transferred by excited electrons moving within semiconducting matrices[93]. The next 30 years saw the accumulation of evidence for the solid state properties of biological systems[94].

The foundation of nonlinear optics was laid in 1950 when French physicist Alfred Kastler discovered optical pumping. Charles H.Townes, soon after, developed the principle of maser, and in 1960, Theodore H.Maiman built the first laser. In 1968, Herbert Fröhlich[95] showed that long-range coherent EM oscillations arising from cooperative behaviour of dipoles may exist in biological systems. Finally, in 1970, the English translation of the groundbreaking work on *Electromagnetic Fields and Life*[96] by the Russian biophysicist Alexander S.Presman made known to the West, the results of the work on the biological effects of non-ionizing EM radiation, undertaken in the 1960s mainly in Russia. It not only made clear that the assumption that these frequencies had no influence on living systems was no longer tenable, but also proposed the first modern hypothesis on the biological role of EM fields. Presman argued that environmental EM fields have played an important role in the evolution of life and are also involved in the regulation of the vital activity of organisms. He suggested that living beings behave as specialized and highly sensitive antenna systems for diverse parameters of weak fields of the order of the ambient natural fields. According to Presman, EM fields serve as mediators for connecting the organism to its environment as well as for the communication between organisms. EM fields produced by the organisms themselves are involved in the coordination and communication of physiological systems within living organisms. He also suggested that informational, as opposed to energetic, interactions play a significant, if not the main, role in electromagnetic biocommunication. Presman's book aroused a great deal of interest in the West, fuelling the impetus for biophysical electromagnetics in this present era.

Bibliography

[1] M.A.B.Brazier, *A History of Neurophysiology in the 17th & 18th Centuries* (Raven Press, New York 1984).

[2] M.A.B.Brazier, *A History of Neurophysiology in the 19th Century* (Raven Press, New York 1988).

[3] K.E.Rothschuh, *Aus der Frühzeit der Elektrobiologie, Elektromedizin* **4** (1959) 201.

[4] M.Rowbottom and Ch.Susskind, *Electricity and Medicine - History of their Interaction* (San Francisco Press, San Francisco 1984).

[5] Ch.C.Gillispie, *Dictionary of Scientific Biography* (Scribner's, New York 1981).

[6] Wellcome Institute for the History of Medicine and Related Sciences (ed.), *Subject Catalogue of the History of Medicine and Related Sciences* (Kraus International Publishers, Munich 1980).

[7] U.Warnke and U.Warnke, Geschichte der therapeutischen Anwendung von Magnetfeldern, *Biophysics and Medicine Report* **4** (1983) 1.

[8] A.Schmid: Zur Geschichte der Elektrotherapie vom Altertum bis zum Beginn des 19.Jahrhunderts, in *Festschrift für Jacques Brodbeck-Sandreuter*, ed. K.Reucker (Basel 1942), p.73.

[9] H.E. Hoff, Galvani and the Pre-Galvanian Electrophysiologists,*Annals of Science* bf 1 (2) (1936) 157.

[10] M.Pera, The Ambiguous Frog. *The Galvani-Volta Controversy on Animal Electricity* (Princeton University Press, Princeton 1992).

[11] D.Stillings, A survey of the history of electrical stimulation for pain to 1900, *Medical Instrumentation* bf 9 (1975) 257.

[12] K.E.Rothschuh, Du Bois-Reymond, Emil Heinrich, in *Dictionary of Scientific Biography*, ed. Ch.C.Gillispie (New York 1981), p.200.

[13] C.A.Culotta, in *Dictionary of Scientific Biography*, Vol.1, ed. Ch.C. Gillispie (Scribner's, New York 1981) p.302.

[14] W. Nernst and J.O.W. Barratt, Elektrische Nervenreizung durch Wechselstrome, *Zeitschrift für Elektrochemie* **10** 1904 664.

[15] W.Nernst, Zur Theorie des elektrischen Reizes, Pflüger's *Archiv für die gesamte Physiologie des Menschen und der Tiere* **122** (1908) 275.

[16] W.Nernst, Zur Theorie der elektrischen Nervenreizung, *Zeitschrift für Elektrochemie* **14** (1908) 545.

[17] Zeynek, *Gottinger Nachrichten, Mathematisch-Physikalische Klasse*, Heft 1 (1899) 94.

[18] J.Kowarschik, Die Behandlung mit Hochfrequenzströmen (Arsonvalisation, Diathermie), in *Neue Erfahrungen auf dem Gebiet der medizinischen Elektrizittslehre* ed. L.Mann (Georg Thieme, Leipzig 1928) p.316.

[19] K.F.Nagelschmidt, Über Diathermie (Transthermie, Thermopenetration), *Münchener Medizinische Wochenschrift* **56** (1909) 2575.

[20] K.F.Nagelschmidt, *Lehrbuch der Diathermie* (Springer, Berlin 1913).

[21] G.Lakhovsky, *L'oscillation cellulàire (ensemble des récherches expérimentales)* (Doin, Paris 1931).

[22] J.W.Schereschewsky, The physiological effects of currents of very high frequency (135,000,000 to 8,300,000 cycles per sec.) *U.S. Public Health Reports* **41** (1926) 1930.

[23] E.Schliephake, *Short wave therapy* (Actinic Press, London 1936).

[24] Cl.A.Neymann and S.L.Osborne, Artificial fever produced by high frequency currents - Preliminary report, *Illinois Medical Journal* bf 56 (1929) 199.

[25] F.H.Krusen, Medical applications of microwave diathermy, *Proceedings of the Royal Society of Medicine* **40** (1950) 638.

[26] J.Ross, Biological effects of pulsed high peak power electromagnetic energy using Diapulse, in *Emerging Electromagnetic Medicine*, ed. M.E.O'Connor, R.H.C.Bentall, and J.C.Monahan (Springer, New York 1990) p.269.

[27] E.Neumann and R.Blanton, The early history of electrodermal research, *Psychophysiology* **6** (4) (1970) 453.

[28] W.Boucsein, *Elektrodermale Aktivitt* (Springer, Berlin 1988) p.5.

[29] E.K.Müller, Das elektrische Leitverrmogen des menschlichen Körpers als Masstab für seine Nervositt, *Schweizerische Blätter fur Elektrotechnik* **9** (1904) 321.

[30] O.Veraguth, *Das psychogalvanische Reflex-Phänomen* (Berlin 1909).

[31] C.G.Jung, On psychophysical relations of the associative experiment, *J. Abnorm. Psychol.* **7** (1907) 247.

[32] W.Boucsein, Elektrodermale Aktivitt (Springer, Berlin 1988).

[33] V.J.Walter and W.G.Walter, The central effect of rhythmic sensory stimulation, *Electroenceph. clin. Neurophysiol.* **1** (1949) 57.

[34] E.D.Adrian and D.W.Bronk, The discharge of impulses in motor nerve fibres, *Journal of Physiology* **67** (1929) 119.

[35] H.E.Piper, *Elektrophysiologie menschlicher Müskeln* Springer, Berlin 1912).

[36] O.Bernhard, in *Handbuch der Lichttherapie*, eds. W.Hausmann and R.Volk (Springer, Vienna 1927) p.3.

[37] Z.R.Kime, *Sunlight* (World Health Publications, Penryn CA 1980).

[38] F.Birren, *Color Psychology and Color Therapy* (University Books, New Hyde Park N.Y. 1961).

[39] F.Birren, *The Symbolism of Color* (Citadel Press/Carol Publishing, New York 1989).

[40] J.Liberman, *Light - Medicine of the Future* (Bear & Co., Santa Fe New Mexico 1991) p.68.

[41] E.S.Babbitt, *The Principles of Light and Color* (Babbitt, & Co., New York 1878; abridged reprint University Books, New Hyde Park N.Y. 1967).

[42] D.P.Ghadiali, *Spectro-Chromemetry Encyclopedia* (SpectroChrome Institute, Malaga N.J. 1933).

[43] C.Fere, Sensation et mouvement (F.Alcan, Paris 1900).

[44] F.Stefanescu-Goanga, Experrimentelle Untersuchungen zur Gefuhlsbetonung der Farben, *Psychologische Studien* **7** (1911) 284.

[45] A.Metzger, Über das physiologische Substrat der optisch motorischen Erlebnis-inheit, *Bericht Über die 45 Zusammenkunft der Deutschen Ophthalmologischen Gesellschaft in Heidelberg* (J.F.Bergmann, München 1925) 97.

[46] K.Goldstein,
Über induzierte Veranderungen des Tonus, *Acta Oto-Laryngologica* **7** (1924), fasc.1, 17.

[47] K.Goldstein and O.Rosenthal, Zum Problem der Wirkung der Farben auf den Organismus, *Schweizerisches Archiv für Neurologie und Psychiatrie* **26** (1930) 3.

[48] H.Ehrenwald, Über einen photo-dermatischen Tonusreflex auf Bestrahlung mit farbigen Lichtern beim Menschen, *Klinische Wochenschrift* bf 11 (1932) 2142.

[49] R.M.Gerard, Differential effects of colored lights on psychophysiological functions (Unpublished doctoral dissertation, University of California at Los Angeles, 1958).

[50] B.S.Aaronson, Color perception and affect, *American Journal of Clinical Hypnosis* **14** (1971) 38.

[51] J.J.Plack and J.Schick, The effects of color on human behavior, *Journal of the Association for Study in Perception* **9** (1974) 4.

[52] H.Wolfarth and C.Sam, The effect of color psychodynamic environmental modification upon psychophysiological and behavioral reactions of severely handicapped children, *The International Journal of Biosocial Research* **3** (1) (1982) 10.

[53] E.Newton Harvey, *A History of Luminescence from the Earliest Times Until 1900* (American Philosophical Society, Philadelphia 1957).

[54] R.Dubois, Mécanisme intime de la production de la lumière chez les organismes vivants, *Annales de la Societé Linnéenne (Lyon)* **60** (1913) 81.

[55] A.G.Gurwitsch, Die Natür des spezifischen Erregers der Zellteilung, *Roux's Archiv für Entwicklungsmechanik* **100** (1923) 11.

[56] V.M.Inyushin, B.A.Dombrovsky and G.A.Sergeyev (eds.), *Bioenergetic Questions* (Kazakh State University, Alma-Ata 1969).

[57] V.P.Kaznacheev, S.P.Shurin et al., Distant intercellular interactions in a system of two cultures connected by optical contact (in Russian), in *Ultraweak Luminescence in Biology*, ed. B.N.Tarusov *et al*, (Nauka, Moscow 1972) p.224.

[58] L.Colli and U.Facchini, Light emission by germinating plants, *Nuovo Cimento* **12** (1954) 150.

[59] W.S.Metcalf and T.I.Quickenden, *Nature* **216** (1967) 169.

[60] B.Ruth and F.A.Popp, Experimentelle Untersuchungen zur ultraschwachen Photonenemission biologischer Systeme, *Zeitschrift für Naturforschung* **31c** (1976) 741.

[61] various authors, in *Recent Advances in Biophoton Research and its Applications*, ed. F.A.Popp, K.H.Li, and Q.Gu (World Scientific, Singapore 1992).

[62] E.E.Crane, Bioelectric potentials, their maintenance and function, *Progress in Biophysics and Biophysical Chemistry* **1** (1950) 85.

[63] R.O.Becker, The significance of bioelectric potentials, *Bioelectrochemistry and Bioenergetics* **1** (1974) 187.

[64] A.P.Mathews, Electrical polarity in the hydroids, *American Journal of Physiology* **8** (4) (1903) 294.

[65] I.H.Hyde, Difference in electrical potentials in developing eggs,*American Journal of Physiology* **12** (1904) 241.

[66] R.O.Becker and G.Selden, *The Body Electric* (W.Morrow, New York 1985) p.74.

[67] S.Ingvar, Reaction of cells to the galvanic current in tissue cultures, *Proceedings of the Society for Experimental Biology and Medicine (New York)* **17** (1920), 105. Meeting, 198.

[68] E.J.Lund, *Bioelectric Fields and Growth* (University of Texas Press, Austin 1947). Contains extensive bibliography on "Continuous bioelectric currents and bioelectric fields in animals and plants", by H.F.Rosene.

[69] H.S.Burr, *Blueprint for Immortality - The Electric Patterns of Life* (Neville Spearman, London 1972); American edition: *The Fields of Life* (Ballantine, New York 1973).

[70] R.O.Becker and G.Selden, *The Body Electric* (W.Morrow, New York 1985).

[71] R.O.Becker, *Cross Currents* (J.P.Tarcher, Los Angeles 1990).

[72] C.A.L.Bassett and R.O.Becker, Generation of electric potentials in bone in response to mechanical stress, *Science* **137** (1962) 1063.

[73] R.Croon, *Elektroneural-Diagnostik und Therapie* (Konkordia, BGhl-Baden 1959).

[74] R.Voll, Der heutige Stand der Elektroakupunktur (*Medizinisch Literarische Verlagsgesellschaft*, Uelzen 1962).

[75] Y.Nakatani, Skin electrical resistance and Ryodoraku, *J. Autonomic Nerve* **6** (1956) 52.

[76] X.Zhang et al., The principle of diagnosis and the method of application of the meridian detector, *Journal of the Institute of Traditional Chinese Medicine* **9** (1958) 579.

[77] Zh.Zeng et al. A study on the determination of the normal value of cutaneous conductance and temperature of acupuncture points, *Journal of Traditional Chinese Medicine and Pharmacology of Shanghai* **12** (1958) 33.

[78] V.G.Adamenko, The tobiscope: its use in hypnosis, *Journal of Paraphysics* **6** (1972) 12.

[79] V.G.Adamenko, V.H.Kirlian and S.D.Kirlian, Biometer: Detection of acupuncture points, *Journal of Paraphysics* **6** (1972) 14.

[80] K. Kane and A. Taub, A history of local analgesia, *Pain* **1** (1975) 125.

[81] R.A.Herin, Electroanesthesia: A review on the literature (1819-1965), *Activitas Nervosa Superior (Prague)* **10** (1968) 439.

[82] S.Leduc, Production of sleep and general and local anesthesia by intermittent current of low voltage, *Arch. Elect. med.* **10** (1902) 617.

[83] P.D.Wall and W.H.Sweet, Temporary abolition of pain in man, *Science* **155** (1967) 108.

[84] W.H.Sweet and J.G.Wepsic, Treatment of chronic pain by stimulation of fibres of primary afferent neurons, *Transactions of the American Neurological Association* **93** (1968) 103.

[85] R.Melzack and P.D.Wall, Pain mechanisms: a new theory, *Science* **150** (1965) 971.

[86] F.M.Wageneder, A.Iwanovsky and C.H.Dodge: Electrosleep (Cerebral Electrotherapy) and electroanesthesia - The international effort at evaluation, *Foreign Science Bulletin (Library of Congress)* **5** (4) (1969) 1.

[87] S.H.Rosenthal and N.L.Wulfsohn, Studies of electrosleep with active and simulated treatment, *Current Therapeutic Research* **12** (1970) 126.

[88] *The Principles and Practical Use of Acupuncture Anesthesia* (Medicine and Health Publishing Co., Hong Kong 1974).

[89] B.Sjölund and M.Ericksson, Electroacupuncture and endogenous morphines, *The Lancet* **13** (1976) 1085.

[90] M.Patterson, Electro-acupuncture in alcohol and drug addictions, *Clin.Med.* **81** (1974) 9.

[91] E.Gomez and A.R.Mikhail, Treatment of methadone withdrawal with cerebral electrotherapy (electrosleep), *British Journal of Psychiatry* **134** (1978) 111.

[92] M.Patterson, *Hooked ? NET - The New Approach to Drug Cure* (Faber & Faber, London 1986).

[93] A.Szent-Györgyi, The study of energy-levels in biochemistry, *Nature* **148** (1941) 157.

[94] F.W.Cope, A review of the applications of solid state physics concepts to biological systems, *Journal of Biological Physics* **3** (1975) 1.

[95] H. Fröhlich, see T.M. Wu, this volume.

[96] A.S.Presman, *Electromagnetic Fields and Life* (Plenum Press, New York-London 1970).

Chapter 2

Electromagnetism and Living Systems

F.A. Popp

2.1 Introduction

The physics of living systems is based on (1) electrodynamics, (2) thermodynamics and (3) quantum theory.

Electrodynamics is a basic field for understanding "life" simply because of the nature of interactions in biological systems (b.s.). Nuclear forces (strong and weak interactions within atoms) are excluded because they have too short a range for acting within the biological dimensions of some Angstrom units up to, say, some meters. Gravitational forces, on the other hand, are long-range-interactions, but many orders of magnitude weaker than the electromagnetic forces within the dimensions of cells and organisms. Consequently, only electromagnetic interactions (which include, according to the modern view of electromagnetism, both virtual and real photons) remain the primary candidate for the interactions in biological systems.

Thermodynamics cannot be disregarded for understanding the whole physics of life. Actually, every cell is a many-body system. Although it displays a typical volume of only about 10^{-9} cm^3, the number of molecules interacting there is certainly high enough to follow the laws of statistical physics, that is, thermodynamics. Thermodynamics provides the general macroscopic regulation principles in biological systems.

Finally, quantum theory is a necessary fundamental discipline for the science of life. For instance, it is generally accepted that the essential electromagnetic binding forces of macromolecules follow quantum laws simply because of the small number of bosons in the field under consideration which means that the product of time and energy transferred within this time remains in the order of the Planck's constant - the least quantum of action. Typical examples are the van der Waals-forces,

London-dispersion forces, exchange forces, virtual forces, quasi-particles and photons in biological systems.

Actually, conventional understanding of biological systems is already based largely on the disciplines mentioned. Biochemistry deals with the Coulomb interactions of the valence electrons of organic molecules within the cells, but it takes account, in addition, of thermodynamics by evaluating the rate equations of biochemical reactions as well as by explaining the heat production of biological systems. Tacitly, biochemistry includes also quantum theory, as the discrete electronic spectra cannot be described by classical potentials. The highly selective biochemical reactivity is then certainly a topic of quantum chemistry rather than of electrodynamics or classical thermodynamics. Models for ligand-receptor binding or lock-and-key principle, so common in biochemical reactions, can be regarded as very simplified outlines of the eigenstates of the Schrödinger equation.

The many-body Hamiltonian of the biochemical reaction is based on Coulombic interactions between all the valence electrons (the energy operator of the system) of all the molecules under consideration. The system is very often implicitly regarded as a closed one, subject to a definite temperature which determines the actual occupation of all the possible energy eigenstates of the system's Hamiltonian. We shall see that this approach is rather crucial for understanding life.

In conclusion, there is no doubt at present that a complete physical description of biological systems has to be given at least in terms of electrodynamics, thermodynamics and quantum theory. The question remains open as to the particular emphases which are relevant within the different disciplines and whether the latter are sufficient for understanding biological systems. The situation is summarized in Table 2.1.

It is obvious that if biological systems are made up of a "noble gas" of its compounds, the usual biochemical model would provide the most appropriate description. The doubts about its validity originate from the extraordinary physical properties of biological tissues, e.g.the extremely high polarizability, the high degree of cooperativity (which corresponds to an exceptionally low degree of freedom) and at the same time the rather high flexibility. All these features cannot be understood by simple equilibrium physics. Rather, it becomes more and more evident that by turning to non-equilibrium quantum electrodynamics, one gets as dramatic a change in perception about biological matter as when one turns from high temperature to low temperature physics. For example, superconductivity changes the properties of matter often to such an extent that the best insulators at high temperature turn into the very best superconductors below the critical temperature. Hence, the choice of the physical approach determines crucially the results which can be obtained and at the same time, the applicability, validity and accuracy of the pictures which we can obtain about "life".

The different approaches fall into two general classes according as to whether they

Table 2.1: Usual Description of Living Systems

	Electrodynamics	Thermodynamics	Quantum Theory
Most simple approach	Coulomb-potential of electrons and cores $V(r_{ij})$	System has a definite temperature T	Hamiltonian is a function of $V(r_{ij})$, linear
Necessity:	Electromagnetic forces are the strongest ones in cells	Living systems are exposed to an external heat bath	Biomolecules interact as single units
Questions:	Are Coulomb-forces sufficient?	Is the organism a closed system?	Are classical models sufficient?

involve classical or quantum models (see Table 2.2).

We shall discuss these models which give a good representation of the current status of "Bio-Electromagnetism" - the study of endogenous fields within the biological systems and "Electro-Magneto-Biology" - the effect of exposure of biological systems to external electromagnetic fields.

At first glance, the classical description seems to be sufficient, since macroscopic matter which is used for electrotechnical devices, as, for instance, resistors, capacitors, coils, etc. are usually understandable in terms of classical electrodynamics and can serve as the circuit elements for understanding macroscopic living tissues. There are actually some striking analogies which justifies this point of view. One can get some valuable estimates of the magnitudes of electric currents within the tissues, the resistance, capacitance and inductance of cells and organelles, as well as of cell populations and tissue layers. It is reasonable to develop models of nerve conduction, including the group velocity of nerve impulses, the phenomena of electroencephalogram EEG and magnetoencephalogram MEG, the response of biological tissues to external electromagnetic waves, etc. solely in terms of classical electrodynamics, that is, Maxwell's equations.

Table 2.2(a)

CLASSICAL MODELS

Interactions	Thermodynamics	Physical Consequences
Coulomb's law	closed system	membrane potential
+ Ampere's law	"	"bioplasma"-concept
		Nordenström's model
		R.O. Becker's model
		nerve conduction
+ Faraday's law		bio-electrochemistry
		matter and radiation
		resonant cavities and wave-guides
	heat bath and chemical potential	Fröhlich's postulate
essential: non-linearity	open system	Prigogine's postulate
		Chaos-models

Table 2.2(b)

QUANTUM MODELS

Interactions	Thermodynamics	Physical Consequences
Linear Hamiltonian (Coulomb force) interaction with radiation	closed system	biochemistry photobiology
Nonlinear Hamiltonian		Davydov's postulate
Hamiltonian commutative with annihilation operator	ideal open system	generalized coherence postulate

The very fruitful relaxation theory of Debye and Wagner[1,2] has been deduced from this classical approach, which does not (and cannot, of course) distinguish between the non-living and the living state of matter. Apart from the passive elements like resistors, capacitors and coils, one can certainly introduce active elements like batteries, transistors and so on, in order to understand living matter as an open instead of a closed system. Even more refined models may be based on cavity resonators, waveguides or plasma waves, in order to describe characteristic phenomena like pattern formation, biological rhythms or wave propagation in biological systems.

However, as Fröhlich[3,4] pointed out, it is impossible to explain long-range order in living systems in terms of classical electrodynamics under the constraint of equilibrium thermodynamics. Fröhlich introduced therefore a chemical potential and showed that a Bose condensation-like amplification of definite modes can occur. He took the cell membrane as an example where the supply of metabolic energy can amplify microwaves in a narrow frequency range to such an extent that they become the origin of coherent long-range oscillations with stable phase relations.

This new aspect of living order arises by introducing a chemical potential which effectively identifies the biological system as an open one instead of one at thermal equilibrium. A topologically equivalent approach was developed by Prigogine[5] under the more general consideration of the non-linearity of interactions as well as the special thermodynamic boundary condition of a phase transition far away from equlibrium. His so-called "dissipative structures" describe, as in Fröhlich's model, supermolecular, coherent spatio-temporal pattern-formation as transitions between disordered and ordered regimes. It is important to note that the nature of the interaction plays no role for understanding the essential features. It is only necessary that the interactions are nonlinear. Consequently, Thom's catastrophe theory[6], limit cycle models[7] or the most fashionable chaos theory[8] which are essentially based on non-linearity of differential equations or iteration equations are logical progressions of Fröhlich's and Prigogine's approach. None of these models require a quantum description.

However, a complete theory of the living state has to be based also on quantum theory. Biological systems react rather specifically to molecular compounds, sometimes to single molecules. Such high specifity and sensitivity already require a quantum description, as molecular structure cannot be understood in terms of classical physics alone. However, while ordinary biochemistry is based only on Coulombic interaction of the valence electrons, photobiology has, in addition, to take into account also the interaction of biological systems with single photons. Consequently, a complete description includes non-classical light, which is the subject of a rather new discipline of quantum optics[9]. Whether the Hamiltonian has to be bound or unbound[10], time-independent or even time-dependent for a full understanding of biological phenomena, remain at present secondary problems compared to the question of whether linear or non-linear Hamiltonians should apply. The approaches of biochemistry and photochemistry for example are based on linear Hamiltonians, whereas the soliton concept depends on non-linear Hamiltonians. As Davydov and others have shown[11,12], the non-linearity of the Hamiltonian may induce "self-focussing" in matter in much the same way as the "classical" concept of non-linearity does. The difference between Fröhlich's and Prigogine's concept and Davydov's solitons lies mainly in the fact that "dissipative structures" need sufficient distance from equilibrium conditions in order to stabilize, while solitons do not interact with the thermal heat bath considerably such that even at equilibrium a rather high stability may arise.

It appears reasonable therefore, to propose a more general concept which includes both the stability at equilibrium as well as at non-equilibrium conditions, essentially based on the formation of coherent states in the most general sense, e.g. eigenstates of the annihilation operator. In this generalized coherence model, organisms are systems which satisfy all the laws of physics as well as the optimization condition of the highest possible signal to noise ratio under the permanent exposure to an external heat bath[13].

2.2 Classical Models

2.2.1 Maxwell's Equations

Let us start with the observation that resting charges (i.e. electrons) either repell each other or, in case that they have opposite charges, like electrons and protons, are mutually attractive. The force decreases in inversely proportion to the distance between the charges under consideration. By introducing the force per unit of charge, called *electric field strength* $\vec{E}(\vec{r},t)$, which depends on space point $\vec{r}$ and time t, *Coulomb's law* is most elegantly written in the differential form:

$$\vec{\nabla} \cdot \vec{E} = 4\pi\rho$$

where $\vec{\nabla}$ is a vector operator with the components $\left(\vec{i}\frac{\partial}{\partial x}, \vec{j}\frac{\partial}{\partial y}, \vec{k}\frac{\partial}{\partial z}\right)$ and ρ is the charge density (amount of charge per unit of volume at $\vec{r}$ and t). This law holds for the vacuum. In matter, however, the force $\vec{E}$ gives rise to a polarization $\vec{P}$ which, as a force oppositely directed to $\vec{E}$, originates from an opposite charge density ρ_{ind} that is induced by the external electric field. This polarization of matter is described by

$$\vec{\nabla} \cdot \vec{E} = 4\pi\rho - 4\pi\vec{\nabla} \cdot \vec{P}$$

where $\vec{\nabla} \cdot \vec{P} = \rho_{ind}$.

Note that the factor 4π reflects a convention and not a real physical effect. After introducing the mathematical (and not actually measurable) quantity $\vec{D}$, called "displacement":

$$\vec{D} = \vec{E} + 4\pi\vec{P}$$

we arrive at the first Maxwell's equation:

$$\vec{\nabla} \cdot \vec{D} = 4\pi\rho$$

As the polarization in a passive medium can, to first approximation, only linearly increase with increasing $\vec{E}$, we get the macroscopic "matter equation"

$$\vec{D} = \varepsilon\vec{E} \ ,$$

where ε as the *dielectric constant* ($\varepsilon \geq 1$) is introduced empirically. Note that the determination of ε is not a question of classical physics. It can be calculated only by quantum theory. This is a further argument for quantum description as a necessary basis of understanding biological systems, in particular the extraordinary high values of ε at low frequencies for an alternating E-field. For vacuum we have by definition $\vec{P} = 0$ and therefore $\varepsilon = 1$.

Coulomb's law describes the force $\vec{K}_e(\vec{r})$, acting on a resting charge q as soon as $\vec{E}(\vec{r})$ is known. By definition of $\vec{E}(\vec{r})$, we have consequently

$$\vec{K}_e = q\vec{E}$$

As soon as the charges can move, the electric force $\vec{K}_e(\vec{r})$ will shift them away from their resting position. While ε describes the capacity of $\vec{E}(\vec{r})$ to separate positive and negative charges in the medium under consideration, a further matter equation defines the mobility of the charges under the influence of an external electric field. The quantity σ which is introduced is called the *conductivity* of matter and defined by the Ohm's law according to

$$\vec{j} = \sigma\vec{E}$$

where $\vec{j}$ is the current density and σ, as a specific property of matter, is taken as a constant in the linear approach to the problem. Once again, σ can be explained only in terms of quantum theory.

Besides the electric force $\vec{K}_e(\vec{r})$ there arises a magnetic force always (and only) in case that a charged particle moves. This additional force per unit of charge is described analogous to the electric field strength $\vec{E}(\vec{r}, t)$ by a *magnetic induction* $\vec{B}(\vec{r}, t)$ according to

$$\vec{K}_m = q(\vec{v} \times \vec{B}) \ ,$$

where $\vec{v}$ is the velocity of the charge q and $\vec{B}(\vec{r}, t)$ the magnetic induction at the position of this charge at time t. The description in terms of vector analysis is necessary since the magnetic force $\vec{K}_m$ always acts perpendicular to the direction of the velocity $\vec{v}$ of the charge as well as perpendicular to the magnetic induction $\vec{B}(\vec{r}, t)$. This is the reason why static magnetic fields can never transfer energy to the charges on which they act. The induction law for magnetic fields which are induced by moving charges (currents i) has been found by Ampere:

$$\vec{\nabla} \times \vec{B} = \frac{4\pi}{c}\vec{j}$$

where $\vec{j}$ is the current density (current per unit area) at $\vec{r}, t$ and c the velocity of light. This *Ampere's law* is for the creation of magnetic fields by moving charges what Coulomb's law is for the creation of electric fields by resting charges.

Like electric polarization in matter, magnetic forces induce changes of the magnetic-moment densities (*magnetization M*) according to

$$\vec{j}_m = c(\vec{\nabla} \times \vec{M})$$

where $\vec{j}_m$ is the additional current density corresponding to the magnetization $\vec{M}$ of the matter ($\vec{M} = 0$ for vacuum). As the total effective current in matter is then $\vec{j}_{eff} = \vec{j} + \vec{j}_m$, we arrive at the general *Ampere's law*

$$\vec{\nabla} \cdot \vec{B} = \frac{4\pi}{c}\vec{j} + 4\pi\vec{\nabla} \times \vec{M}$$

In analogy to the displacement $\vec{D}(\vec{r})$ in case of an electric field, a vector field $\vec{H}(\vec{r})$ is introduced which is called magnetic field and defined by

$$\vec{H} = \vec{B} - 4\pi\vec{M}$$

Note that only $\vec{E}$ and $\vec{B}$ are the real physically measurable quantities, while $\vec{D}$ and $\vec{H}$ are purely mathematical terms which serve for simplifying the formalism. By insertion of $\vec{H}$ we then write Ampere's law in the form:

$$\vec{\nabla} \times \vec{H} = \frac{4\pi}{c}\vec{j}$$

Since $\vec{M}$ is proportional to $\vec{B}$ in first order expansion, the analogous equation to $\vec{D} = \varepsilon\vec{E}$ becomes,

$$\vec{H} = \frac{1}{\mu}\vec{B}$$

where μ is a constant characteristic of the material under study. It is called *permeability*. For vacuum ($\vec{M} = 0$) we have by definition $\mu = 1$, while in case of matter only quantum theory provides an accurate calculation of μ. *Diamagnetic* substances are by definition those without remnant magnetization, where $\mu < 1$ describes just the effective weakening of the external field by its magnetization. For *paramagnetic* substances which display a remnant magnetization (by non-compensated rotating microscopic currents originating from orbital angular momenta and/or spins of charges) we have $\mu > 1$, indicating an ordering of the remnant magnetization to the direction of the external field $\vec{B}(\vec{r})$. For many other substances including ferromagnetic ones, μ becomes a complicated function of $\vec{B}$, losing its character as a constant.

The total force on a charge q is then

$$\vec{K} = \vec{K}_{el} + \vec{K}_{magn} = q\vec{E} + q(\vec{v} \times \vec{B})$$

which is called Lorentz force. As in every coordinate system moving with constant velocity $\vec{v}$ the total force on a charge must remain constant when $\vec{v}$ itself may change, there cannot be a principle difference between electric and magnetic forces. Rather, they have to be transformable into each other. While in a moving coordinate system the transformation reflects the different relative velocities of the reference systems, in a fixed coordinate system the change of $\vec{v}$ describes the time dependence of $\vec{E}$ and $\vec{B}$, which "induce" each other mutually by means of the transformation law between $\vec{E}$ and $\vec{B}$. This physical necessity originating from the equality of all coordinate systems moving with constant velocities relatively to each other is expressed by *Faraday's law*:

$$\vec{\nabla} \times \vec{E} + \frac{1}{c}\frac{\partial \vec{B}}{\partial t} = 0$$

which is at the same time the physical origin of electromagnetic induction, e.g. the stability of electromagnetic waves and wave propagation in vacuum.

According to Coulomb's law, the sources of electric fields are charges representing electric monopoles. It seems reasonable therefore to look for corresponding magnetic monopoles as the sources of $\vec{B}$-fields. However, no one has ever discovered any hint of magnetic monopoles. This leads to the strange but rather remarkable asymmetry of electric and magnetic fields in our universe. The lack of sources of $\vec{B}$-fields is expressed in the differential form

$$\vec{\nabla} \cdot \vec{B} = 0$$

Now all the fundamental equations of the classical electrodynamics are complete. They are the famous *Maxwell's equations*, describing all the classical phenomena of electromagnetism. Table 2.3 summarizes these relevant equations and displays, in addition, an outline of the dimensions and units which are generally used at present.

2.2.2 Classification of Bio-Electromagnetism and Electromagnetobiology

The variety of parameters which may be varied in an electromagnetic field is so great that only a systematic classification allows us to survey all the possible influences.

Electromagnetic fields may

(a) originate in living systems and function there (bio - electromagnetism),

(b) be produced in the surroundings of biological systems, but have influences on them (electro-magneto-biology),

(c) connect isolated biological systems irrespective of whether they are produced within the systems or not (electromagnetic bio-information).

The fields may consist of electric, magnetic or both electric and magnetic components. The field amplitudes may vary over many orders of magnitude without losing their biological relevance. They may be created in living tissues or be influencing them. They include electric fields from 10^{-6} V/m up to at least 10^{8} V/m, and magnetic induction between nano-Teslas up to some thousands of Teslas locally. The frequency ranges from ultralow (ULF, 0-30 Hz) including direct current (DC) to extremely high frequencies (ELF, 30 to 300 GHz) up to the optical range, and beyond: UV-radiation, X- and gamma-rays. Biological functions are affected over the whole spectrum. A rough classification of the frequencies is given in Table 2.4.

Table 2.3(a): Maxwell's Equations

$$\begin{aligned} \text{Lorentz force} &: \vec{K} = \vec{K}_{el} + \vec{K}_{magn} = q\vec{E} + q(\vec{V} \times \vec{B}) \\ \text{Matter equations} &: \vec{D} = \varepsilon\vec{E}\,;\ \vec{B} = \mu\vec{H} \\ \text{Ohm's law} &: \vec{j} = \sigma\vec{E} \end{aligned}$$

Maxwell's Equations:

$$\begin{aligned} \text{Coulomb's law} &: \vec{\nabla} \cdot \vec{D} = 4\pi\rho \\ \text{Ampere's law} &: \vec{\nabla} \times \vec{H} = \frac{4\pi}{c}\vec{j} \\ \text{Faraday's law} &: \vec{\nabla} \times \vec{E} + \frac{1}{c}\frac{\partial B}{\partial t} = 0 \end{aligned}$$

Absence of magnetic sources: $\vec{\nabla} \cdot \vec{B} = 0$

Very often, it is convenient to introduce for electric as well as for magnetic fields a potential, from which $\vec{E}$ and $\vec{B}$ can be derived.

They are defined according to

$$\begin{aligned} \vec{E} &= -\vec{\nabla}\varphi \\ \vec{B} &= \vec{\nabla} \times \vec{A} \end{aligned}$$

where φ is a scalar potential, $\vec{A}$ a vector potential.

Table 2.3(b): Units of Measurements (Standard SI)

Length :	l	meter (m)
Mass :	m	kilogram (kg)
Time :	t	second (s)
Frequency :	f	Hertz (Hz) 1 cycle per second

Abbreviations:

1 Coulomb (C)	=	1 As
1 Weber (Wb)	=	1 Vs
1 Tesla (T)	=	1 Vs/m^2
1 Gauss (Gs)	=	10^{-4} Tesla
1 Watt (W)	=	1 VA
1 Ohm (Omega)	=	1 V/A
1 Siemens (S)	=	1 A/V
1 Farad (F)	=	1 As/V
1 Henry (H)	=	1 Vs/A

Power P Joule (J) = 1 kg m^2/s^2
Power P Watt (W) 1 Ws = 1 J

Electric potential: U Volt (V)
Electric current: i Ampere (A)

Electric field strength: E(V/m)
Magnetic induction (magnetic flux density): B (Tesla,T)

Displacement: D (As/m^2)
Magnetic field strength: H (A-turn/m)

Conductivity: σ Siemens/m (S/m)
Dielectric constant: ε (dielectric permittivity) dimensionless
Electric field constant: $\varepsilon_0(F/m)(\varepsilon_0\mu_0 = c^2)$
Magnetic permeability: μ (magnetic induction/magnetic field strength) (H/m)
Magnetic field constant: $\mu_0 = 4\pi 10^{-7}$ Wb/(Am)
Relative permeability: dimensionless

Table 2.4: Frequency Ranges in Electromagnetism

Band designation	Typical uses	Frequency	Wavelength
Extremely High Frequency (EHF)	Satellite communications, radar, radar microwave relay, radio navigation, amateur radio	30-300 GHz	1-10 mm
Super High Frequency (SHF)	Satellite communications, radar, amateur, taxi, police, fire, airborne, weather radar, ISM	3-30 GHz	1-10 cm
Ultra High Frequency (UHF)	Microwave point to point, amateur, taxi, police, fire, radar, citizen's band, radio-navigation, UHF-TV, micro-wave ovens, medical diathermy, ISM	0.3-3 GHz	10-100 cm
Very High Frequency (VHF)	Police, fire, amateur, FM, VHF TV, industrial RF equipment, diathermy, emergency medical radio, air traffic control	30-300 MHz	1-10 m
High Frequency (HF)	Citizen band, amateur, medical, diathermy, Voice of America, broadcast, international communications, industrial equipment	3-30 MHz	10-100 m

(according to J.T. Zimmerman and V.J. Rogers: in: 19a)

Table 2.4

Band designation	Typical uses	Frequency	Wavelength
Medium Frequency (MF)	Communications, radio-navigation, marine radio-phone, amateur, industrial RF equipment, AM broadcast	0.3-3 MHz	100-10,000 m
Low Frequency (LF)	Radio navigation, marine communications, long-range communications	30-300 kHz	1-10 km
Very low Frequency (VLF)	Very long range communications, audio frequencies, navigation	3-30 kHz	10-100 km
Voice Frequency (VF)	Voice, audio frequencies	0.3-3 kHz	100-10,000 km
Extremely low Frequency (ELF)	Power lines, audio frequencies, submarine communications	30-300 Hz	1000-10,000 km
Ultra Low Frequencies (ULF)	Direct-current (DC), 50 or 60 Hz power, usual range of brainwave frequencies	0-30 Hz	10,000 km to infinity

Although it is useful to discuss the different effects for different frequencies separately, it would be a mistake to believe that the frequency domains are working independently. Rather, over the whole frequency range there is a strong mode coupling[13], which is responsible for the synergetic effects of different frequencies. Living systems may be playing an unimaginably huge concert of all the modes, creating a completely new category of phenomena outside classical electrodynamics.

The important parameters for the biological effects of electromagnetic fields include: (1) the location of the biological system with respect to the field, (2) whether the electric or magnetic component is acting, (3) the field amplitudes, (4) the frequency spectrum and timing, (5) field gradients and (6) directions of polarization. A small change in any one of these parameters may dramatically alter the biological effects. This extremely broad range of possibilities make electromagnetic fields a most informative source for regulation within and between living systems, and at the same time, increases the difficulties in examining electromagnetic interactions systematically and with sufficiently high accuracy and reproducibility.

As far as classical electrodynamics is concerned, one expects that the same laws which are valid for dead materials should hold also for living tissues. Nevertheless, we should quard against the uncritical application of the usual thermodynamics, electrodynamics and quantum theory to explain the properties of living systems. At least two factors should make us cautious: the strong inhomogeneity of biological matter giving rise to complex gradients of electromagnetic fields, and the very important fact that organisms are never at thermal equilibrium conditions, not even locally in biologically relevant time-intervals. By contrast, they are permanently excited and regulated by electromagnetic waves from ULF even up to the optical range inside the body, and furthermore, they are permanently subject to the electromagnetic ocean in the environment, to which they react sensitively. This leads to significant fluctuations of the electromagnetic properties of the tissues. As an example, excitation in the optical range by exposure to light or by the remnant "biophoton field"[14] may dramatically change the permittivity, conductivity and even the permeability of the tissue under consideration.

2.2.3 Basic Classical Understanding of Electric Properties of Living Tissues and Heat Loss

Most fundamental work has been done in order to understand the electric properties of tissues, e.g. conductivity and permittivity, which can be investigated, as soon as instruments for measuring of electrical resistance and capacitance became available. Systematic work in measuring the frequency dependence of $\sigma(\nu)$ and $\varepsilon(\nu)$ performed during the last hundred years, forms a solid basis for understanding the electric properties of living tissues, which are useful for further investigations.

The results can be classically described in terms of a general relaxation theory[1,2] which, together with the most simple damped-oscillator models of Drude and Lorentz[15] provide a basic description of conductivity and permittivity of living tissues which, at times, agree surprisingly well with observations, while at other times, show rather dubious correspondence to experimental results. This dialectic between theoretical description and experimental observations enables us to get an increasingly profound understanding of what happens when electromagnetic fields interact with living tissues.

Relaxation theory is based on the characteristic time period of separation between opposite charges by external electric fields, following Coulomb's law, and giving rise to damped oscillations as soon as the charges are elastically bound to their resting positions. Depending on the mass, the strength, and the nature of the reaction forces, the partially bound or free charges need different relaxation times to return to their equilibrium positions, and thus contribute in accordance with this relaxation behaviour to the conductivity and permittivity of the matter under investigation. This is approximated in the following general relations:

$$\sigma(t) = \sigma_\infty + (\sigma(\nu) - \sigma_\infty)\left(1 - \exp\left(-\frac{t}{\tau}\right)\right)$$

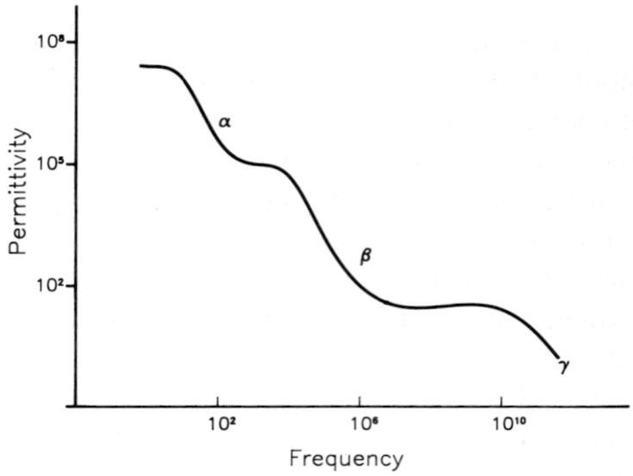

Figure 2.1: *The permittivity (and conductivity) of tissues explained in terms of the general relaxation theory of Debye and Maxwell-Wagner. It explains at least the three major dispersion regions (alpha, beta und gamma) found in tissues.*

$$\varepsilon(t) = \varepsilon_\infty + (\varepsilon(\nu) - \varepsilon_\infty)\left(1 - \exp\left(-\frac{t}{\tau}\right)\right),$$

where $\sigma(t)$ and $\varepsilon(t)$ are the conductivity and the permittivity at time t, respectively, σ_∞ and ε_∞ describe the residual values at $\nu \to \infty$, $\sigma(\nu)$ and $\varepsilon(\nu)$, the final values after applying an electric field of frequency ν, and τ is the characteristic relaxation time. Since σ and ε are connected by the same relaxation time, they are not independent of each other but linked together by the famous Kramers-Kronig relation. The relation applies to dielectric materials with an arbitrary distribution of relaxation times, but with linear response and time-independent properties of the system under study. However, it is just the deviations from the Kramers-Kronig relations which may provide a useful tool for examining the important range where these relations become invalid.

Nevertheless, the dispersion equations (p. 16) can explain in general terms the electric properties of living tissues, that is

(1) the increase of $\sigma(\nu)$ and the decrease of $\varepsilon(\nu)$ with increasing frequency ν of the applied electric field,

(2) the $\alpha-, \beta-, \gamma-$ dispersions at about 10-100 Hz, 10-100 MHz and 10-100 GHz, respectively (Fig. 2.1).

As charges which form currents cannot at the same time contribute to local dielectric concentration, the opposite dependence of $\sigma(\nu)$ and $\varepsilon(\nu)$ is understandable even without tracing it back to the Kramers-Kronig relation. For $\sigma(\infty) > \sigma(\nu)$ and $\varepsilon(\infty) < \varepsilon(\nu)$, σ increases and ε decreases with increasing frequency ν. Because of the inertia of charged particles, however, $\sigma(\nu)$ and $\varepsilon(\nu)$, becomes less and less able to follow the external field as its frequency increases. Rather, an increasing phase shift between field amplitude and the oscillation of the charged particles will take place. As a consequence, the conduction loss (σE^2) turns more and more into dielectric loss, where the relaxation of the polarized charges heats up the surrounding medium. As an example, the loss due to the relaxation of water molecules at a frequency of a GHz is about 50% of the total loss. At a frequency of 10 GHz it increases at 90% and at 30 GHz it becomes 98% of the total loss[16].

The regions of dispersion, where σ and ε display the strongest increase and decrease, respectively, with increasing frequency, are at present, attributed to the following relaxation processes: α-dispersion at about 10-100 Hz is probably due to ionic diffusion processes in micron- and larger sized objects, where the tissue can be considered as an electrolyte. In view of the high (and remarkably constant and stable) surface capacity of cell membranes, the currents in the LF-range can flow only through the extracellular space. With increasing frequency, however, the capacitance of the cell contributes more and more to the conductivity. Consequently, at a sufficiently high frequency, the cell membranes become short-circuited and the relaxation of capacitive charging of proteins becomes the determining factor of the dispersion region. This range has been estimated to begin at some MHz, forming the β-dispersion region. Finally, the γ-dispersion at about 25 GHz can be explained in terms of the dipolar relaxation of water and its interactions with biomolecules.

The penetration of e.m. fields into the body is then considered at present, as follows:

(1) For static fields and at low frequencies, there is no thermal absorption. While magnetic fields penetrate the tissue practically without any hindrance, electric fields are strongly attenuated due to the jumps in the σ- and ε -values between air and skin. The boundary conditions of Coulomb's law require a corresponding gradient in the electric field components between outside and inside such that only a very small field amplitude can penetrate in practice perpendicular to the surface of the body.

(2) With increasing frequency, eddy currents becomes more and more significant, which means that time varying magnetic field- components penetrating into the tissues induce electric fields of the same frequency in view of Faraday's law. Consequently, electric currents of about this frequency are induced as well which becomes subject to conductivity loss.

(3) However, also on account of Faraday's law and, in addition, the boundary conditions, the penetration depth of electromagnetic fields is inversely proportional to $\sqrt{\nu\sigma\mu}$. External electromagnetic waves thus become more and more excluded by the skin with increasing frequency ν. At about 10 MHz the penetration depth into the body is only some cm, and it reduces to about 1 mm at a frequency of 10 GHz.

(4) A further effect which inhibits the penetration of electromagnetic waves is the reflectance at the interface of materials with different σ, ε and μ. The wave impedance of a medium is the ratio of the E- to the B-field in a plane wave travelling through the medium:

$$\gamma = \sqrt{\frac{j\nu\mu}{\sigma + j\nu\varepsilon}}$$

For perpendicular incidence, the reflection coefficient is then:

$$\Gamma = \frac{E_r}{E_i} = \frac{\gamma_2 - \gamma_1}{\gamma_2 + \gamma_1} \ ,$$

where the indices 1 and 2 refer to the different media.

(5) From frequencies of about some MHz onwards, electric field amplitudes and magnetic ones, which are higher than about 10 T, cannot be separated seriously in view of the Faraday's law. Rather, the magnetic component induces a sufficiently high electric field amplitude which converts energy into heat by conduction loss. Heat production increases with the square of the frequency and field amplitude as well as with increasing resistance of the medium.

So far, we have discussed only the "classical effects" - relaxation, dispersion, boundary conditions at low and high frequencies, heat loss due to conductivity loss at low frequencies and eddy currents for magnetic fields as well as the significance of Faraday's law (for the skin effect and the conversion of magnetic energy into heat) and heat loss due to dielectric relaxation at very high frequencies. They explain approximately the observed results when living tissues are exposed to external electric and magnetic fields of different frequencies. Table 2.5 summarizes these effects.

2.2.4 Nonthermal Effects

Of course, there is no limitation to the classical effects which can occur in biological tissues, as electric, magnetic and electromagnetic fields over wide ranges of amplitudes and frequencies are always present there, including the availability of charges, currents, electric and magnetic dipoles and multipoles. However, whether and to what extent most of the possible effects actually play an important biological role is not clear at present. Even the question, whether any one of the possible effects following Maxwell's equations will certainly not become significant in biological systems cannot be answered with certainty.

Table 2.5: Most important Effects: Classical Bio-Electromagnetism and Electro-Magneto-Biology

Frequency Range	Field	Mechanism	Formalism	Biological Consequence
All frequencies	$\vec{E}$	conduction loss	$\vec{j} = \sigma \vec{E}$	Loss by free charges
		dielectric loss	$\vec{D} = \varepsilon \vec{E}$	Loss by bound charges
	$\vec{B}$	eddy currents	$\left\{ \begin{array}{c} \vec{\nabla} \times \vec{E} = -\frac{1}{c}\frac{\partial \vec{B}}{\partial t} \\ \vec{j} = \sigma \vec{E} \end{array} \right\}$	Loss by penetrating alternating magnetic fields in conducting zones
Static Fields	$\vec{E}$	Ionic current (extracellular medium)	$\vec{j} = \sigma \vec{E}$	Polarization of biological tissue Cells act as insulators
		Boundary conditions forbid pentration of field components parallel to the surface		Protection against DC- and LF- electric fields
	$\vec{B}$	Field penetrates practically without any absorbance		Organisms may make use of external fields (like magnetostatic bacteria) Organisms may react sensitively to weak external magnetic fields

Table 2.5

Frequency Range	Field	Mechanism	Formalism	Biological Consequence
Low Frequency Range ($\simeq$ 10 Hz)	$\vec{E}$	α-dispersion ε decreases, σ increases with frequency Cell membranes become integrated in the processes (impedance decreases, walls become charged).	Debye-equations, Oscillator - models	Frequency-region, where a strong coupling occurs between the whole organism and the cells
	$\vec{B}$	Loss possible by eddy currents. Effect increases with increasing frequency		Loss is mainly due to conductivity loss
Radio-Frequency Range $\simeq$ 10 -100 MHz	$\vec{E}$	β-dispersion ε decreases, σ increases with frequency Same reasons as for α-dispersion, continuing and reaches saturation Cell membranes relaxation processes Cavity models of substructure of cell	Maxwell-Wagner equations	Electromagnetic "Compartimentation" of cells and substructures of cells by relaxation processes over the whole range from LF to RF and higher Loss changes from conductivity to dielectric loss

Table 2.5

Frequency Range ν	Field	Mechanism	Formalism	Biological Consequence
10 kHz to β-dispersion region ($\geq$ 100 MHz)	$\vec{E}$ and $\vec{B}$	Wavecharacter of the Fields Faraday's law	$Q \propto \nu^2 E^2$ $\underbrace{\propto \nu^2 B^2}$ heat production in the body	Heat production due to conductivity loss, time-varying $\vec{B}$ induces $\vec{E}$-field, which transfers energy
$>$ 10 MHz	$\vec{E}$ and $\vec{B}$	Skin-effect Penetration dephts decreases $\propto \frac{1}{\sqrt{\nu\sigma\mu}}$	Skin-Effect arises by including Faraday's law and boundary conditions	Protection by skin at high frequencies
10 - 10 kHz to β-dispersion region and higher	$\vec{E}$ and $\vec{B}$	Reflection of Waves from the surface Dielectric losses become more and more important compared to conduction losses (at 1 GHz conduction and dielectric loss are about equal)	Maxwell's equation and boundary conditions Maxwell-Wagner equations	Complicated, resonance-like interactions of electromagnetic waves in biological tissues

Table 2.5

Frequency Range ν	Field	Mechanism	Formalism	Biological Consequences
100 MHz to 30 GHz		Increasing amount of incident energy absorbed in skin. Skin-Effect	Debye-dispersion equations	Dielectric loss of water becomes more and more significant
$\simeq$ 30 GHz		γ-dispersion	Debye and Maxwell-Wagner equations	Biomolecules and their polarization and water are contributing

Actually, there are a lot of electric, magnetic or electromagnetic interactions which are likely candidates for explaining as yet unexplained biological phenomena, e.g. orientation, movement and pattern formation of biomolecules within the cells or in the extracellular space, energy transfer to biomolecules as, for instance, rotational energy for DNA replication and - transcription, spatio-temporal organization of biomolecules and their reactions, and the huge variety of possible resonance effects which may be responsible for regulation and control over the whole electromagnetic spectrum, from ULF involved in biological rhytms and action potentials up to the resonance-like phenomena of photobiology in the optical range. There is now a diversity of literature on possible nonthermal electromagnetic effects[17–20]. Let us discuss some typical examples, i.e. orientation, rotation, and resonance absorption.

In order to understand orientational effects in cells, one may take account, for instance, of

- the rather high electric or magnetic, permanent or induced dipoles or multipoles and their field orientation due to the minimization of the potential energy,
- Lorentz forces due to moving charges (e.g. in the blood stream) in a magnetic field or resting charges which are exposed to electromagnetic or magnetic wave propagation, i.e. in a cavity resonator or in a wave guide,
- coupled interactions of electric fields and photons which excite molecules electronically so that they get high enough polarization for autocatalytic chain reactions and chain formation within the electric or electromagnetic field. Biopolymerization is one candidate of such a mechanism *in vivo.* The orientation as well as the high reactivity rate are explicable even quantitatively using such an approach.

A further example is the unsolved problem of where the angular momentum of DNA comes from as this macromolecule rotates with a frequency of about 100 Hz when it is replicated or when transcription takes place.

However, by taking account of the angular momentum density

$$l = \frac{\omega}{V} \int r^2 dM \approx 10^{-12} g cm^{-1} s^{-1}$$

of the DNA, where ω is the rotation frequency, V the volume element of DNA, and r the radius over which the mass-distribution dM is integrated, and provided that this angular momentum originates from the Poynting vector of an electromagnetic field within the cell such that

$$l = \frac{1}{4\pi c} |\vec{r} \times \vec{E} \times \vec{H}| \ ,$$

one obtains rather reasonable values of $\vec{E}$ and $\vec{B}$, e.g. $|E| = 10^6$ V/m and $|B| \simeq 0.3$ T. The transferred energy per base pair is then about 3×10^{-4} eV, the Poynting vector provides a radiation intensity of about 10^{14} eV/s, and the dispersion relation of the DNA lattice system with the rotation frequency of about 100 Hz leads to an effective mass of quasi-particles of the order of 10^{-28} g, moving with a group velocity of about 10^8 cm/s. These data agree surprisingly well with spin waves (bogolons, magnons or even Cooper pairs). The broadening velocity of these wave packets perpendicular to their movement along the helix axis is about 10^4 cm/s much lower than the group velocity[21].

A further example of excited states of DNA, or exciplexes with quasistable equilibrium positions between the strands of around 3.4$\AA$ has been discussed in an other paper[22].

There is no experimental evidence for the correctness of these models. However, these examples demonstrate that nonthermal electromagnetic interactions may explain a variety of as yet unsolved problems of molecular biology, including, for instance, biopolymerization, orientation and movement of macromolecules, as well as the mitotic figures in terms of cavity resonator waves in a cell[23,24].

The whole story of bio-electromagnetism is then the consideration of resonance-like effects like those of the simple Oscillator-(Drude-model[15] p. 16), but extended to the manifold of possible couplings, not only under equilibrium conditions, but in particular by taking account of non-equilibrium in biological systems. A brief survey of those effects is displayed in Table 2.6.

Tables 2.6: Systematic Classification of Electromagnetism in Biology

Frequency-Range	Location	E-Field	B-Field	Typical Biological Phenomena
ULF-ELF 0 to about kHz	Internal	$> 10^8 V/m$	-	Extraordinary high charge density of proteins (about 1 free charge per 100 $\dot{A}^2$ surface area of monomolecular layers).
		$\geq 10^6 V/m$	-	Membrane-potential Electric properties of membranes at equilibrium
		$\geq 10^6 V/m$		Active transport
		$\simeq 10^6 V/m$ $10^{-6} V/m$		Nerve impulse action potentials (EKG, EEG, ERG, CNS, ANS) electric brain activities muscle activities Electro-Navigation Electroreception

Maxwell's Equation	Further Equations	Explanation Content	Open Questions	Models
Coulomb's equ. Matter equation $\vec{D} = \varepsilon\vec{E}$		formation of electrical double-layers tendency of compartimen-tation	Micro-Structures (e.g. Biopolymers) cannot be ex-plained in this frame. They are subject of Quant-tum Chemistry.	Macroscopic Double-Layers Membrane-Models
Coulomb's equ. Matter equ: $\vec{D} = \varepsilon\vec{E}$	Thermo-dymamics: Nernst-Planck equ. Diffusion (Fick's laws)	Equilibrium conditions Order of amplitude of the membrane potential	Stabilization of potential Source(s) and nonlinear pro-perties, interlinkage of all membrane-transport phenomena	electro-chemistry
Coulomb's equ. Matter equ: $\vec{D} = \varepsilon\vec{E}$	Introduction of a chemical potential (ATP hydrolysis)	Chemically driven battery	Source(s) triggering the reactions, in vivo	electrochemistry and alternatively: - electrodiffusion - irreversible thermo-dynamics - external pump
$\vec{j} = \sigma\vec{E}$	Wave equation: mechanical theory of elastic media Non-equilibrium conditions	Possibility of Oscillations	Source(s) and interactions of stimulations, distribution and complete biological significance of the electric waves	Cable Theory Electro-Diffusion

Maxwell's equation	Further equations	Explanation content	Open Questions	Models
$\vec{\nabla} \cdot \vec{B} = 0$ $\vec{\nabla} \times \vec{H} = \frac{4\pi}{c}\vec{j}$ Matter equation $\vec{B} = \mu\vec{H}$		Magneto-statics Current as the sources of magnetic fields Diamagnetism ($\mu < 1$) Paramagnetism ($\mu > 1$) Ferromagnetism ($\mu >> 1$, $\mu = \mu(\vec{H})$).	Theory of Matter Equation which is subject of Quantum Theory	Human Body as penetrated by external magnetic fields. Biomagnetism (MCG, MEG, interaction of dia-, para- or ferromagnetic substances in the body with magnetic field). Magneto-Biology

Frequency-Range	Location	E-Field	B-Field	Typical Biological Phenomena
ULF-ELF 0 to about 1 kHz	Internal	-	$\geq 10^3$ Tesla (locally) to 10^{-16} Tesla	Magnetotaxic behavior of certain bacteria, Magneto-navigation MCG, MEG
ULF-ELF	External	up to about 10^5 V/m	-	Resultant internal field is by several orders of magnitude smaller than the external applied field Boundary conditions require perpendicular penetration of E-Field at the body surface
		-	up to 10^4 Tesla	Magnetic field penetrates the body practically without change of B-Field ($\mu \approx 1$). Changes of biochemical reactivity, mutations disorders, change of reproduction and growth locomotion, embryogenesis, aging, behavior, biological clocks CNS, ANS, MCG, MEG, navigation

Maxwell's Equation	Further Equations	Explanation Content	Open Questions	Models
Coulomb's equation Matter equation: $\vec{D}(w) = \varepsilon(w)\vec{E}(w)$ $\vec{j}(w) = \sigma(w)\vec{E}(w)$		Penetration of electric fields into tissue, distribution over tissues	Do resonance-like effects appear?	Relaxation-time models: Debye equations Maxwell-Wagner equations etc.
Faraday's law (Wave equation) + Boundary Conditions Matter Equation		Penetration and distribution of electromagnetic waves	How Important are Standing Waves and similar resonance-like wave phenomena?	Wave-guides, Cavity-modes, Plasma-Concepts etc.

Frequency Range	Location	E-Field	B-Field	Typical Biological Phenomena
RF (> 300 Hz and ≤ 20 MHz)	Internal or External	up to 10^6 V/m		Conductivity of most tissues gets nearly independent of frequency Biological matter is modelled as consisting of poorly conducting protein plus its water of hydration
up to EHF (≤ 300 GHz)	Internal or External	up to 10^6 V/m	up to 10^4 Tesla	Mitotic figures, Wave propagation in biological tissues standing-wave-phenomena, resonance-like response
RF $>$ MHz	Internal or External	$> 10^5$ V/m	$> 10^3$ Tesla	Heat production in biological tissues

2.3 Fröhlich's Model and Related Topics

One of the most important questions of modern physics is the problem how biological systems overcome thermal dissipation to such an extent that a high spatio-temporal order arises and remains stable over a relatively long time. The extraordinary high sensitivity may be a further consequence of this capacity to optimize (and probably minimize) thermal dissipation.

While on the one hand some scientists do not believe that biological systems should display unusual thermodynamic properties, as they are looked upon as systems near thermal equilibrium, other physicists have focussed precisely on the problem of the extraordinary thermodynamical features of the living state:

(1) Szent-Györgyi stressed that the electronic excitation of biomolecules is an essential factor of life[25],

(2) Schrödinger showed that biological systems are not closed but open systems, built up by the "order" (negentropy) of their food[26],

(3) Prigogine directed attention to "dissipative structures" far away from thermal equilibrium[5].

There are further possibilities for minimizing thermal loss, i.e. the avoidance of interactions between excited states of the biological systems and the surrounding heat bath, for instance, by completely different relaxation times, or by special properties of coherent states like Davydov solitons which do not interact with phonons.

A special solution of the problem of stability away from equilibrium, enunciated by Fröhlich, contains just the most necessary elements of thermodynamics. It has been originally based on the rate equations of a pumped system connected to a heat bath. As it describes a mechanism which can be subjected to experimental investigation it has the importance of a paradigm for understanding life from a physical point of view.

Fröhlich's model includes principally all the significant suggestions of other scientists who tried to solve the problem:

(1) energy supply s by pumping the biological system with an external food source,

(2) the possibility of thermal dissipation by a dissipation interaction factor Φ between biological systems and the heat bath,

(3) permanent deviation from equilibrium,

(4) Bose condensation-like formation of coherent states.

The most simple formulation of the solution has been written by Fröhlich[27] as:

$$z + \frac{s}{\Phi} \simeq \sum_{j=1}^{z} \frac{(\exp(\beta w_j) - 1)}{(\exp[\beta(w_j - \mu)] - 1)}$$

where z is the number of modes in a small range of frequencies where energy is supplied to the system with the rate s. For $s = 0$, the chemical potential $\mu = 0$ provides an equilibrium system where the identity $z = \sum_{j=1}^{z} 1 = z$ holds. With increasing s the chemical potential μ increases such that $\mu \to w_j$ for a small number of modes within the narrow frequency band. A more accurate description takes into account the coupling between z and Φ (which accounts for thermal dissipation). However, it is obvious that a possible solution of this mechanism provides a Bose condensation-like narrowing of modes with

$$z \to 1, \quad \Phi \to 0 \text{ for } s > s_0, \quad \mu \to w_l$$

for $\hbar w_l$ describing the quantum energy of one excited mode.

As soon as the energy supply s increases to such an amount that a threshold s_0 is exceeded, μ for at least this one mode l within the narrow frequency band approaches the value w_l, which means that the expression on the r.h.s. of Fröhlichs formula above approaches ∞. This requires that on the l.h.s., $\Phi \rightarrow 0$, which means that no thermal dissipation takes place any more, narrowing the frequency band to the overoccupation of a single mode l.

Further analysis of Fröhlich's theory has been carried out by Bhaumik et.al.[28,29], Wu (see this volume) and others (see references in paper Wu).

The most essential points in Fröhlich's model are

(1) to connect the problem of open biological systems to the occupation of modes in a Bose-Einstein-like distribution of modes with a chemical potential μ,

(2) to describe this fundamental thermodynamical model on the basis of necessary conditions of coherent excitations in biological systems with typical threshold behavior of phase transition phenomena ($s > s_0$, $\mu \rightarrow w_l$) and Bose condensation-like processes,

(3) to make the model available for experimental investigations.

Fröhlich considered in particular, the cell membranes, which display an electric potential of about 100 mV across a thickness of about 10^{-6} cm.

This gives a field strength of about 10^5 V/cm, and is of the order of the break-through amplitude of an electric field in air. Fröhlich suggested that by means of strong fluctuations of this field and its coupling with the highly polarizable membrane, typical oscillations with frequencies of the order of $\frac{d}{v}$ will occur, where d is the thickness of the membrane (about 10^{-6} cm) and v the velocity of sound waves which represent the elasticity and polarizability of the medium under consideration. With $v = 10^5$ to 10^6 cm/s, a typical frequency of the optical phonons would lie in the region between 10^{11} to 10^{12} Hz. Fröhlich was convinced that very striking effects of 6-7 mm coherent microwaves on the behaviour of a variety of biological systems which are strongly frequency selective at very low power supply, are evidence for his theory[30–33].

However, systematic investigations in this field, in particular, the examination of the growth rate of yeast cells under the influence of microwave exposure, did not lead either to convincing and consistent experimental results or to a valuable or new theoretical concept. Nevertheless, this does not diminish the importance of Fröhlich's work, who pointed out that

(1) coherent state formation is decisive for understanding biological phenomena,

(2) it is necessary to base the life sciences on non-equilibrium thermodynamics,

(3) these physical hypotheses of biological organization should be subjects for experimental investigation.

Models similar to that of Fröhlich originate from Prigogine's "dissipative structures", where again, the deviation from thermal equilibrium is necessary (but, as in Fröhlich's concept, not sufficient) for the formation of spatio-temporal order, including coherent state-formation. The decisive point is always a nonlinear autocatalytic coupling within the system which, as soon as it is supplied with enough energy, lead to phase transitions between regimes of different order.

Fröhlich's concept can therefore be applied to a variety of biological problems. It is not surprising that Fröhlich[33] and Prigogine propose to understand biological phenomena such as chain formation of proteins[34], cell growth, enzymatic activity, brain activity and cancer on the basis of different orders of nonlinear equations which can be formulated within their models.

A further branch of theoretical biology which can be traced back to models like that of Fröhlich, Prigogine, is the "catastrophe theory" of Thom or limit cycle models based on nonlinear equations which give rise to "deterministic chaos". Within this mathematical concept it is possible to describe a manifold of spatio-temporal patterns with impressive similarity to the real structures of plants or organisms. As these equations contain different elements of certainties and uncertainties as well as stabilities and instabilities, they become more and more basic tools for understanding the complex behaviour of biological systems in terms of regulation principles rather than of the elementary mechanisms.

2.4 Towards Quantum Biology

All the models discussed so far do not require quantum descriptions as they are based on classical electrodynamics and thermodynamics. Even if "deterministic chaos" comes into play, the non-linearity of interactions and the uncertainty of boundary conditions or the complexity of the couplings are not reasons for departing from classical physics. However, there are some phenomena in biology which cannot be explained in purely classical terms: (1) the extraordinary physical properties of biological matter as, for instance, the extremely high polarizability, giving rise to strong gradients of electric and magnetic fields, (2) the exceptionally high efficiency of actions like muscle contraction or nerve conduction, as well as of biochemical reactivity, (3) the rather high stability of wave propagation in biological tissues and, last but not least, (4) the extremely high sensitivity of biological systems to electromagnetic waves over the whole range of frequencies. It is precisely the last point that requires a quantum description, as it is well accepted now that biological systems work at the quantum limit, i.e. at the lowest possible uncertainty of signal processing. In order to elucidate the problems and possible solutions, let us discuss some striking examples and models which have been proposed for solving some crucial problems.

2.4.1 Extraordinary Polarizability of Biological Tissue

It is not surprising that macromolecules may display rather high dipole moments. The surprising fact is that in the noncrystalline, impure, particulate and wet biological system which should be subject of quasi-equilibrium thermodynamics at about room temperature such an extremely high and dense ordering and loss-less energy transfer dynamics of the macromolecules takes place. One of the first careful and systematic attempts to throw light on this problem has been made by Cope[35], who proposed that within a new discipline of "solid state biology", the concepts of superconductivity should be applied to research on the extraordinary properties of tissues and biological macromolecules. In line with these proposals are both, Fröhlich's idea of Bose condensation-like energy storage in biological systems and, Little's attempts to evaluate the conditions of Cooper-pairing along the DNA chain by taking into account the quantum description of Coulomb interactions at room temperature[36]. However, despite the recent discovery of "high-temperature superconductivity" which makes those ideas rather reasonable in retrospect, there is as yet no convincing experimental evidence for its existence in biological systems nor have there been enough investigations about the alternative of "non-equilibrium-superconductivity". The latter would contribute to the understanding of those effects in biological systems which are subject to both equilibrium and non-equilibrium optical excitation of electronic states. Dicke's subradiance regime[37] would enable superconductive-like processes even at high temperatures.

2.4.2 Extraordinary High Efficiency of Biological Activity

The role of heat conversion in case of exposure of the biological system to electromagnetic radiation is generally not as simple as in an equilibrium system. An active biological system may be subject to homoeostasis which means that instead of increasing its temperature after exposure to external radiation (or to heat) it

- may become transparent to the incident energy,
- may change its structure and/or its dynamics without production of free energy,
- may produce heat and/or radiation.

Only in rare cases will it react in a predictable way. These determinable events are generally due to the unphysiological exposure to rather high incident power, where after the biological system is killed, it is returned to balanced equilibrium with the external heat bath. The temperature will hence increase by the conversion of incident energy into heat. However, as the investigation of biophoton emission from living systems after exposure to external increase of temperature or to external electromagnetic radiation has shown, hysteresis-like reactions arise, showing evidence of cooperative effects which are based on nonlinear interactions of the system with weak external electromagnetic waves[14]. One of the reasons is certainly that the biological system is not in a thermal equilibrium state. Within optical frequencies, the deviation from equilibrium in terms of the Boltzmann factor may amount to many orders of magnitude, depending on the wavelength under consideration. The excited biological matter can be looked upon as a dissipative structure in the sense of Prigogine, which displays a phase transition between the two antagonistic states of Bose condensation on the one hand, where coherent storage of Bosons induce a permanent ordering and formation of coherent states, and thermal dissipation on the other hand. The photon field increases its entropy by metabolism and the maintenance homoeostasis of the biological system against the external heat bath. This free part of the photon field is responsible for the heat production of the organism which, for a living system, cannot be considered as black body radiation, but rather as the necessary consequence of entropy production originating from ordering processes feeding on "negentropy" in Schrödinger's sense.

It is evident that high cooperativity or a high degree of coherence is sufficient for the extremely high efficiency of biological response, e.g. muscle contraction, biochemical reactivity or bioluminescence reactions. However, because the system is poised at the threshold between two completely antagonistic phases, it is not possible to predict whether the system will react

- by lowering its own entropy and produce of heat,
- by keeping its entropy constant, becoming transparent by means of compensation mechanisms, or
- by increase of its entropy and functioning like a heat bath.

These possibilities of completely different reactions to the same external influences may provide an essential reason for the frequently observed irreproducibility of experimental results concerning the exposure of biological systems to external electromagnetic fields.

2.4.3 Stability of Wave Propagation

The extraordinary electric properties of living tissues, i.e., its high polarizability which has been taken into account also by Fröhlich, gives rise to further suggestions and models for understanding biological phenomena. For some striking examples, let us discuss briefly (1) the proposal of Inyushin, Sedlak and Zon[38–40] that living tissues form some kind of "bioplasma", then (2) the suggestions that cavity resonator and wave-guide properties are responsible for pattern formation as, for instance, the mitotic figures[18,23,24], and (3) the theory of Davydov that solitons play an important role for macromolecular organization and energy transport phenomena[41].

The plasma concept is a consequence of local charge fluids which may become coupled by Faraday's law to such an extent that collective oscillations with typical frequencies (the plasma frequencies) will arise. Take $\vec{v}$ as the typical velocity of the charge fluid, n as the charge density exceeding the background density n_0, the Maxwell's equations can be written

$$\begin{aligned}
\vec{\nabla} \cdot \vec{E} &= 4\pi e(n - n_0) \\
\vec{\nabla} \cdot \vec{B} &= 0 \\
\vec{\nabla} \times \vec{E} &= -\frac{1}{c}\frac{\partial \vec{B}}{\partial t} \\
\vec{\nabla} \times \vec{B} &= \frac{1}{c}\frac{\partial \vec{E}}{\partial t} + 4\pi \frac{en\vec{v}}{c}
\end{aligned}$$

Linearization of these equations around $n = n_0$, $\vec{v} = 0$ and neglecting strong magnetic forces, one gets solutions where the charge density, the velocity and the electric field all oscillate around the "plasma frequency" Ω:

$$\Omega^2 = 4\pi n_0 \frac{e^2}{m}$$

where m is the mass of the charge. For high enough frequencies one may consider electrons with the lowest mass. Quickenden and Tilbury rejected the bioplasma model[42]. While their indications to the lack of experimental evidence are clearly justified, their theoretical arguments based on (1) equilibrium thermodynamics, (2) the permettivity of water, and (3) linear first-order approach to the calculation of the plasma relaxation time are probably none of them valid. Rather, this questionable approach to living tissues may account for the one to two orders of magnitude difference from that required to meet the conditions of plasma formation.

A second approach to understanding wavelike phenomena and pattern formation in biological tissues in terms of superpositions of low- and high-frequency fields has been stimulated by the similarity of cavity resonator waves and the mitotic figures in a cell. The whole process of pattern formation during mitosis can be described by the slow change of superpositions of cavity resonator waves which are selected according to the boundary conditions within the cell. This means that a part of the excess electromagnetic energy of the non-equilibrium cell is distributed over the cell due to the boundary conditions of the cell walls, this leads to the cell slowly changing, by means of own activity, the shape and probably the physical parameters of the membranes which in turn alter the boundary conditions, thus acting back on the field, and so on. Fig. 2.2 demonstrates an impressive agreement between the mitotic figures and one of the selected cavity resonator waves.

Cell division is a logical consequence of such a model, as the different solutions of the mode equations favour the doubling of the available space by the doubling of

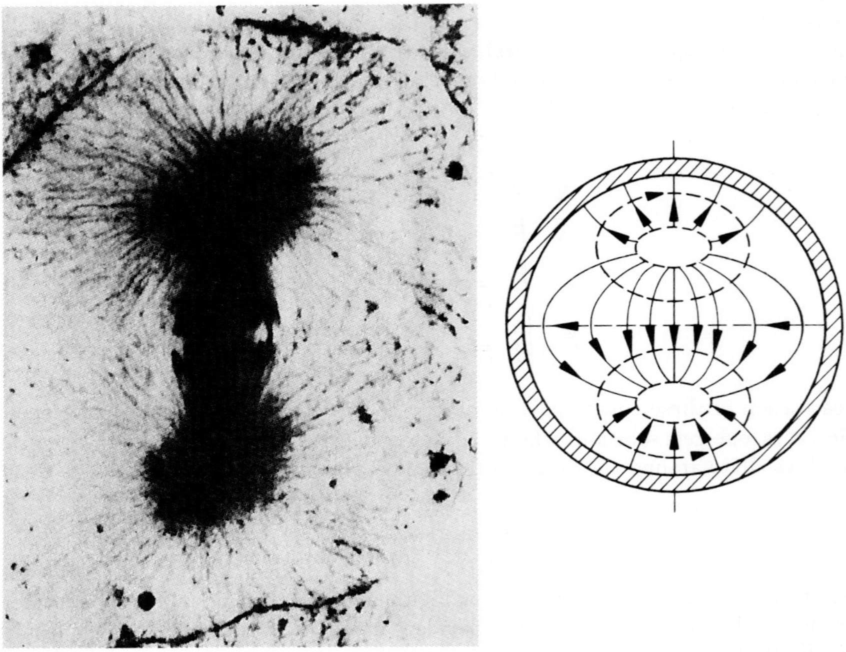

a. b.

Figure 2.2: *a. Completely developed spindle apparatus of a fish (Corregonus) in mitosis. (From: Darlington, C.D.; Lacour, L.F.: The Handling of Chromosomes. Allen and Unwin, London, 1960). b. Electric field of* TM_{11} *cavity modes in a right circular cylindrical cavity. Comparison with Fig. 2a shows that mitotic figures are striking examples of long-lasting photon storage within biological systems (From: Popp, F.A.: Photon Storage in Biological Systems, In: Electromagnetic Bio-Information, Urban & Schwarzenberg, Muenchen - Wien - Baltimore 1979)*

modes under just the same boundary conditions of electric and magnetic fields. The Maxwell's equations provide for those cases of wave guides and resonant cavities all the possible solutions of the Faraday's law which satisfy the boundary conditions with enough energy available. As in an atom or molecule, higher excited modes need more energy and form more complicated patterns. The Maxwell's equations which describe the Faradays law under the constraint of a single-frequency solution with a sinusoidal time dependence, $\exp(-i\Omega t)$, then take the simple form:

$$\begin{aligned} \vec{\nabla} \times \vec{E} &= i\frac{\Omega}{c}\vec{B} \\ \vec{\nabla} \cdot \vec{B} &= 0 \\ \vec{\nabla} \times \vec{B} &= -i\mu\varepsilon\frac{\Omega}{c}\vec{E} \\ \vec{\nabla} \cdot \vec{E} &= 0 \end{aligned}$$

The boundary conditions, i.e. geometry and matter properties such as ε and μ, are decisive for whether the particular solutions are due to conducting or dielectric wave guides or resonant cavities. It seems very likely that solutions of this kind play some role in biological sytems. In order to give an explanation for the avoidance of rapid thermal dissipation of these possible solutions we like to mention again that the extraordinary electric and magnetic properties of the membranes as well as the deviation from thermal equilibrium and the availability of energy supply even in the optical region which may well support the excitation and stability of these modes.

A most elegant proposal for overcoming the crucial problem of thermal dissipation has been enunciated by Davydov. Very stable solitary waves, observed in 1834 on the water surface, were originally described by the equation

$$\left(\frac{\partial}{\partial t} + u\frac{\partial}{\partial z} + \beta\frac{\partial^3}{\partial z^3}\right) u = 0, \quad u = u(z,t)\,.$$

This "Korteweg and De Vries"-equation[43] describes waves with exceptionally great stability and self-organizing character.

The first step to introduce this concept into modern physics was proposed by Sagdeev (1958)[44] and in 1973 by Kadomtsev and Karpman[45], who suggested that in a plasma subjected to strong magnetic fields "solitons" similar to those on the water surface may propagate. Instead of the classical Korteweg-De Vries equation the nonlinear one-dimensional Schrödinger equation (NSE) was proposed as the fundamental mechanism behind soliton formation:

$$\left(i\hbar\frac{\partial}{\partial t} + \frac{\hbar^2}{2m}\frac{\partial^2}{\partial z^2} + G|\psi|^2\right)\psi(z,t) = 0\,,$$

where G describes the nonlinearity parameter. This NSE also has the advantage that it can be solved exactly. Davydov extended this concept to any solution of a general NSE which shows, after autolocalized excitation, wave propagation without significant change in the form and velocity, where the dynamical balance between nonlinearity and dispersion provides the mechanism behind the extraordinary stability of this wave. Although exact solutions cannot be given in general, in real systems even unstable solitons may become significant as soon as their lifetime is comparable to the relaxation time of the system under investigation. Davydov shows that there are three reasons for the exceptionally high stability of solitons:

(1) The presence of an energy gap: in order to destroy a soliton, energy is required for splitting the soliton into its decay-products exciton and deformation of the lattice system. This energy gap increases with the chain deformation energy, and may therefore amount to considerable high values in biological systems which provide just the appropriate "matter" in their biopolymers.

(2) Solitons do not interact considerably with thermal noise: as their velocity is less than sound, they do not emit phonons. Their kinetic energy cannot become thermalized.

(3) Solitons display the self-focussing property of stabilizing their shape: as soon as a deformation of the solitary wave takes place, a displacement in the lattice system couples back in such a way that the original displacement is restored. This self-consistent compensation of the dispersion is the reason that such waves may penetrate each other without colliding.

Davydov proposes solitons for explaining, for instance, high temperature superconductivity, and in biological systems energy transport and information transmission along molecular chains and fibers, muscle contraction, and the role of microwaves as an alternative to Fröhlich's approach.

2.4.4 Extremely High Sensitivity of Biological Systems

There are well known and well-documented cases of high sensitivity of biological systems, e.g. electroreception, electronavigation, sensitivity to changes in the weather, or to seismic waves, as well as allergic reactions. There are reasons to believe that the sensitivity achieves just the quantum limit which means that the highest signal/noise ratio which is theoretically possible, limited only by the uncertainty relation (including squeezed states).

There is no doubt that all the different models which explain non-thermal effects, like Fröhlich's approach, Prigogine's dissipative structures, chaos-theory as well as Davydov's solitons, provide essential elements for explaining the high sensitivity of biological systems, at least qualitatively. They are all more or less based on the assumption that the signal/noise-ratio in biological systems may exceed considerably the value 1. However, a quantitative model has not yet been developed. In order to understand the growing sensitivity of living systems with inceasing size and development stage, Smith suggested a very fruitful idea. He compared the electric and magnetic energy density with the thermal energy density of one degree of freedom which has to be at least exceeded in order to arrive at a signal/noise ratio higher than one. This allows us to determine a lowest threshold of sensitivity of an antenna-system. Equality of electric, magnetic and lowest possible thermal energy density leads to the formulae[46]:

$$E_{min} = \sqrt{\frac{kT}{\varepsilon V}}$$

$$B_{min} = \sqrt{\frac{\mu kT}{V}}$$

V is the volume of the biological system under consideration. In considerably good agreement to observations he gets threshold values of E between 10^7 V/cm for molecular dimensions to about 10^{-4} V/cm for the whale, while B_{min} lies between 10 Tesla for molecules and 10^{-16} Tesla for whales. If he takes into account quantum effects and coherent states, he gets even lower values. However, at present, it is not possible to decide whether this fits the experimental results. Nevertheless, the dependency of the sensitivity to electromagnetic interactions on the dimensions of the system seems to follow qualitatively the model proposed by Smith. This means that the living state should minimize its noise by freezing the number of degrees of freedom due to mode coupling.

A similar result originates from the signal/noise-ratio SNR of an antenna within a heat bath:

$$SNR = \frac{\dot{n}}{4\Delta\nu}\left(\exp\left(\frac{h\nu}{kT}\right) - 1\right),$$

where ν is the frequency, $\dot{n}$ the number of incident quanta per unit of time, and $\Delta\nu$ the width of the frequency band.

For radiowaves and microwaves we have:

$$SNR \approx \frac{\dot{n}h}{4kT}\left(\frac{\nu}{\Delta\nu}\right)$$

In the optical range the SNR is considerably high such that a protection against damage by UV-radiation seems to become more important than the improvement of the SNR. However, the optimization of the SNR in a non-equilibrium system requires the independence of the Boltzmann-factor on the wavelength:

$$\delta \left(\exp \left(\frac{h\nu}{kT} \right) \right) = 0$$

Just this has been established in the optical range of the electromagnetic spectrum (see Fig. 2.3). There are three cases possible:

(1) The excitation temperature approaches infinite high values which describes the case of protection against external perturbations. Any "signal" including cosmic radiation incidence is then transformed into thermal noise: SNR = 0 in this case.

(2) The excitation temperature approaches 0 which corresponds to the Fröhlich model: the chemical potential arrives just at the quantum energy and induces a Bose-condensation of the mode under consideration. Instead of thermal dissipation the system uses the condensed bosons as amplifiers, arriving at a negative SNR:

$$SNR = -\frac{\dot{n}}{4\Delta\nu}$$

This model has been used to describe biophoton emission[19b] in fairly high agreement to the measurements.

(3) The excitation temperature remains proportional to the frequency of the mode under investigation which corresponds to a generalized Prigogine model of an "ideal open" system just at the phase transition between a chaotic and an ordered regime, analogously to the laser threshold. This f_ν=constant rule has been described and discussed in detail elsewhere[13].

However, it should be noted that the f_ν=constant rule provides an optimization of the signal/noise-ratio which fits the quantum aspects as well as the classical criteria.

2.5 Some Basic Problems and Open Questions

A lot of biological problems finds its answers in taking account of Maxwell's equations, for instance,

1. the membrane potential,

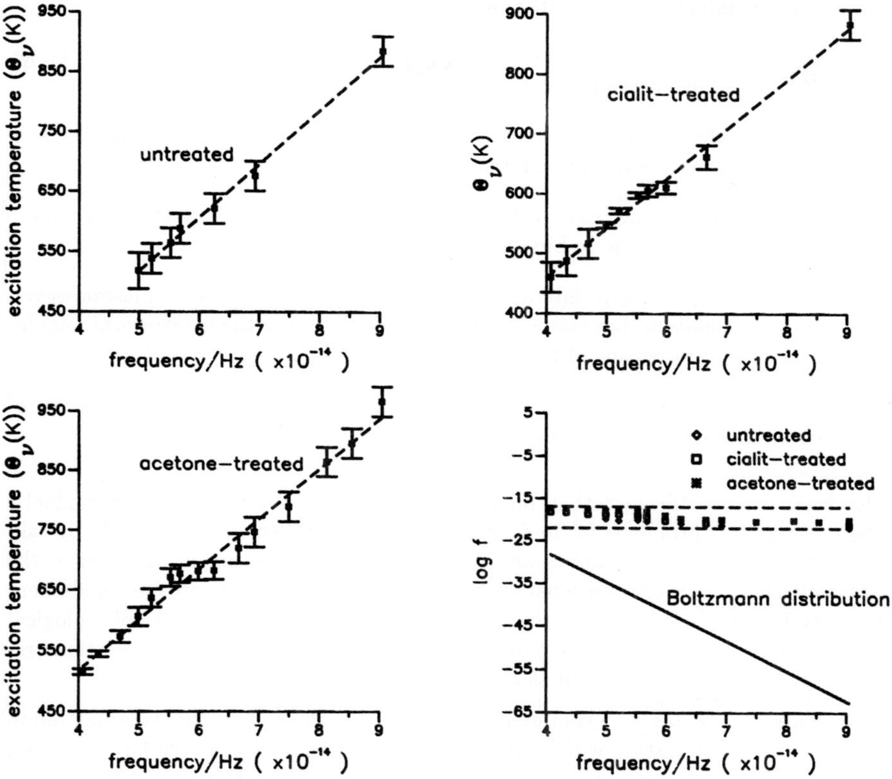

Figure 2.3: *Measurements on cucumber seedlings with the aid of interference filters show that $\Theta(\nu)$ increases proportional to the frequency ν. Left upper side: Untreated outgrown cucumber seedlings. Right upper side: Outgrown cucumber seedlings, treated with cialit. Left lower side: Outgrown cucumber seedlings after treatment with acetone. From the $\Theta(\nu)$-values, one can calculate lower limits of the occupation probabilities $f(\nu)$. Right lower side: compared to the Boltzmann distribution $f(\nu) = \exp\left(-\frac{h\nu}{kT}\right)$ at physiological temperatures T the $f(\nu)$-values of biological systems are (1) many orders of magnitude higher and (2) display no distinct frequency dependence.*

2. nerve conduction and related phenomena such as action potentials,

3. the response to external electromagnetic fields in the entire frequency range, but mainly concerned with the domain of high field amplitudes,

4. thermal effects, based on conductivity loss or dielectric loss including the skin effect and protection by boundary conditions,

5. probably also some basic principles of electroreception, electromagnetic navigation, electromagnetic bio-information,

6. some very speculative interactions in biological tissues such as plasma waves, cavity resonators, Lorentz forces on the blood stream, the transport of angular momentum by means of the Poyinting vector, etc.,

7. a general antenna theory for understanding bio-communication in terms of communication engineering[47].

However, despite the fact that biological systems are essentially electromagnetic the most serious problems for understanding "life" can become solved only by taking account of thermodynamics and/or quantum theory, either in addition to the Maxwell's equations or even separately. Roughly speaking, there are two fundamental questions wich fall into this category, e.g. the problem of stability of biological systems and the extraordinary properties of biological matter. Both questions require thermodynamical as well as quantum description.

Thermodynamics has to explain, how biological systems use the external pump (food) to establish a state which can least partially can remain stable far away from equilibrium. This answer has to involve the energy distribution over the spatio-temporal structure (its evolution in the phase space) including the spectral distribution over all the possible modes (its Fourier transform). Fröhlich's proposal of a Bose condensation-like energy storage, Prigogine's dissipative structures and also the f_ν=constant-rule provide a first step in the solution of this crucial problem, which has to be looked upon as the key question for understanding biological systems. The reason why it is not electrodynamics but thermodynamics that has to be used to solve this problem is simply that the stability of living systems should follow the usual optimization principles of physics, e.g. the first and the second law of thermodynamics when applied to the closed system of both biological systems and its surroundings. Fröhlich suggested the heat bath as the "idealization" of the external world and the chemical potential as the interaction between the surroundings and the biological systems. However, it is necessary to describe the nature and the dependence of the chemical potential on the frequency and the particular conditions, in order to understand the stability and the sensitivity of biological systems. One of the possible solutions of this problem which has been deduced from the optimization of the signal/noise-ratio concerns the description of the excitation temperature of the non-equilibrium system in terms of the chemical potential. After taking account of Brillouin theorem[19,48], we come to the following phenomenological description of the excitation temperature.

$$\Theta(\nu) = \frac{h\nu T}{h\nu - \mu - q(\mu)kT}$$

where q represents the number of stored photons. In view of Brillouin theorem $q = n_B$ for a living system under the constraint of Bose condensation-like energy storage, where n_B is the number of coherent bosons (photons) which are "condensed", as well as a measure of the resonator value of the store under consideration[19a]. On the other hand, for μ we can write

$$\mu = h\nu - n_D kT$$

where n_D is the number of photons which are thermally dissipated by means of the interaction with the external heat bath. After insertion of μ into Θ-equation we then have

$$\Theta(\nu) = \frac{h\nu}{(n_D - n_B)k}$$

This formula provides a phenomenological description of the basic mechanisms which may help to stabilize the non-equilibrium state of a biological system: At first it establishes the linear increase of the excitation temperature with increasing frequency in case of $n_D - n_B$ = constant. Secondly, it integrates the Fröhlich-mechanism to the "dissipative structures" far away from equilibrium as for $n_D = n_B$ we get the phase transition from a chaotic state ($n_D > n_B$) to an ordered regime by means of occupation inversion ($n_B > n_D$), see Fig. 2.4.

Despite the rather simple basic description of the most important mechanisms for stability of biological systems, including the proposals of Fröhlich and Prigogine, it should be noted that for $n_D - n_B$=constant the "driving force" is the maximation of the entropy of an "ideal open system", governing thus the stability of a biological system. This means that according to this most general model, biological systems are special physical systems which follow the optimization of the signal/noise-ratio by integration of longer and longer wavelengths into the bose condensation-like formation of coherent states. However, it is precisely in the mechanisms of how to establish coherent states (in the strong sense of eigenstates of the annihilation operator) that neither thermodynamics nor electrodynamics can give an answer, for it concerns quantum theory. Davydov's soliton concept is one example. Another possibility is the creation of coherence by means of "subradiance" or "superradiance" in terms of the Dicke's theory[38]. We do not know at present whether all the possible mechanisms

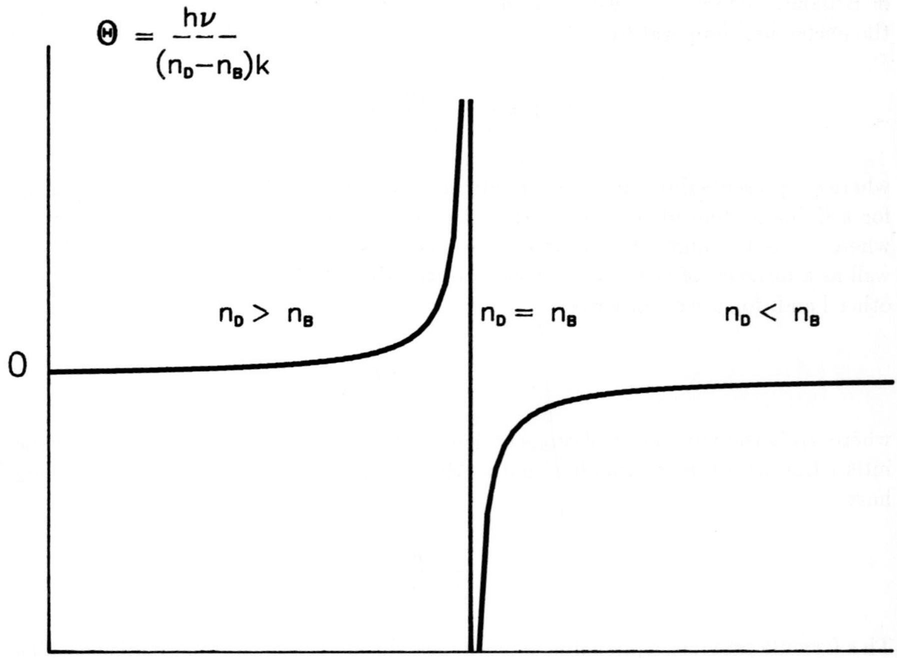

Figure 2.4: *A general Bose-condensation process, described in terms of the excitation temperature Θ of the tissue. This defines the regions below, at, and above threshold into three corresponding deviations from thermal equilibrium, i.e. the zone of positive excitation temperature, where Bose-condensation is outnumbered by thermal dissipation, the zone of negative excitation temperature above threshold, and the very sensitive change of the excitation temperature at threshold. Biological systems seem to work in the threshold region.*

for establishing coherent states under the optimization procedure of entropy of an ideal open system with mode coupling are actually taking place, for creating and establishing living states. This may become the most essential question for future research.

Bibliography

[1] *Handbook of Biological Effects of Electromagnetic Fields.* CRC Press, Boca Raton 1986.

[2] K.W. Wagner: *Arch. Electrochem.* **2** (1914), 371 (see also [1]).

[3] H.Fröhlich: Long-Range Coherence and Energy Storage in Biological Systems. *Int. J. Quantum Chem.* **2** (1968), 641-649.

[4] T.M. Wu, this volume.

[5] I. Prigogine, G. Nicolis and A. Babloyantz: Thermodynamics of Evolution. *Physics Today* **11** (1972), 23-28.

[6] R. Thom: *Stabilite Structurelle et Morphogenese.* Reading, Mass., Benjamin 1972.

[7] F. Kaiser: Specific Effects in Externally Driven Self-sustained Oscillating Biophysical Model Systems. In: *Coherent Excitations in Biological Systems* (H. Fröhlich and F. Kremer eds.), Springer, Berlin 1983.

[8] B.L.Hao: *Chaos.* World Scientific, Singapore 1985.

[9] E.Wolf (ed.): *Progress in Optics Vol. XXX*, North Holland, Amsterdam 1992.

[10] F.A. Popp: Weak quantization. *J. Math. Phys.* **14** (1973), 604-608.

[11] A.S. Davydov and V.N. Ermakov: Soliton Generation at the Boundary of a Molecular Chain. *Physica D 32* (1988), 318-323.

[12] A.S. Davydov: this volume.

[13] F.A. Popp, K.H. Li and Q. Gu: *Recent Advances in Biophoton Research*, World Scientific, Singapore 1992.

[14] W.P. Mei: this volume.

[15] J.D. Jackson: *Classical Electrodynamics.* Wiley, New York 1963.

[16] T. S. England: Dielectric Properties of the Human Body for Wave-lengths in the 1-10 cm Range. *Nature* **166** (1950), 480-481.

[17] M.F. Barnothy: *Biological Effects of Magnetic Fields*, Plenum Press, New York 1969.

[18] A.S. Presman: *Electromagnetic Fields and Life,* Plenum Press, New York 1970.

[19a] F.A. Popp, G. Becker, H.L. König and W. Peschka (eds.), Electromagnetic Bio-Information, Urban & Schwarzenberg, München 1979.

[19b] F.A. Popp, U. Warnke, H.L. König and W. Peschka (eds.), Electromagnetic Bio-Information, Urban & Schwarzenberg, München 1989.

[20] W. Beier: *Einführung in die theoretische Biophysik.* G.Fischer, Stuttgart 1965.

[21] F.A. Popp: *Biophotonen.* Verlag für Medizin, Heidelberg 1976.

[22] F.A. Popp and W. Nagl: DNA Conformations. Towards an Understanding of Stacked Base Interactions: Non-Equilibrium Phase Transitions as a Probable Model. *Polymer Bull.* **15** (1986), 89-91.

[23] M. Rattemeyer: Diplomarbeit, Experimentalphysik, Marburg 1980.

[24] F.A. Popp and V.E. Strauss: *So könnte Krebs entstehen*, Deutsche Verlagsanstalt, Stuttgart 1976.

[25] A. Szent Györgyi: *Electronic Biology and Cancer.* M.Dekker, New York 1976.

[26] E. Schrödinger: *What is Life?*, Cambridge University Press, Cambridge 1955.

[27] H. Fröhlich: The Biological Effects of Microwaves and Related Questions. *Advances in Electronics and Electron Physics* **53** (1980), 85-152.

[28] D. Bhaumik, K. Bhaumik and B. Dutta-Roy: On the Possibility of Bose Condensation in the Excitation of Coherent Modes in Biological Systems. *Phys. Lett.* **56A** (1976), 145-148.

[29] D. Bhaumik, K. Bhaumik and B. Dutta-Roy: A Microscopic Approach to the Fröhlich Model of Bose Condensation of Phonons in Biological Systems. *Phys. Lett.* **59A** (1976), 77-80.

[30] N.D. Deryatkov: Influence of Millimeter-Band Electromagnetic Radiation on Biological Objects. *Soviet Physics USPEKHI (Translation)* **16** (1974), 568-569.

[31] L.A. Sevast'yanova and R.L. Vilenskaya: A study of the Effects of Millimeter-Band Microwaves on the Bone Marrow of Mice. *Soviet Physics USPEKHI (Translation)* **16** (1974), 570-571.

[32] A.Z. Smoloyanskaya and R.L. Vilenskaya: Effects of Millimeter-Band Electromagnetic Radiation on the Functional Activity of certain Genetic Elements of Bacterial Cells. *Soviet Physics USPEKHI (Translation)* **16** (1974), 571-572.

[33] H. Fröhlich: Evidence for Bose Condensation-like Excitation of Coherent Modes in Biological Systems. *Phys. Lett.* **51A** (1975), 21-22.

[34] L. Lauck, A.R. Vasconcellos and R. Luzzi: On Fröhlich's Coherent Effects in Biological Systems: Influence of Carriers and High Order Dissipative Effects. *J. theor. Biol.* **158** (1992) 1-13.

[35] F.W. Cope: A Review of the Applications of Solid State Physics Concepts to Biological Systems. *J. Biol. Phys.* **3** (1975), 1-41.

[36] W.A. Little: Fluxoid Quantization in a Multiple Connected Superconductor. *Phys. Rev.* **133** (1964) 1A, 97-103.

[37] R.H. Dicke: Coherence in spontaneous radiation processes. *Phys. Rev.* **93** (1954), 99-110.

[38] V.M. Inyushin and P.R. Chekerov: *Biostimulation through Laser Radiation and Bioplasma.* Engl. transl., Danish Society for Physical Research, Copenhagen 1976.

[39] W. Sedlak: Bioplasma - the new state of matter, pp. 13-30. In: *Bioplasma.* Proceedings of the first Conference on Bioplasma, 9 May 1973, The Catholic University of Lublin, (W.Sedlak ed.), 1976.

[40] J.R. Zon: The living cell as a plasma physical system, *Physiol. Chem. Phys.* **12** (1980), 357-364.

[41] A.S. Davydov: Solitons in Biology. In: *Solitons* (S.E. Trullinger, V.E. Zakharov and V.L. Pokrovsky eds.), Elsevier Science Publishers B.V. 1986, 1-51.

[42] T.I. Quickenden and R.N. Tilbury: A Critical Examination of the Bioplasma Hypothesis. *Physiological Chemistry and Physics and Medical NMR* **18** (1986), 89-101.

[43] D.J. Korteweg and G.De Vries: *Philos. Mag.* **39** (1985), 422, see 42).

[44] R.Z. Sagdeev: *The Question of Plasma Theory and of the Problem of Controlled Thermonuclear Reactions*, Vol. 4 (Acad. Sci. USSR Moscow), 384.

[45] B.B. Kadomtsev and V.I. Karpman: *Usp. Fiz. Nauk* **109** (1973), 193, see 42).

[46] C.W. Smith: Coherent Electromagnetic Fields and Bio-Communication. In: 19)

[47] P.S. Callahan: Nonlinear Infrared Coherent Radiation as an Energy Coupling Mechanism in Living Systems. In: *Molecular and Biological Physics of Living Systems* (R.K. Mishra, ed.), Kluwer, Dordrecht 1990.

[48] F.J. Zucker: Information, Entropie, Komplementarität und Zeit. In: *Offene Systeme I* (E. von Weizsäcker, ed.), Klett, Stuttgart 1974.

Chapter 3

Biological Effects of Weak Electromagnetic Fields

Cyril W. Smith

3.1 Introduction

Textbooks of electrophysiology[1] make the explicit or implicit assumption that no frequencies greater than the highest components of the cellular action-potential waveform (10 kHz) are of biological significance. If this were indeed the case, this Chapter would finish here. Twenty years ago, Frey and Bowers[2] realised that electromagnetic spectrum allocations and health hazards were matters of public concern as the availability of inexpensive solid-state power sources engendered new microwave applications. Relatively large numbers of microwave systems might then come under the unsupervised control of private individuals or organizations. Safety standards established when microwave systems were uncommon and when the average citizen was unlikely to be irradiated by a microwave beam could become inadequate if microwave beams were to be emitted from cars, as well as traffic signals, telecommunications towers and airport radars.

As a measure of the magnitude of the problem, they considered the case of radar-equipped cars emitting 50 mW within a beam of 2° producing a power density of 100 μ W/cm^2 at a distance of 5 m. Although the beam might be scanned and the car in motion, thereby reducing greatly the radiation incident on any given individual, that same person could be subject to simultaneous irradiation from many cars similarly equipped. Such power levels would be inconsequential according to the current U.S. safety standards but, would be significant relative to standards in Eastern Europe.

The microwave region of the electromagnetic spectrum formally begins at 300 MHz (free-space wavelength 1 m), but in general, electromagnetic radiation begins to be appreciably absorbed by humans at frequencies greater than about 15 MHz. The eye lens and the testes in particular have limited ability to dissipate heat and hence are especially vulnerable to microwave irradiation. The threshold power densities required to produce thermal effects in animals and humans are well-known. Non-thermal effects were not included in the considerations when the U.S. radiation safety level was set at an average of 10 mW/cm^2 (100 W/m^2) for long term exposures. However, non-thermal effects were apparently influential in the establishment of the maximum standard of 10 $\mu W/cm^2$ (100 mW/m^2) in the former Soviet Union[3].

The U.S. radiation safety standard was confirmed by their Tri-Service Program of research on biomedical aspects of microwave radiation[4], although some workers felt this research was "largely irrelevant"[5] to the subject of non-thermal non-ionizing radiation because it rejected the East European data and did not adequately consider the possibility of low-power radiation hazards.

"Incomplete understanding of the effects of microwave radiation on biological systems, combined perhaps with personal and institutional prejudice, has led to the current anomalous situation in which different countries have adopted safety standards that differ by several orders of magnitude"[2].

When in 1989, Simon Best and I wrote *Electromagnetic Man*[6], the situation was much the same. We noted that,

"The general public - particularly in America - got their first real idea of the potential military applications of electromagnetic fields in 1976, when the United States claimed that their Embassy in Moscow was being irradiated with microwave beams by the Russians...However, what was perhaps equally if not more disturbing was the eventual disclosure that the irradiation had been happening since 1953 and that the State Department and five previous governments had known about it since it started.......several diplomats developed leukaemia; two of the U.S. Ambassadors serving during the period (1953-77) died of cancer and the third died in 1986 of leukaemia which was first discovered in 1975."

Recently, it was suggested to me while I was on a visit to the Commonwealth of Independent States (C.I.S.), that this episode represented no great malevolent plot against the U.S.A. but, rather, was the result of incompetence in the use of microwaves to recharge the batteries of Soviet listening devices planted within the U.S. Moscow Embassy.

By the mid-1970's it was already apparent to many that microwave radiation could be biologically harmful. Dr. Milton Zaret, of Scarsdale, New York, had identified a particular form of posterior subcapsular cataract in the eye as being the signature of radiofrequency and microwave exposure; this became apparent while he was carrying out a health study of radar maintenance men for the U.S. Air Force in 1964[7].

In our Conclusion to *Electromagnetic Man*[6], we posed the question, "What level of electromagnetic radiation is safe?" and continued:

"The present U.S. and U.K. guideline of 100 W/m^2 (10 mW/cm^2) as the upper limiting exposure to microwave radiation represents a fraction of the power density of tropical sunlight. But it is only reasonable from the point of view of avoiding widespread thermal injury, and represents enforceable legislation. The level of non-ionising radiation which will produce no effects of any kind in any person could not be lived with from an engineering stand-point in view of the expectations society has from modern electronics. It is even below the level of non-ionising radiation which people themselves emit and which can affect other hypersensitive persons in their vicinity."

3.2 Review of Literature

In his introduction to *Interactions between Electromagnetic Fields and Cells*[8], Schwan pointed out that in connection with microwave radiation:

"The medically oriented work before the second world war led almost immediately to a split between two different schools of thought. One had it that whatever therapeutic or other effect was observed was caused by a noticiable temperature increase and that, therefore, the results observed were due to heat and not caused by electrical fields per se. The other school of thought believed that direct field interactions termed nonthermal or athermal are important. This controversy led to some hundred publications of variable quality. The debate was never settled and these papers are largely forgotten."

The radar developments during the second world war were rapidly applied to therapeutic techniques from which the interest in the dielectric properties of biological materials had its resurgence, this because even by the late 1920's, Fricke, at the Cleveland Clinic, U.S.A., had investigated the dielectric properties of cells and found that the capacitance of tumour tissue differed from that of normal tissue, a fact which has yet to penetrate conventional medical awareness.

Dielectric properties of materials relate to their interaction with electric fields (steady and alternating), as for example when they are placed between the plates of a capacitor. The dielectric properties arise from the patterns and motions of electric charges associated with atoms and their chemical bonds. At very high frequencies, the dielectric properties merge with those of optical refraction.

The dielectric properties of biological materials, fluids and water have been thoroughly measured over a wide frequency range[9,10,11]. They do not show indications of sharp resonance phenomena. The classical Debye resonance is broad and extends over several decades in frequency[9], not characteristic of the observed non-thermal biological responses to microwaves and millimeter waves which will be discussed later.

Until recently, it was very necessary to look at the 'Acknowledgments' section of publications on the effects of non-ionizing radiation to see who had funded the work, in order to be able to assess its significance. Increased openess with regard to the electrical environment came partly as the result of the judicially instituted 'New York Power Lines Project'[12] and partly as a result of the recession whereby laboratories which had ceased to be funded by vested interests chiefly concerned with minimizing the risk of legal action being taken against them, were no longer inhibited regarding the publication of "inconvenient" results. Becker[13] and Brodeur[14] discuss the background to this manipulation of information and research projects.

The best sources to scan for updating information on the biomedical effects of non-ionizing radiation are to be found in the three newsletters: *Bioelectromagnetics Society Newsletter*[15], past issues of which contain many articles relevant to microwave effects in biological systems[16]; *Electromagnetics News*[17] and *Microwave News*[18] both of which have published various articles on the effects of non-ionizing radiation as well as reports, updates and correspondence relating to the electromagnetic environment, current research studies and safety standards in different countries.

The specialist journals for this area include: *Bioelectromagnetics*, and *Electro- and Magnetobiology* (formerly *Journal of Bioelectricity*).

The Office of Naval Research, U.S.A., has, for many years, published a quarterly digest of current literature on the "Biological Effects of Nonionizing Electromagnetic Radiation". This has now resumed publication as a scientific abstracting journal under the name, *BENER Digest Update*[19] There is also an electronically searchable extended set of abstracts *EMF Database* available. Many of the leading scientific and engineering journals carry the occasional article of relevance.

There are also a whole series of conferences whose proceedings represent valuable sources of information. These include: The Annual Conferences of the Institute of Electrical and Electronics Engineers (IEEE) Engineering in Medicine and Biology Society (EMBS)[20] ; the various meetings of The Bioelectromagnetics Society (BEMS)[15], the European Bioelectromagnetics Association (EBEA)[21] and the Bioelectrical Repair and Growth Society (BRAGS)[22].

For those who want an introduction to the biological effects of microwaves and millimeter waves rather than to be updated in the subject, I offer the following summary.

At a lecture given in the 1990 General Assembly of the International Union of Radio Science (URSI) held in Prague[23], Ross Adey concluded that,

"It is no longer a matter of speculation that biomolecular systems are responsive to low level, low frequency electromagnetic fields. Not only is tissue heating not the basis of these interactions, but the many instances of responses windowed with respect to field, frequency and intensity set a rubric for their consideration in physical mechanisms involving long range ordering at the atomic level".

"From theoretical consideration of the collisional basis of molecular interactions with microwave and far-infrared fields, there is no compelling evidence for resonant absorption in ordinary molecular fluids below 3,000 GHz. This model is supported by the virtual absence of experimental evidence for interaction with CW fields at frequencies below the GHz range other than by heating. On the other hand, RF fields that are sinusoidally amplitude-modulated at ELF frequencies produce a wide range of biological interactions...."

Among the interactions listed were, the entrainment of brain rhythms and the conditioning of brain responses to imposed fields, and the modulation of brain states and behavioral states[24]; strong effects on cell membrane functions including the modulation of intercellular communication through gap-junction mechanisms[25]; the reduction of cell mediated cytolytic immune responses[26]; and the modulation of intracellular enzymes that are markers of signals arising at cell membranes which are then coupled intracellularly.

For more than two decades, Adey and his co-workers have been studying the effects of ELF electric fields and amplitude modulated RF and microwave fields on cerebral tissue Ca^{2+} efflux[23]. They found that the maximum effect occurred for the ELF or modulation frequency of 16 Hz, but the effects exhibited narrow windows in frequency and amplitude and were biphasic in respect of applied electric field gradients, differing by a factor of 10^6. Adey discussed these effects in terms of dissipative processes and cooperative phenomena in cell membranes containing receptor proteins.

With the discovery of intracellular enzymes that respond to signals initiated at cell membranes as a response to electromagnetic field exposure, Adey and his co-workers[23] also found intramembranous particles inserted into the lipid bilayer membrane. Their outer tips are negatively charged glycoprotein strands which attract calcium and hydrogen ions and form receptor sites for chemical stimulation of the molecules. These form a calcium-mediated direct path for inwardly directed biosignals between the cell surface and intracellular enzymes and organelles. Again there are windows in frequency and amplitude as well as differing sensitivities for various cell functions (see Tsong and Gross, this volume). The presence of Ca^{2+} ions is essential for microwave effects on the binding of a ligand to β-adrenergic receptors of rat erythrocyte membranes.

The effects of low frequency modulation on the microwave carrier represent a further complication to the assessment of microwave and millimeter wave effects on living systems. It is now possible to buy microwave oscillators at frequencies going above 100 GHz, well into the millimeter wave region, which are coherent to a fraction of a Hertz. However, oscillators with this coherence were not in general used for most of the experiments reported in the literature. One must presume that in the majority of experiments, frequency components which included the power supply frequency and its harmonics were present as amplitude or frequency modulation of the microwaves. Such frequencies are within that band of frequencies extending from below 1 Hz to a few hundred Hz that are known to be particularly effective biologically and which include the all permeating Schumann Bands of ionospheric radiation within which all evolution has taken place.

Blackman[27] has reviewed the highly reproducible biological influences of low frequency sinusoidal electromagnetic signals, both alone and superimposed on RF carrier waves. The association of calcium ions with brain tissue was selected as the biochemical marker because calcium-ion efflux from chick brain tissue *in vitro* occurs at non-thermal levels. Combined with the results of studies of brain biochemistry and EEG in animals, synaptosomes and human neuroblastoma cells in culture, this provides evidence that CNS tissue from several species, including humans, are affected by low intensity RF-fields modulated by specific low frequencies. The biological systems which have so far been investigated are quite diverse, nevertheless, the consistent features are the specificity of the ELF and the involvement of the geomagnetic field.

In 1968, Webb first reported the frequency dependent inhibition of bacterial growth by 136 GHz millimeter radiation[28] and, in 1974, Devyatkov and co-workers[29] reported the existence of frequency-dependent low-intensity microwave effects in biological systems over the frequency range 39 GHz to 60 GHz.

Devyatkov's report contained results of experiments by Sevastyanova and Villenskaya showing that millimeter waves at certain frequencies around 42 GHz and at intensities above a certain threshold, exerted a protective effect on mouse bone marrow cells pre-exposed to X-radiation (see Wu, this volume). Subsequently this work was repeated with better frequency resolution and showed indications of a 60 MHz periodicity within the band of millimeter wave frequencies that were effective. Another study, this time investigating the decrease in synthesis of β-lactamase in penicillin-resistant strains of *E. coli* and *Staphylococcus aureus*, showed frequency selective effects in *S. aureus* at frequencies which did not have any effect on the *E. coli*.

Grundler *et al*[30,31,32] have described experiments which showed a resonant response in the growth rate of yeast cells irradiated with millimeter waves, they used the region of 42 GHz as suggested by the work reported by Devyatkov[29] on *Rhodotorula ruba*. They found an exponential growth rate reproducible within $\pm4\%$ and that this could be influenced by continuous millimeter wave fields corresponding to a power flux density of a few mW/cm^2. There was a fine frequency structure which had a periodicity of 8 MHz. Initially, the cell concentrations were measured by photometry, which only gave the results of the average reaction to millimeter radiation of 10^6 cells in a stirred suspension.

To overcome the disadvantages inherent in photometry, Grundler and co-workers[31] developed a method for studying the kinetics of single cell growth during microwave irradiation. With this, they clearly demonstrated a direct microwave influence. In comparison with the controls, there were significant radiation induced asymmetries and double peaks in the cell cycle time distributions for both the first and second cell division cycles indicating the presence of subgroups of cells. In the case of the double peaked distributions, one peak remained close to the control value while the other showed a displacement due to the irradiation. Even though all the cells were contained in a monolayer within an area of only 0.2 mm^2, the effect of the microwaves was to produce a sub-culture of cells. These microwave frequencies were stabilised to ± 1 MHz of the 42 GHz used. The results again showed a similar resonant frequency dependence with the same resonance width of 8 MHz to that observed in the photometric measurements.

Experiments at 84 GHz — double the previous frequency — also gave significant radiation effects, although these were mainly during the second cell division cycle. By means of a computer controlled cell counter, tests could be made over one complete cell cycle, a lower temperature (25°C) was used to extend the cell cycle to 5 hours. These results showed that the microwave influence is restricted to the G_1-phase of the cell cycle at the frequency used.

Cycles of cell division are divided into four phases.The first phase (G_1)and the third phase (G_2) are periods of cell nuclear activity during which most of the significant metabolic activities occur. The DNA replication takes place during the second phase (the so-called S-phase, or synthesis period). By the time that the brief fourth mitotic (or M-phase) begins, the chromosomes will have completed replication. The first signs of mitosis are condensations of chromosomes (prophase), which then move apart towards the cell equatorial plane with a microtubule spindle between them (prometaphase), they align at opposite ends of the cell (metaphase), the chromatids separate to become independent chromasomes (anaphase), after which nuclear re-organization takes place and nuclear membranes form (telophase). If cell division (cytokinesis) is to occur this may or may not synchronise with mitosis.

Webb[33] has investigated the effects of millimeter waves for over 25 years. In experiments from 59 GHz - 143 GHz using cell cultures of the bacterium *E. coli*, he found effects on the growth rate, and on DNA and RNA and protein synthesis, which had sharp frequency windows and as well as windows within the cell division cycle. Two sets of sub-frequencies were effective, one set went in integer multiples of 7 GHz, the other in integer multiples of 5 GHz. Raman spectroscopy revealed no resonant activity in the nutrients or in resting cells, but sharp peaks were observed when the cells were activated by a nutrient containing an oxidizable carbon energy source (see Wu, this volume).

Resonant microwave absorption in solutions of *E. coli* DNA was first observed by Edwards, Swicord and co-workers[23,34]. The DNA was randomly nicked by low concentrations of the endonuclease DNase. It was supposed that this produced a dynamic length distribution whose mean length decreased with time. Enhanced microwave absorption occurred as the sample length distribution changed to correspond to the region giving significant absorption in the experimental frequency range, which extended from 0.4 to 12 GHz. The frequency f of the observed resonances and the DNA length l as determined by the number of base pairs, were related to the acoustic velocity v by

$$v_{circular} = (n+1)fl$$

or

$$v_{linear} = \left(n + \frac{1}{2}\right) fl$$

where the subscript distinguishes linear and circular DNA molecules, and $n = 0, 1, 2 \ldots$.

These resonances appear to be sharper than the frequency stability of the experiments since the standard deviations at the peaks are seen to be generally worse than at the skirts of the resonances[34].

With these as with many other bioelectromagnetics experiments, replication in other laboratories did not come easily, sometimes it did not come at all. It must be realised that in such work, one is dealing with a multi-variable experimental system involving a living system. Pickard[16] has listed criteria for the reproducibility of bioelectromagnetics experiments[36]. Marino[35] has made some pertinent remarks on, "Negative studies and common sense" in an editorial of this title. He points out that, "The careful student of bioelectricity quickly learns to separate poison-pill experiments and sophistry from facts and rational analysis, and to determine which individuals and groups are truly interested in building bioelectricity into a useful and important science, and which are interested in burying the subject under a mountain of innuendo, doubt and disdain. The bad news is that judges and other generalist laymen, unfamiliar with the concept of the null hypothesis, may be susceptible to the Siren call of the negative study."

The scanning-tunnelling-microscope (STM) has been applied to the visualization of biological materials[37]. Recently, Michel *et al* at I.B.M., Zurich, have devised a scanning surface harmonic microscope which is capable of operation at microwave frequencies[38,39]. It essentially consists of a conventional STM inside a tuneable microwave resonator positioned so that the microwave electric field at the tip is maximized and normal to the sample surface. The resonant cavity is tuned to a harmonic of the applied fundamental frequency. Non-linearities in the electron movement on the surface generate harmonics in the resonator which are detected by a spectrum analyser and fed-back to the tip-to-surface spacing control servo. The resulting surface images contain contrast related to local changes in the non-linearity of electron movement in the sample.

The operation of the scanning tunnelling microscope at microwave frequencies permits the study of surface processes at frequencies up to about 10 GHz. The use of third-harmonic imaging, which is possible due to the non-linearity induced in the surface electrons, avoids the ubiquitous problem of tip contact when there are thin insulating layers on the sample surface. Promising areas of application include the identification and properties of molecular absorbates with resolutions of 0.3nm, self-assembled monolayers, single organic molecules, biological macromolecules and biological membranes.

3.3 Characteristic Response to Microwaves and Millimeter Waves

The response to microwave irradiation which is common to all biological systems, alive and dead, and to biomaterials and water, is the microwave heating effect. The initial bioengineering work on this came largely from Schwan's laboratory[8] at the University of Pennsylvania. One result of which was the invention of the microwave cooker for preparing food in submarines. The penetration depth of microwaves into tissues, the reflection effects at interfaces, and the specific absorption rate (SAR) in skin, fat and muscle were all evaluated for a range of frequencies. When the organism, organ or biostructure is a half-wavelength in size (for humans that is in the range 30-300 MHz or just below the microwave region), there is resonance absorption and the absorbed energy is a maximum. From 300 MHz to 3 GHz, there can be focussing of the microwaves by curved surfaces such as the bone of the skull; this can lead to local hot-spots. Beyond 3 GHz, the penetration depth of the radiation decreases and surface heating effects predominate and resemble those of infrared or sunlight. In the frequency range 500 MHz - 2 GHz, an incident energy flux density of 10 mW/cm^2 (100 W/m^2) will produce an average temperature rise of about 0.5°C in the human body.

There is a fundamental relation between the incident power density of electromagnetic radiation and the electric and magnetic field components of the radiation. There is a generalised form of Ohm's Law relating the electric field $E(V/m)$ to the magnetic field $H(A/m)$ through the characteristic impedance Z (ohms) of the medium within which the radiation propagates. This is valid so long as measurements are made far enough away (usually the order of a wavelength) from the antenna or radiation source, thus:

$$E = H.Z$$

or, in terms of power density W (Watts/m^2)

$$W = E^2/Z = H^2.Z$$

In free space, the characteristic impedance has a fundamental value which is approximately Z = 377 ohms. The various microwave exposure limit tables cited by Grandolfo[40] have been rounded off and implicitly assume values for Z in the range 372 - 400 ohms.

Hence, an incident power density of 100 W/m^2 corresponds to an electric field $E =$ 194 V/m or a magnetic field $H = 0.5$ A/m in free-space.

The magnetic flux density B is related to H by

$$B = \mu_o H$$

where μ_o is the permittivity of free space and has the value 1.26 x 10^{-6} H/m. Hence 100W/m^2 corresponds to a $B = 0.65\mu$T. This would be superimposed on a static geomagnetic field of about 50μT.

Within water or a biological fluid, the effective values of permittivity must be taken into account and the calculations become more complicated. The magnetic permeability in bio-materials is effectively that for free-space. Hasted[9] gives the dielectric parameters for water up to optical frequencies. For biological materials, data can be found in books by Hasted[9], Pethig[10] and Grant *et al*[11]. However, if one is dealing with a coherent system, as is very likely the case, then the interaction of the incoming radiation will not be with individual molecules, but with domains of coherence as considered by Del Guidice *et al*[41]. The result is that the velocity of propagation in the coherent medium may fall from the 300 Mm/s of free-space value to as little as 1-10 m/s.

Other possible mechanisms of interactions between microwaves and biological systems include[42] the effect of electric fields on chemical equilibrium, chemical rate constants and conformational transformations; interaction-forces between microscopic particles in an external electric field leading to the formation of pearl-chains; torque forces inducing cell rotation; effects on ligand-binding in the cell membrane.

The absorbed microwave energy also gives induced conduction currents and their associated magnetic fields in irradiated tissues. These can affect ion motion and membrane potentials. Non-linearities of cells and tissues may rectify alternating currents to give D.C. which has ion-transport properties and can lead to electrolysis.

However, all these phenomena are within the realms of classical physics. The real issue is whether living systems are completely described by classical physics, or whether quantum physics is needed in addition. This boundary is crossed as soon as energy gap phenomena become involved. Pethig observed many years ago[43] that since the energy gap of a protein is of the order of 5 eV, and all the redox reactions going on in the human body represent an electric current of 200A, this corresponds to a power of 1 kW, of the correct order for the metabolic rate.

People have repeatedly reported being able to 'hear' the presence of a microwave beam[81,82], such as from a radar transmitter. This was finally tracked down to a thermal effect mediated by electromechanical interaction distal to the cochlea[44]. The threshold for a response from single auditory neurons in the cat to pulsed microwave radiation is as low as 4 μJ/g per pulse. Although this is regarded as a thermal effect, it corresponds to about 3 phonons per pulse of microwaves, emphasising that the sensitivity of the ear is as close to the phonon threshold as the eye is to the photon threshold.

One biological effect which does involve energy levels is magnetic resonance (NMR). The nuclear magnetic moment, even in a strong magnetic field, does not give a resonance up into the microwave region (e.g. proton-NMR = 42.6 MHz/Tesla), but the electron has its resonance at 28 GHz/Tesla and this is applied to electron spin resonance (ESR) techniques which are important for the detection of free radicals[45].

In my laboratory, we have shown that microwaves and radiofrequencies modulated at the particular frequency which satisfies the proton-NMR condition in the geomagnetic field (approx. 2 kHz) are particularly efficient in the production of cataracts in bovine eye lenses *in vitro*[46].

These hazardous conditions could arise in a subject moving routinely about in the non-uniform steady magnetic field of a typical laboratory in the presence of suitably modulated microwaves. These NMR-resonances are so sharp (ppm) that it is not practicable to set the frequency of an ordinary oscillator to obtain NMR conditions given the nominal value of the magnetic field. The resonance condition is realised by sweeping slowly through the appropriate frequency or magnetic field holding the other parameter constant. The relaxation time for NMR in biological tissues is in the range 0.5-3 seconds.

In the early literature on biological effects of microwaves, there are cases where reported effects increased as the incident microwave power densities were reduced, although there was usually an associated increase in the scatter of the results, so they were dismissed as unreliable. No one seems to have asked the questions, "Do the effects extrapolate to infinity as the power is reduced to zero? If not, at what incident power density does the turn-down occur?"

In experiments involving low level microwave radiation, the biological effects were initially characterised by being small and difficult to reproduce by other workers in other laboratories, or even failing to reproduce at all. Grundler[31], while noting the theoretical conjectures of Kaiser[47] on reproducibility, considers several criteria in respect of low power microwave experiments.

A stressed biological system is more sensitive to microwaves (synergism), as for example microwaves combined with X-rays[29,48]. The latter would increase the free radical concentration, the former might prolong their lifetimes. Incubation in an alternating magnetic field has been reported to enhance the effects of ionizing radiation[49]. Chemical or other known environmental stressors might also act synergistically with microwaves.

Frequency specific biological reactions may depend on the biological cell system used. When the same specific enzymatic reaction was tested in different cell systems, microwave irradiation did have an effect on *Staphylococcus aureus*, but had no effect on *E. coli*[50].

The biological parameters to be tested should not be of too specific a nature. Cell growth, which is dependent on many factors, is affected by microwaves in a frequency specific manner but, microwaves produced no lethal mutagenic or chromosomal effects in micro-organisms.

Microwave effects may only occur in certain phases of the cell cycle. A yeast was only sensitive to microwave irradiation in the G_1-phase at the frequency used[31].

The temperature of cells in pre-experiment storage may affect subsequent microwave interactions[31]. Many experimental techniques rely on synchronised cell cultures which are notoriously difficult to achieve[51]. These techniques may involve pre-experiment storage of the culture at a low temperature while in the resting phase to bring all cells to the same condition. Growth is then promoted by raising the temperature to that for incubation, possibly accompanied by an osmotic shock. Grundler[31] reports preliminary results whereby cells stored at 4^oC gave no microwave effects, whereas cells stored at 30^oC did give microwave effects.

The existence of microwave modulation effects, particularly modulations within the ELF range, which covers the most biologically active frequencies, has not been considered or controlled in many experiments. Grundler was stabilising frequency to only ± 1MHz (in 42 GHz) so the spectral composition of the microwaves used would not have been known or controlled from 0.1 Hz to 100 Hz. Mechanical vibration is used by insect systems to generate highly coherent sub-millimeter wave effects[52]. Vibration is difficult to eliminate particularly in urban buildings, and experiments are unlikely to have been made in vibration-free laboratories.

Finally, I could add to this list the possibility that the experimenters themselves may affect the outcome of their experiments. This could happen if the experimenter were radiating biologically active frequencies as has been observed with electrically hypersensitive persons[53] and healers[54]. Robert Jahn, of Princeton University[55], has carried out many experiments with a wide range of mechanical and electrical systems and found that ordinary subjects can, by intention, influence a system in some way and consistently at highly significant levels of probability.

As evidence for operator effects, I have demonstrated in cooperation with Wekroma at Brione, Switzerland[6], that a beaker of water held in the hand for one minute showed changes in its optical spectrum, particularly at wavelengths below 420 nm. It is necessary to use 10 cm optical path length cuvettes in a differential spectrometer, suggesting the possibility of some resonance interaction between the water and the cuvette. Similar results have been independently obtained by Kurick, Institute of Physics, Kiev, Ukraine, and reported at the 1st. International Conference on "Water Systems and Information", Kiev, May 12-17, 1992.

3.4 Threshold Effects

The interesting things in physics happen at thresholds, yet it is characteristic of research into biological effects of microwaves that workers have been more concerned with accurately reproducing power levels, measuring SAR's and building anaechoic chambers than in looking for thresholds. The reason is tied to concepts of ionizing radiation dosimetry and the power levels permitted by regulatory bodies on the basis of thermal effects.

Rea and co-workers have developed an effective protocol for demonstrating electromagnetic field sensitivity in human subjects[56]. The study was carried out in four phases. The first phase devloped criteria for controlled testing using an environment low in chemical, particulate, and EMF pollution. The second phase involved a single blind challenge of 100 patients who complained of EMF sensitivity. The 25 of these who were found to be EMF sensitive were compared to 25 healthy naive volunteer controls, none of whom reacted to any challenge on test. Of the 25 EMF sensitive patients, 16 had positive signs and symptoms plus objectively determined autonomic nervous system changes. In the fourth phase, the 16 EMF-sensitive patients were re-challenged double-blind only to frequencies to which they were found to be most sensitive. These were inserted randomly into 5 placebo challenges. The active challenges were 100% positive and all placebo tests were negative. The frequency range of these tests was from 0.1 Hz to 5 MHz. I have found[53] that some patients show a continuum of sensitivities extending well into the microwave region as seen in Figure 3.1, so this protocol would be applicable to tests at microwave and millimeter wave frequencies.

The approach to the detection of microwave effects used in the former Soviet Union and certain East European countries was based on changes shown by functional states such as reversible changes in nervous and cardiovascular systems and behavioral or psychological changes. Boris Savin of the Institute of Industrial Hygiene and Occupational Diseases, Moscow, has discussed the importance of experimental data on the effect of radio and microwaves on the higher nervous activity for the determination of safety standards[57]. Yuri Kholodov heads the Group of Electromagnetic Neurophysiology at the Institute of Higher Nervous Studies, Moscow, which has studied the reactions of the human nervous system to electromagnetic fields from steady fields to microwaves using the EEG as an indicator[58].

The consequence of these different approaches has been that for many years, safety standards were a thousand times higher in the West than in the East. Hasted wrote in 1973,

"Although it is very difficult to understand in physical terms just how non-thermal effects can arise, it is worth mentioning that there are many responsible scientists in the U.S.A. who do not discount the Soviet reports, and make use of the lower permissible levels in their own laboratories."[9].

Lack of adequate microwave technique has been the basis of much of the Western criticism of work from the former Soviet Union, where workers have on the other hand considered other aspects to be of greater importance. Increasing the incident power density above a threshold does not produce a proportional increase in the

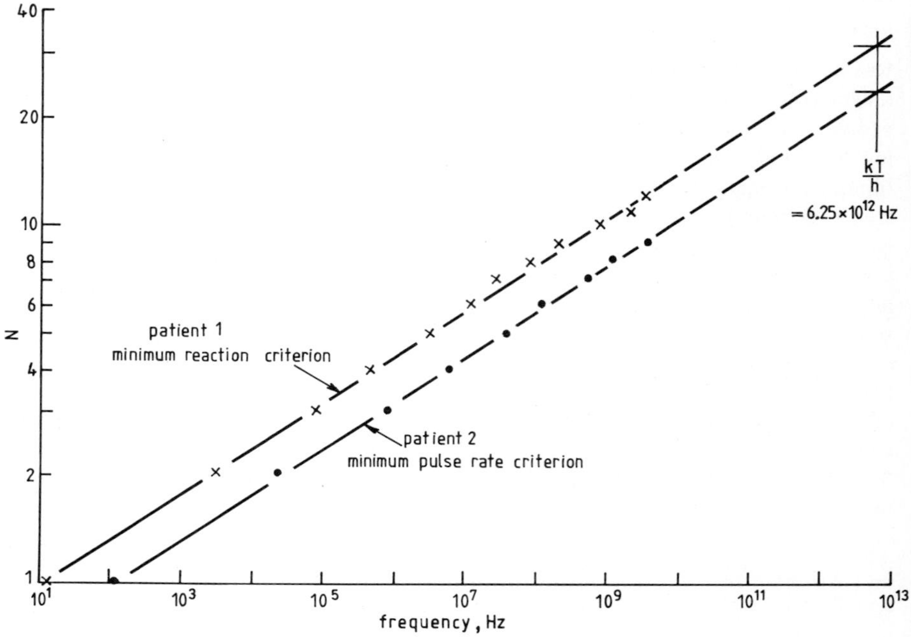

Figure 3.1: *Diagram showing the range of continuous electromagnetic hypersensitivities for two patients. The sections of solid lines represent frequencies which provoked patient specific reactions. Patient 1 had symptoms consistent with non-specific disturbances to the autonomic nervous system. Patient 2 had increased heart rate (tachycardia). The points indicate frequencies which provided neutralization of the respective symptoms. The measurements in the case of Patient 2 were done 'blind' to the patient. The dotted lines are extrapolations to the frequency corresponding to the mean thermal energy (kT). The ordinates are integer numbers representing the order of successive harmonics. The mathematical relationship is characteristic of a fractal equation.*

biological effect so, knowing the power density at the threshold is more important than elsewhere.

To investigate long term irradiation effects, the Institute of Clinical and Experimental Medicine, Novosibirsk, has had single cell lines in culture for 25 years, a feat of which any brewery could be proud. The Institute has found that live and dying cells can communicate very precise information using only optical (ultra-violet) means, and that a laser beam can take the imprint of bio-information and transmit it to other cell cultures.

Workers at the environmental research A. N. Marzeev Institute, Kiev[59], have stimulated autoimmune reactions in rats by microwave irradiation (2375 MHz) getting a destabilization of functional activity of the immune system humoral factors at 500 $\mu W/cm^2$. Cytochrome was affected at these levels. Autoimmunity could be stimulated by immunization of intact animals with brain tissue of rats exposed at 50 and 500 $\mu W/cm^2$, the process was dependent on the microwave intensity[60].

Chronic exposure of domestic fowl to very low intensity microwave radiation was reported by Tanner and Romero-Sierra[61] to give increased mortality rate and profound deterioration in health of the survivors.

A report from The People's Republic of China by Chiang *et al*[62] of investigations into the effects of the environmental microwave exposure of humans, concluded that in the highest exposed groups the visual reaction time and short term memory were worse and that there may have been effects on the CNS and immune systems.

Lester and Moore[63,64] found that cancer tended to occur on leading terrain crests relative to radar transmissions and was less frequent in the valleys, and also that counties in the U.S.A. with an Air Force Base had significantly higher incidences of cancer mortality.

Szmigielski *et al*[65] have carried out a retrospective study from 1971-1980 of immunologic and cancer related aspects of exposure to low-level microwave and radiofrequency fields. It showed a clear increase in the risk of cancer among subjects occupationally exposed to microwaves and radiofrequencies.

Some years ago, Andreyev and co-workers[66] published their first results of indications of special characteristic frequencies in humans, and subsequently a comprehensive report of research into the physical mechanisms of low-intensity radiation on biological systems[67]. The experiments showed that a human organism with functional disorders can distinguish insignificant frequency changes of external electromagnetic radiation in the millimeter frequency band, with a resonance half-width of 20 MHz. Low intensity radiation at specific frequencies in the range 50-70 GHz and at power flux densities from mW/cm^2 to $\mu W/cm^2$ incident on acupuncture zones connected with 'malfunctioning' organs, produces a sensory response accompanied by a strongly pronounced therapeutic effect. The method was tested on more than 4000 patients for some pathologies. When visiting Kiev, I enquired how the choice of millimeter wave frequency was made, and was referred to the microwave oxygen lines in this region all of which have therapeutic applications. Millimeter wave applicators have been designed for various clinical applications. These authors conjecture that,

"An informational connection with an external field and energy transport along limit cycle space trajectories may be conditioned by protein spin states..... Electromagnetic waves in the range 45-65 GHz arising in the organism due to transitions between sub-levels of a triplet spin-spin splitting, provide a universal long-acting coherence which is not limited by nonuniformities of real living structures. The role of short-acting activators may be played by enzyme complexes, as their activity is known to depend on spin orientation of the external electrons in active centres in a trigger way... Thus we consider the living organism to be a quantum system and a dissipative structure formed as a result of a non-equilibrium phase transition which constantly reproduces itself due to self-organization processes."

Devyatkov and co-workers have published a study of millimetric wave interactions with biological objects and water[68].

From the point of view of physics, Fröhlich[51] characterised thresholds and active biological systems by three properties. First, they are relatively stable but far from equilibrium, which require that the various excitations are stabilized, pointing to the existence of metastable states. Second, they exhibit a non-trivial order, which requires a motional order as is found in the existence of macro-wavefunctions in superconductors and superfluids, but which also exists in non-equilibrium systems such as lasers or in maintained particular excitations such as sound waves; its generalisation leads to coherence. Third, they have extraordinary dielectric properties, which arise from the high electric fields maintained in membranes in conjunction with the sensitivity to very weak electromagnetic fields with sharp frequency response.

The coherent excitation of a single polar mode depends in a step-like manner on the rate of energy supply but with a time-lag while coherence becomes established. This type of excitation requires a strong non-linear interaction with the "heat bath" which attempts to impose its temperature on the particle distributions. While in a Bose gas the number of particles is fixed so that Einstein condensation and superconductivity arise only when the temperature is sufficiently lowered, in the biological situation the temperature is fixed and the number of quanta is increased by the rate of energy supply until the threshold is reached.

The ultimate threshold for magnetic effects is the linkage of a single quantum of magnetic flux with the cross-sectional area of a coherent cell or organism. The integral of the magnetic field over a cross-section perpendicular to it is defined as the magnetic flux, Φ, which is shown by Fröhlich to be an integer multiple of the flux quantum, $\Phi_o = h/2e = 2.07 \times 10^{-15}$ Wb. Although this originated in superconductivity, it is not restricted to it but is a completely general phenomenon. The possibility and experimental evidence that living systems are sensitive to, and use magnetic flux quanta, has been presented by Del Giudice *et al*[69]. If a system is able to respond to the magnetic flux quantum, then it has the Josephson effect available.

The Josephson effect is a macroscopic quantum phenomenon of superconductivity in which the current flow between two regions of superconducting long-range order, separated by a barrier which is a non-superconductor, is dependent , not on the voltage between the regions, but on the phase difference, Ψ, of order parameters, which have properties similar to those of wave functions in elementary quantum mechanics. The phenomena are divided into stationary (DC) and non-stationary (AC) effects depending upon whether the variables change with time. The most important of the ac effects are Josephson oscillations which take place if the voltage has a dc component, V, at a frequency, f related to the voltage by (approximately) 500MHz/μV, or exactly:

$$2\pi f = d\Psi/dt = \frac{2\epsilon}{\hbar} \cdot V$$

where e is the electron charge and $\hbar$ is Planck's constant divided by 2π.

3.5 Cellular Basis

The cellular basis for bioelectromagnetic effects in the microwave and millimeter wave region requires a physical model for the biological cell[70]. The frequencies for non-linearly excited coherent oscillations may be based on any of the possible modes of resonance as shown in Figure 3.2, but they may also be much lower as in the case of limit cycles. They may also not be determined by structures but, by hyperfine energy levels.

Any structure, whether a biological system or a musical instrument, will have some natural resonance frequencies determined by its dimensions and the velocity with which waves travel within it. The fact that a cell is visible against its surroundings means that it has a different refractive index and hence waves will be, at least, partially reflected at its boundaries. These waves may acquire energy as they travel and build up to become sustained oscillations.

There are of the order of 3000 enzymes controlling the chemistry of a biological cell. The high catalytic power of an enzyme requires a reduction of the activation energy. A metastable state with a high internal electric field may be nature's way of activating an enzyme catalysed reaction. Frequency selective long-range interactions may arise from the excitation of coherent vibrations and give the selective attraction of enzymatic substrates[71].

The importance of the frequency 100 GHz assumed by Fröhlich for fundamental biological activities was strongly supported by the publication of research from the former Soviet Union[29]. Physically, this frequency corresponds to a an acoustic mode

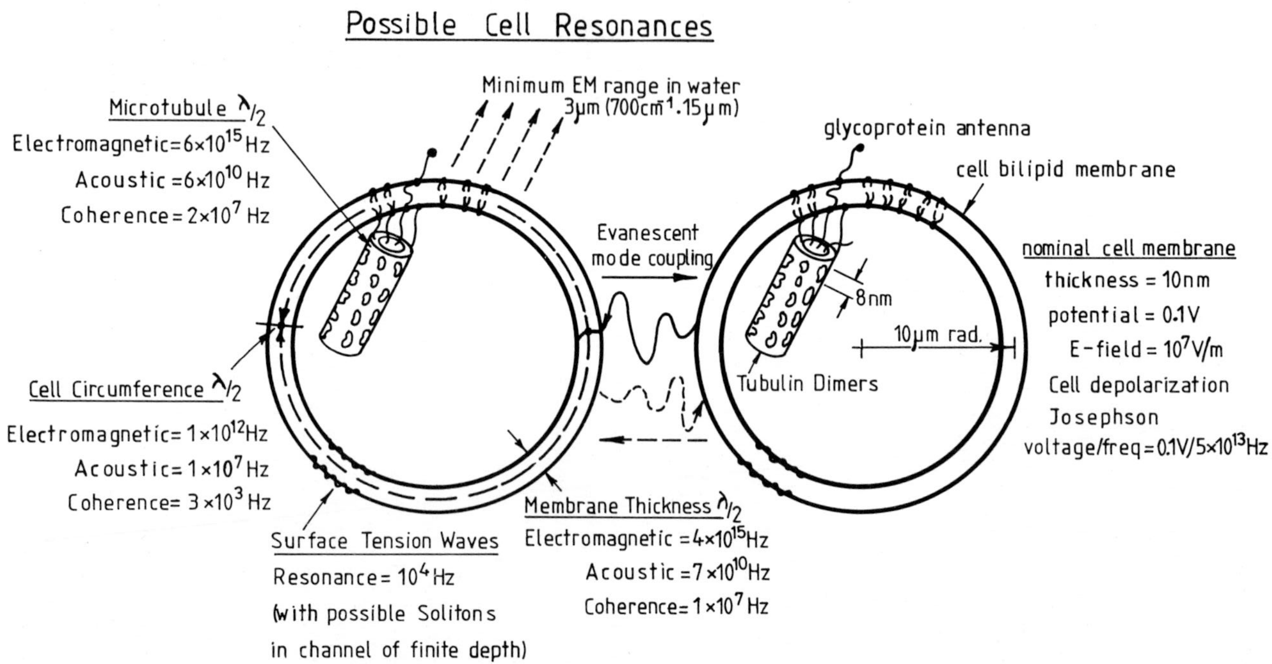

Figure 3.2: *Diagram showing possible resonances relating to the sizes of features of an idealized biological cell. There will also be resonances corresponding to hyperfine energy levels not related to the cell geometry.*

resonance in the thickness of the cell membrane for which its dimension is a half-wavelength (Fig. 3.2).

The stimulated Raman effect has been used to investigate the possibility of far-infrared sub-millimeter coherent oscillations in biological molecules and cells. The effect depends on the coherent pumping of molecules into excited vibrational or rotational states by the electric field of a laser beam which is above some critical intensity threshold. The molecules relax and emit isotropically the inelastic scattered radiation. This relaxation may be coherently stimulated. The intensity of the Raman scattered radiation is three orders of magnitude below the Rayleigh elastically scattered radiation which is at the laser frequency. In the inelastic collisions, the laser photons may lose or gain energy. If they lose energy to the molecules, the Raman scattering appears at a lower frequency giving a Stokes line, if they gain energy from the molecules, the Raman scattering appears at a higher frequency giving an anti-Stokes line. The freqency shift is given by the energy change divided by Planck's constant. The ratio of the anti-Stokes to Stokes intensities compares the number of molecules donating and accepting energy. In 1977, Webb *et al* published a series of Stokes and anti-Stokes laser-Raman spectra of synchronized active cells of *E. coli* bacteria[72]. The average cell size of an *E. coli* cell ranges from 1.1 μm - 1.5 μm wide by 2 μm - 6 μm long over the growth cycle. For the circumferential half-wavelength mode in the lipid bilayer shown in Figure 3.2, the resonances should come within the band of wave numbers 320 cm^{-1} to 100 cm^{-1} according to the above range of cell sizes. The spectra taken by Webb 40 minutes after incubation showed Raman lines at 150 cm^{-1}, at 50 minutes they were at 120 cm^{-1}, and at 60 minutes they were close to 100 cm^{-1}. The corresponding frequencies are from 4.5 THz to 3 THz (1 THz = 10^{12} Hz).

The mean frequency corresponding to thermal energy (kT_{300}) at biological temperatures is 6.25 THz, so that these highly coherent frequencies could be thermally pumped resonances in which a Bose condensation had occurred.

My laboratory was probably the first to demonstrate coherent oscillations in the radiofrequency region of around 8 MHz from yeast cells at the time of cytokinesis[73]. This frequency is the same as the periodicity found in the millimeter wave experiments by Grundler *et al*[31,32]. Webb has pointed out that this is also the frequency that one would expect from the rate constant for the hydrolysis of ATP.

It was only possible to find these coherent 8 MHz oscillations experimentally because a series of steps had been found in voltage-current characteristics of thin films of yeast cells[69]. On previous occasions, cell systems had been found to react to magnetic fields at levels corresponding to a single magnetic flux quantum linking the measured cell cross- section. One of the consequences of magnetic flux quantization is the Josephson effect which gives a frequency to voltage interconversion of 500 MHz/μV. It was possible to adjust the experimental conditions while observing the voltage steps until the expected frequency came within the range of a radio-frequency spectrum analyser. The coherent oscillations were then sought and found. They lasted for only for a few minutes one mean generation time (4 hours) after starting the synchronous cell incubation[73]. Under other experimental conditions, voltage steps corresponding to frequencies up to about 2 GHz were observed[69].

In earlier experiments in my laboratory[74], the enzyme, lysozyme was used as the dielectric in a point-plane electrode configuration within a section of X-band to wave-guide. It was found that on irradiation with 9 GHz, steps appeared in the voltage-current characteristic. Applying the Josephson effect conversion gave the corresponding frequencies of 20 MHz and 40 MHz. The coupling-in of weak signals at either of these frequencies gave a 15-fold increase in the DC conductivity and the steps increased so that they then corresponded to 300 MHz.

All chemical reactions involve energy change and involve quantum physics; there is no chemistry in classical physics. The energy of a chemical reaction is of the order of electron-volts (1 eV = a chemical energy of approx. 100 kJ/mole = a photon wavelength of 1.24 μm). Biochemical reactions catalysed by enzymes require an activation energy to switch on their remarkable effectiveness. That living systems use coherent optical frequencies for biocommunication has been demonstrated by Popp[75] (see also Popp et al, this volume). There are biophotons available in living systems at frequencies high enough for single quantum photochemistry to occur.

Ordinary organic (non-bio-) chemical reactions are affected by an external magnetic fields, static and alternating, from the low audio up to microwave frequencies. McLauchlan[76] has described how weak magnetic fields can affect chemical reactions involving free radicals, which also includes much of essential bioenergetics[17]. The major effect of the magnetic field is to remove degeneracies of the sub-levels of triplet radical pairs whose energies are equal in zero field, but differ progressively and linearly as an applied field is increased, according to the Zeeman effect. The Zeeman separation of the states exceeds the magnitude of the hyperfine interaction at a few milliTesla. Vanag and Kuznetsov[77] have considered how the radical reactions of lipid peroxidation, the enzyme reactions involving paramagnetic molecules, and photochemical reactions could be influenced by magnetic fields through the mechanism of spin exclusion.

Evidence that living systems are making use of the Josephson effect has already been referred to[69]. This offers a postulate for a procedure whereby the power density threshold for the interaction of microwaves with living systems might be determined.

Grundler *et al*[31] arranged their microwave exposure system so that the cells were exposed to the electric field component. If a microwave electric field across a biological structure (e.g. cell, synapse, membrane) is considered as a Josephson weak link experiment and, if the applied microwave radiation is sufficient to give a field related voltage across the weak-link equal to, or greater than the Josephson voltage corresponding to its frequency, then this combination will try simultaneously to force the phase difference of the order parameter and the potential across the weak-link junction, leaving no degrees of freedom to the biological system.

The result might be the paralysis of bio-signals trying to use that resonance to trigger a chemical reaction at the site of the weak-link junction. The range of frequencies over which this could happen will depend on the extent of available resonances or hyperfine energy level structure.

The 42 GHz experiments on yeast cells by Grundler *et al*[31,32] were reported at power flux densities of the order of 1 mW/cm^2 (10 W/m^2). This must have exceeded all the thresholds since highly resonant effects were observed. The corresponding fields over the diameter of the yeast cells cells of mean volume 30 μm^3 would not have been greater than 30 V/m, assuming that the effective dielectric constant was that of water (approximately 16 at 42 GHz). This would give about 900 μV across the cell diameter, the largest likely weak-link, or less, if the weak-link is a sub-cellular structure and is more than 10-times the Josephson voltage of 84 μV corresponding to the frequency 42 GHz. This suggests that the ultimate power density threshold at this frequency might be 100 times less, as little as 10 μW/cm^2 (0.1 W/m^2). This is nearer to the safety levels adopted by the East European countries[40] and suggested effects at still lower power densities[78].

At the 1st. International Conference on "Water Systems and Information" held in Kiev, Ukraine, May 12-17, 1992, I presented evidence that living systems can make use of a low frequency alternating magnetic vector potential. The magnetic flux density B (Tesla) is of mathematical necessity related to a vector **A** such that **B**=curl**A**.

It was long thought that **A** was merely a mathematical convenience, but it was eventually shown experimentally by a number of workers that **A**, the magnetic vector potential, has a physical reality. The most thorough demonstration[79] came from the Hitachi Research Laboratories, near Tokyo. A toroid contains the magnetic field within its volume but the magnetic vector potential spreads into the surrounding space and can influence a beam of electrons in a diffraction experiment by interacting with their wavefunction. This is the Aharanov-Bohm effect, the production of a relative phase shift between two electron beams enclosing a magnetic flux even if they do not experience the magnetic flux[80].

The magnetic vector potential is a vector directed in the same direction as the current giving rise to the magnetic field. A changing magnetic vector potential gives rise to an electric field E thus,

$$\mathbf{E} = -d\mathbf{A}/dt$$

E is not affected by magnetic screening materials so there is the possibility of magnetic effects at near zero values of the magnetic flux density **B** if the magnetic flux to which they are subjected is attenuated by magnetic shielding rather than distance from the current source.

From these experiments, it appears that bio-information might be carried on the magnetic vector potential while the magnetic field formats the medium. If magnetic vector potential effects in living systems extend to the microwave and millimeter wave region, this would provide a possible mechanism for interaction with the long-range order of the postulated Josephson weak-link junctions in living systems through their interaction with the order parameter, which has properties analogous to the wave function in elementary quantum mechanics.

What is now urgently needed is to be able to read the language of electromagnetic bio-communication to complement our understanding of the genetic code.

Bibliography

[1] P. Strong, *Biophysical Measurements* (Tektronix Inc., Beavertron, Oregon, 1970).

[2] J. Frey and R. Bowers,*IEEE Spectrum* **9** (1972) 41.

[3] K. Marha, *IEEE Trans.* **MTT-19** (1971) 165.

[4] S.M. Michaelson, *IEEE Trans.* **MTT-19** (1971) 131.

[5] A.H. Frey,, *IEEE Trans.* **MTT-19** (1971) 153.

[6] C.W. Smith and S. Best, *Electromagnetic Man* (Dent, London, 1989,1990; St. Martin's Press, New York, 1989).

[7] M.M. Zaret, in *Modern Bioelectricity*, ed. A.A. Marino (Marcel Dekker, New York, 1988) p.839.

[8] H.P. Schwan, in *Interactions between Electromagnetic Fields and Cells*, eds. A. Chiabrera *et al* (Plenum, New York, 1985).

[9] J.B. Hasted, *Aqueous Dielectrics* (Chapman and Hall, London, 1973).

[10] R. Pethig, *Dielectric and Electronic Properties of Biological Materials* (Wiley, Chichester, 1979).

[11] E.H. Grant, R.J. Sheppard and G.P. South, *Dielectric Behaviour of Molecules in Solution* (Clarendon, Oxford, 1978).

[12] *New York State Power Lines Project: Biological Effects of Power Line Fields* (N.Y. Dept. of Health, Albany, NY, 1987).

[13] R.O. Becker and G. Selden, *The Body Electric* (Morrow, New York, 1985).

[14] P. Brodeur, *Currents of Death* (Simon and Schuster, New York, 1989).

[15] *Electromagnetics News*, Suite 4, 120 W. Church St., Frederick, MD 21701, U.S.A.

[16] K. Foster and W. Pickard, *Nature* **330** (1987) 531.

[17] S. Best (ed.) *Electromagnetics News* P.O.Box 25, Liphook, Hants GU30 7SE, U.K.

[18] L. Slesin ed., *Microwave News*, P.O. Box 1799, Grand Central Station, New York, NY 10163, U.S.A.

[19] *BENNER Digest Update*, Information Ventures Inc., 1500 Locust St., Suite 3216, Philadelphia, PA 19102, U.S.A.

[20] *Institute of Electrical and Electronics Engineers, Inc. (IEEE)*, IEEE Service Center, 445 Hoes Lane, P.O. Box 1331. Piscataway, NJ 08855-1331, U.S.A.

[21] *European Bioelectromagnetics Association, Proc. 1st. Conference* from: Dr. M. Rooze, Laboratory of Embryology and Human Anatomy, Medical School, Free University of Brussels, 1000 Brussels, Belgium.

[22] *Bioelectrical Repair and Growth Society (BRAGS)*, P.O. Box 64, Dresher, PA 19025, U.S.A.

[23] W.R. Adey in *Modern Radio Science*, ed. J.B. Anderson (URSI/OUP, Oxford, 1990) p.1.

[24] S.M. Bawin, R. Gavalas-Medici and W.R. Adey, *Brain Res* **58** (1973) 365.

[25] W.H. Fletcher, W.W. Shiu, D.A. Haviland, C.F. Ware and W.R. Adey, *Proc. Bioelectromagnetics Soc. 8th Annu. Meeting, Madison WI* (Bioelectromagnetics Society, Frederick MD) p.12.

[26] D.B. Lyle, P. Schechter, W.R. Adey and R.L. Lundak, *Bioelectromagnetics* **4** (1983) 281.

[27] C.F. Blackman, in *Interactions between Electromagnetic Fields and Cells*, eds. A. Chiabrera *et al* (NATO/Plenum, New York, 1985) p.521.

[28] S.J. Webb and D.D. Dodds, *Nature* **218** (1986) 374.

[29] N.D. Devyatkov, *et al, Sov. Phys-Usp* **16(4)** (1974) 568 (Engl. Transl.).

[30] W. Grundler, F. Keilmann, V. Putterlik, L. Santo, D. Strube and I Zimmermann, in *Coherent Excitations in Biological Systems*, eds. H.Fröhlich and F. Kremer (Springer-Verlag, Berlin, 1983) p.21.

[31] W. Grundler, U. Jentzsch, F. Keilmann and V. Putterlik, *Biological Coherence and Response to External Stimuli*, ed. H. Fröhlich (Springer-Verlag, Berlin, 1988) p.65.

[32] W. Grundler, in *Interactions between Electromagnetic Fields and Cells*, eds. A. Chiabrera *et al* (NATO/Plenum, New York, 1985) p.459.

[33] S.J. Webb, in *Nonlinear Electrodynamics in Biological Systems*, eds. W.R. Adey and A.F. Lawrence (Plenum, New York,1984) p.549.

[34] G.S. Edwards, C.C. Davis, J.D. Saffer and M.L. Swicord, *Phys. Rev. Lett.* **53(13)** (1984) 1284.

[35] A.A. Marino, *J. of Bioelectricity* **8(1)** (1989) v.

[36] W.F. Pickard, in *Bioelectromagnetics Soc. Newsletter*, Jan/Feb/Mar (1986).

[37] S. Hameroff, Y.Simickrs, L Vernetti, V.C. Lee, D.Sarid, J.Wiedman, V. Elings, K. Kjoller and R. McCuskey, *J. of Vac. Sci. A* **8(1)** (1990) 687.

[38] M. Welland, *Physics World* **6(1)** (1993) 22.

[39] B. Michel, *Rev. Sci. Instrum.* **63** (1992) 4080.

[40] M. Grandolfo, in *Interactions between Electromagnetic Fields and Cells*, eds. A. Chiabrera *et al* (NATO/Plenum, New York, 1985) p.603.

[41] E. Del Giudice, G. Preparata and G. Vitiello, *Phys. Rev. Lett.* **61(9)** (1988) 1085.

[42] A. Chiabrera, C. Nicolini and H.P. Schwan, eds. *Interactions between Electromagnetic Fields and Cells* (NATO/Plenum, New York, 1985).

[43] R. Pethig, *Electronics and Power* **19** (1973) 445.

[44] R.M. Lebovitz and R.L. Seaman, *Radio Science* **12(6S)** (1977) 229.

[45] S.A. Levine and P.M. Kidd, *Antioxidant Adaptation, Its Role in Free Radical Pathology* (Allergy Research Group, San Leandro CA, 1985).

[46] E. Aarholt, M. Jaberansari, A.H. Jafary-Asl, P.N. Marsh and C.W. Smith, in *Modern Bioelectricity*, ed. A.A. Marino (Marcel Dekker, New York, 1988) p.75.

[47] F. Kaiser, in *Non-linear Electrodynamics in Biological Systems*, eds. W.R. Adey and A.F. Lawrence (Plenum, New York, 1984).

[48] M. Dardalhon, D. Averbeck and A.J. Berteaud, *Radiat. Environ. Biophys.*, **20** (1981) 37.

[49] I.I. Samoilenko, *Prilk. Biokhim. Mikrobiol.* **20(2)** (1984) 230 (Russian with English summary).

[50] A.Z. Smolanskaya, in *Non-thermal Effects of Millimeter Wave Radiation*, ed. N.D. Davyatkov (Acad. Sci. USSR, Inst. Radiotech. Electrotech., Moscow, 1981) p.132 (in Russian).

[51] H. Fröhlich, in *Biological Coherence and Response to External Stimuli*, ed. H. Fröhlich (Springer-Verlag, Berlin, 1988) p.1.

[52] P.S. Callahan *Tuning in to Nature* (Routledge and Kegan Paul, London, 1975).

[53] C.W. Smith, R.Y.S. Choy and J.A. Monro, *Clin. Ecol.* **6(4)** (1989) 119.

[54] M. Usa, T. Maeda, R. Hiratsuka, M. Kobayashi and H. Inaba *Proc. 4th. Symp. on Biological and Physiological Engineering*, (Sendai, Japan, 1989) p.331 (In Japanese with English captions).

[55] R.G. Jahn and B.J. Dunne *Margins of Reality* (Harcourt Brace Jovanovich, Orlando FL, 1987).

[56] W.J. Rea, Y. Pan, E.J. Fenyves, I. Sujisawa, N. Samadi and G.H. Ross, *J. of Bioelectricity* **10(1& 2)** (1991) 241.

[57] B. Savin, *G. Ital. Med. Lav.* **4** (1982) 21.

[58] Yu. A. Kholodov and N.N. Lebedeva, *Problems of Electromagnetic Neurobiology* (Nauka, Moscow, 1988) (In Russian with English table of contents)

[59] G.I. Vinogradova, G.V. Batanov, G.M. Naumenko, A.D. Levin and S.I. Trifonov, *Radiologia* **25(6)** (1985) 840 (In Russian with English abstract).

[60] G.I. Vinogradova and G.M. Naumenko, *Radiologia* **26(5)** (1986) 705 (In Russian with English abstract).

[61] J.A. Tanner and C. Romero-Sierra, *J. of Bioelectricity* **1(2)** (1982) 195.

[62] H. Chiang, G.D. Yao, Q.S. Fang, K.Q. Wang, D.Z. Lu and Y.K. Zhou, *J. of Bioelectricity* **8(1)** (1989) 127.

[63] J.R. Lester and D.F. Moore, *J. of Bioelectricity* **1(1)** (1982) 59.

[64] J.R. Lester and D.F. Moore, *J. of Bioelectricity* **1(1)** (1982) 77.

[65] S. Szmigielski, M. Bielec, S. Lipski and G. Sokolska, in *Modern Bioelectricity*, ed. A.A. Marino (Marcel Dekker, New York, 1988) p.861.

[66] Ye.A. Andreyev, L.N. Christophorov, V.N. Kharkyanen, A.A. Serikov and S.P. Sitko, *Research into the Physical Mechanisms of Action of Low-Intensity on Biological Systems* (Acad. Sci. Ukrainian SSR, Inst. Theor. Phys., Kiev, 1990) ITP-90-49E/50E/79E (In English).

[67] E.A. Andreev, M.U. Beily, S.P. Sitko, *Doklady Akad. Nauk Ukrainsko SSR* **10B** (1984) 60 (In English).

[68] O.V. Betskiy, M.B. Golant and N.D. Devyatkov, *Millimetric Waves in Biology*, (Znanie, Moscow, 1988) (In Russian).

[69] E. Del Giudice, S. Doglia, M. Milani, C.W. Smith and G. Vitiello, *Physica Scripta* **40** (1989) 786.

[70] C.W. Smith, in *Biological Coherence and Response to External Stimuli*, ed. H. Fröhlich (Springer-Verlag, Berlin, 1988) p.205.

[71] H. Fröhlich, in *Adv. in Electronics and Electron Physics*, (Academic, New York, 1980) p.85.

[72] S.J. Webb, K. Lee and H. Fröhlich, *Phys. Lett.* **63A** (1977) 407.

[73] C.W. Smith, A.H. Jafary-Asl, R.Y.S. Choy and J.A. Monro, in *Photon Emission from Biological Systems*, (World Scientific, Singapore, 1987) p.110.

[74] N.A.G. Ahmed and C.W. Smith, *Collective Phenomena* **3** (1978) 25.

[75] F-A Popp, U. Warnke, H.L. König and W. Peschka, eds. *Electromagnetic Bio-Information* (Urban and Schwarzenberg, München, 1989).

[76] K. McLauchlan, *Physics World* **5(1)** (1992) 41.

[77] V.K.Vanag and A.N. Kuznetsov, *Biol. Bull. Akad. Sci. USSR* **24** (1987) 6; **25** (1988) 2.

[78] L.N. Christophorov and A.A. Serikov, in *Problems of Microwave Resonance Therapy*, ed. B.Ugarov (Otlkik, Kiev, 1990).

[79] A. Tonomura, N. Osakabe, T. Matsuda, T. Kawasaki, J. Endo, S. Yano and H. Yamada, *Phys. Rev. Lett.* **56(8)** (1986) 792.

[80] Y. Aharanov and D. Bohm, *Phys. Rev.* **115** (1959) 485.

[81] P.L. Stocklin, *Hearing Device*, U.S. Pat. 4, 858, 612, Aug.22, 1989.

[82] W.B. Brunkan, *Hearing System*, U.S. Pat. 4, 877, 027, Okt.31, 1989.

Chapter 4

Possible Mechanisms for Biological Effects of Weak ELF Electromagnetic Fields

D.T. Edmonds

4.1 Introduction

In this article, I will discuss some of the mechanisms whereby low-amplitude, low-frequency electric and magnetic fields – the so-called ELF fields – may influence biological systems. It is well understood that high frequency electromagnetic radiation such as ultra-violet light or a microwave beam can damage biological tissue by disrupting molecular structure and creating ionization within the tissue. ELF fields do not have the amplitude or frequency to cause heating or ionization and as yet no universally accepted experimental evidence points to their influence upon a biological system. Much of the motivation for investigating this topic has come from epidemiological studies purporting to show that the incidence of some childhood cancers may be increased by proximity to overhead electric power lines. For this reason, I shall concentrate on fields with frequencies from zero (static fields) to a few hundred Hz, electric fields with external amplitudes up to 10kV/m and magnetic fields up to 100 μT (1 gauss) such as are typically found under power lines.

4.2 Background Physics

The fundamental properties of electric and magnetic fields under discussion may be found in undergraduate texts such as Bleaney and Bleaney[1]. The first point to notice is that the energy density of such fields is very small in comparison with that of thermally activated motion. The energy per cubic meter stored in an electric field **E** and in a magnetic field **B** are given below as U_E and U_B, where,

$$U_E = \epsilon_R \epsilon_o E^2/2 \qquad U_B = B^2/(2\mu_R \mu_o) \tag{4.1}$$

and $\epsilon_o = 8.85 \times 10^{-12}$, $\mu_o = 4\pi \times 10^{-7}$, and ϵ_R and μ_R are the relative dielectric constant and relative permeability of the region respectively. For electric fields of 10kV/m and magnetic fields of 100 μT in water, the energy densities becomes,

$$U_E = 3.5 \times 10^{-2} \mathrm{J/m^2} \qquad U_B = 4.0 \times 10^{-3} \mathrm{J/m^2} \tag{4.2}$$

These are indeed very small in comparison with the energy stored in the thermal motion of the water molecules of about 7×10^8 J/m^3 at 20°C.

Another relevant observation is that magnetic fields readily penetrate biological bodies without attenuation but electric fields are often strongly screened from the cell interior. Electrical screening may have two different origins, one associated with the dielectric constant, and the other associated with the motions of free ions in conducting material. If a non-conducting but polarizable parallel-sided slab of material with a relative dielectric constant of ϵ_R is located in an applied electric field $\mathbf{E}_0$ normal to its surfaces as in Figure 4.1a the field within the slab is only $\mathbf{E}_0/\epsilon_R$.

In the applied field, the slab becomes slightly polarized electrically, which leads to the creation of a constant internal reaction field that subtracts from the applied field within the specimen. The reaction field is only uniform for ellipsoids and is maximum for the parallel-sided slab which is the oblate spheroid taken to its limit. For a dielectric sphere, the internal field is parallel to the applied field and has strength $3\mathbf{E}_o/(2+\epsilon_R)$, while for a needle-shaped body aligned along the field, which is the prolate spheroid taken to its limit, the reaction field is zero and the internal field is $\mathbf{E}_o$.

For bodies that conduct electricity, such as the aqueous fluid found within and around biological cells, there exists an even stronger screening effect. Let us consider again the geometry shown in Figure 4.1a, but this time, with a parallel-sided slab of aqueous fluid with dielectric constant ϵ and electrical condutivity σ. At time $t = 0$, let us apply a constant electric field $\mathbf{E}_o$ directed normal to the slab. Let us assume that dielectric polarization is very fast in comparison with ionic diffusion. Initially, no ions have moved and a field $\mathbf{E}_o/\epsilon_R$ exists within the slab. Under the influence of this internal field, positive mobile ions tend to migrate to the right face of the slab and negative mobile ions to the left face. These ions themselves create an internal reaction electric field $\mathbf{E}_r$ in a direction diametrically opposite to the original field which tends to cancel the original internal field.

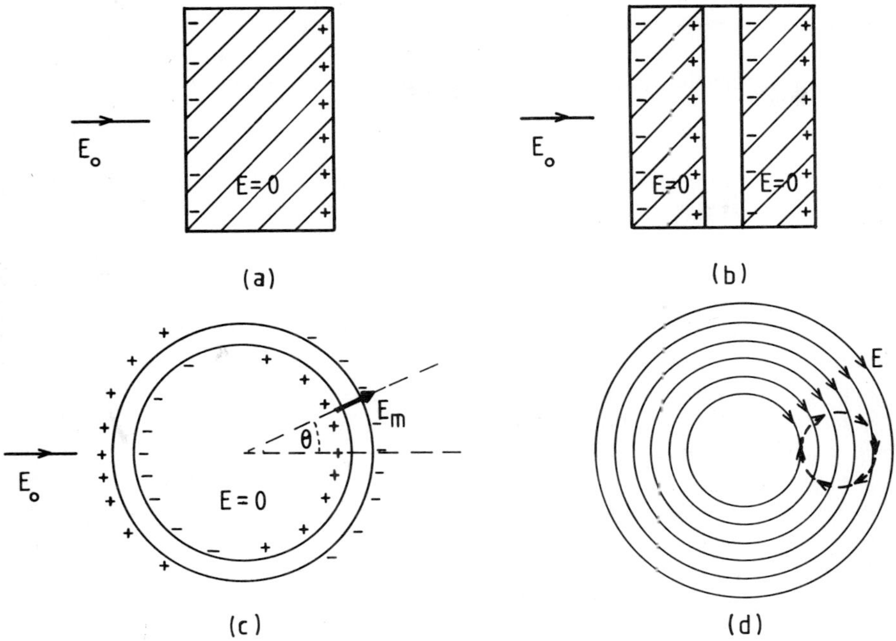

Figure 4.1: *Interaction of electric and magnetic fields with conducting materials of different geometries.*

a. An electric field applied normal to the faces of a parallel-sided slab of permeable or conducting material showing the polarization or free charge buildup that results in a reaction field E_r *opposing the applied field. b. An insulating slab embedded in a conducting slab showing how the interfacial charge that cancels the field in the conducting regions can add to the field across the insulator. c. A cell of radius R with a conducting interior separated from a conducting medium by an insulating membrane of thickness d showing how the field in the interior is cancelled by interfacial charge but the field across the membrane at a position defined by* Θ *is increased to* $1.5 \times E_o(R/d)\cos(\Theta)$. *d. The circular directions of the electric field at time t induced by an axially symmetric alternating magnetic field applied normal to the plane of the diagram. The* emf *taken around the dotted circle is given by the time rate of change of the flux of B which threads the circle even although the electric field takes all angles to the circle from parallel to antiparallel.*

Ions will continue to move so long as any uncompensated electric field exists within the slab and the end result is that the applied field is totally screened from the body of the slab by layers of charge on its two surfaces. It may be shown that the electric field within the slab, $\mathbf{E}_i$, will decay exponentially with a time constant T as in the equation below:

$$\mathbf{E}_i = (\mathbf{E}_o/\epsilon_R)exp(-t/T) \quad \text{where} \quad T = \epsilon_R\epsilon_o/\sigma \tag{4.3}$$

Using a conductivity of 1.5 mho/m, typical of a biological fluid, T becomes 4.7×10^{-10}s. Thus, an abruptly applied electric field only penetrates the slab for a very short time before it is totally neutralized by ionic screening. Alternating electric fields in the frequency range of interest (0-200Hz) are also effectively screened at all times as the response time of the ions is so short in comparison with the period of the applied field. If a sinusoidal electric field of angular frequency ω given by $\mathbf{E}_o \cos \omega t$ is applied normal to the slab, the electric field inside the slab is $(\omega\epsilon_o/\sigma)\mathbf{E}_o \sin \omega t$ at the frequencies ($\omega \ll 1/T$) of interest here. For example, at 100Hz, using the parameters above, the ratio of internal to external fields is 3.7×10^{-9}, giving very severe attenuation. It is interesting to note that when a sinusoidal field is applied, the current density within the slab is given by,

$$\mathbf{J} = \sigma\mathbf{E}_i = (\omega\epsilon_o)\mathbf{E}_o \sin \omega t \tag{4.4}$$

which is 90° out of phase with the applied field and with an amplitude that does not depend upon the conductance σ. At frequencies much higher than our range, the capacitance of the membrane allows the fields to penetrate the cell.

In particular geometries, it is possible for the accumulation of charge at interfaces discussed above to increase the electric field locally. This situation is illustrated in Figure 4.1b.

Here, a parallel-sided slab of insulator is included within the conducting slab. When the steady electric field is applied, the mobile ions in the conducting fluid at both sides of the insulating sheet move such as to cancel any field within the fluids as before. However, the effect now is that the field across the insulating sheet has been enhanced by the sheets of interfacial charge adjacent to it. In a similar manner, if a conducting spherical cell of total radius R, contained within an insulating membrane of thickness d, is placed in a conducting fluid subject to uniform applied electric field $\mathbf{E}_o$, the field within the cell is zero, but the field across the membrane at a position defined by an angle θ as in Figure 4.1c, may be shown to vary as $1.5(R/d)\cos\theta$ (see also Tsong and Gross, this volume).

Thus, the voltage drop across the two membranes at the poles of the cell ($\theta = 0^o$ and 180°) is 1.5 times bigger than the drop in the original field over a distance equal to the diameter of the cell. For a cell of radius 10μm with a membrane thickness of 5nm, the maximum amplification of the field across the membrane at its poles (R/d) is 2000.

A real biological specimen is highly inhomogeneous due not only to its plasma membrane but to other internal membranes and structures with a variety of conductivities. The simple examples of charge accumulation at the boundaries between high and low conductance regions that we have discussed serve as a warning that even if an electric field of high symmetry is applied to a biological specimen, the pattern of internal fields and induced currents may be very complex indeed.

When dealing with ions it is important to realize that the size of the unit charge (1.6×10^{-19}C) is very large at the microscopic level. Thus unlike the diffusion of a quasi-continuous fluid, the electrical charge structure is highly granular so that strong electrical effects are created with very few ions. Let us consider a spherical cell of radius 10μm surrounded by a lipid membrane of thickness 5nm and relative dielectric constant, 2, containing a particular univalent cation at a concentration of 0.1 M and with no voltage across the membrane. If that ion is selectively pumped from the cell until a voltage of -100mV has been created across the membrane, the concentration of the ion within the cell will have dropped by only about 10 parts per million. Thus the migration of a small proportion of any particular ion can have large local electrical effects.

We turn now to the difference in the influence of electric and magnetic fields upon the motion of ions. For an ion of charge q in an electric field **E** the force is qE and acts in the direction of **E**. The same ion travelling with velocity v in a magnetic field **B** experiences a force $qvB\sin\theta$ where θ is the angle between the directions of **B** and v. In this case the force acts in a direction perpendicular to both **B** and v. Many differences between electric and magnetic fields are now apparent. Electric fields act on static and moving ions whereas magnetic fields only act upon moving ions. Because the electric force is in the direction of **E** it may change both the speed and the direction of motion of the ion. However, because the magnetic force upon an ion moving with velocity v is always perpendicular to v, the magnetic force may change the direction of the ion but cannot change its speed. Finally magnetic fields exert no force on ions moving along the direction of the magnetic field whereas electric fields act upon all ions whatever their direction of motion.

To gauge their relative strengths, we will compare the forces exerted by an electric and a magnetic field on a particle with charge q moving at a speed v when the two fields have the same energy density. In this case

$$B/E = [\mu_R\mu_o\epsilon_R\epsilon_o]^{1/2} = n/c = 1/v_c \tag{4.5}$$

where c is the velocity of light in a vacuum and n is the refractive index of the material so that v_c is the velocity of light in that medium. The ratio of magnetic to electric forces is given by

$$F_B/F_E = v(B/E) = v/v_c \tag{4.6}$$

From this it can be seen that the magnetic force on a diffusing ion is much smaller than the electric force when the two fields have comparable energy densities.

Finally I will treat the generation of electric fields by a time varying magnetic field. Electric fields may be generated by static electric charges as in a parallel plate capacitor, but a time varying magnetic field also creates an electric field through the Faraday effect. The equation may be written

$$\oint \mathbf{E}.\mathbf{dl} = -\frac{d}{dt}\int\int \mathbf{B}.\mathbf{dS} \tag{4.7}$$

Stated in words, at any time t, the line integral of the tangential component of the electric field around any closed path is minus the rate of change of the flux of the magnetic field B passing through the area enclosed by that same path. The line integral of the electric field $\mathbf{E}$ is known as the electro-motive force or emf around that circuit and a charge q which moves around the circuit will gain electrical energy given by $q(emf)$.

First, it should be noted that this law in itself is not sufficient to define uniquely the induced electric field. To do that, some extra information of the form of the magnetic field is required. If the magnetic field is generated by a set of coils with axial symmetry about the vertical, the magnetic field also has this axial symmetry and the electric field induced will be directed in circles about the axis as in Figure 4.1d.

Over the central region in which the magnetic field amplitude at any instant may be taken as not changing with the radius R about the axis, the strength of the electric field induced at a radius R from the axis is given by

$$\mathbf{E}(R) = -\frac{R}{2}\frac{d\mathbf{B}}{dt} \tag{4.8}$$

Note that the law of induction holds for any closed path and not only for those that follow the direction of the induced electric field. If, for example, we are interested in the emf about the circular path shown dotted in Figure 4.1d the law still holds. As we pass around the dotted path it can be seen that the direction of the induced electric field changes from parallel to the path to antiparallel to it and back to parallel to it with all angles in between. However because the amplitude of the induced field rises with the radius from the axis there remains a net emf around the path given by the negative rate of change of flux of magnetic field through it. There exists some confusion in the literature on this point where it is sometimes assumed that the existence of an emf around a given path means that the electric field is everywhere directed along the path. It should also be borne in mind that within a biological system the conductivity will be highly inhomogeneous and that even if the electric fields applied have a high symmetry the current induced may form a very complicated pattern.

A characteristic of electric fields created by electro-magnetic induction is that the electric field forms closed loops. This is expressed mathematically by saying that

$$\nabla.\mathbf{E} = 0 \tag{4.9}$$

or that the total flux of the induced electric field through the surface of any volume is zero. This is not true for electric fields generated by static charges where the charges act as sources and sinks of the electric field. This difference is very important in biological systems because electric fields in closed loops cannot be cancelled out by the build up of charged ions as we discussed above for an external electric field applied to a uniform conducting slab. If in Figure 4.1d an axially symmetric time-varying magnetic field is applied normal to a uniform fluid slab, circles of electric field in the plane of the slab will be induced. The induced electric fields will generate circular currents but no cancellation of the field is possible by charge accumulation within a uniform slab. Although the presence of cells in the uniform slab will again lead to screening effects on the microscopic scale, an oscillating magnetic field is often the preferred method of creating internal electric fields. In our discussion of time varying magnetic fields we will distinguish between truly magnetic effects and electrical effects when the electric fields are generated by magnetic fields.

4.3 Effects of Static Magnetic Fields

One area within which the experimental evidence is clear is in the detection of the earth's static magnetic field by animals as an aid to navigation. These range from magnetotactic bacteria[1,2] through fruit flies[3], bees[4] and the newt[5] to migrating birds[6] (see also Warnke, this volume). For example, the rate of firing of some nerves in the head of a pigeon may be shown to depend sensitively upon the direction and amplitude of the applied magnetic field. In fact three apparently independent detectors have been found[7]. One acts upon the pineal gland and is thought to help control melatonin secretion. A second is detected on the optic nerve and depends on light reaching the eye and a third, particularly sensitive detector, does not depend upon light but the location and nature of the transducer is unknown.

In the case of the bacteria the mechanism is clear. They synthesize internally an aligned chain of ferrimagnetic magnetite crystals. The torque exerted by the earth's magnetic field of about 50 μT on the magnets is such that the bacteria have no option but to move along the direction of the field! Very small single crystals of magnetite ($< 0.01\mu$m) are only weakly magnetic (paramagnetic) but larger crystals are magnetically ordered through the exchange interaction and form powerful single-domain ferrimagnets[8]. Still larger crystals ($> 0.5\mu$m) divide into oppositely magnetized domains and lose their powerful magnetic moment. Living creatures synthesize magnetite and can control both the size and the shape of the crystals that form[9]. Since the discovery of the magnetotactic bacteria, single domain magnetite crystals of the optimum size have been found in the heads of birds[10,11] and in fish[12]. A magnet of magnetic moment m experiences no translational force in a uniform magnetic field $\mathbf{B}$ but it is acted on by a couple of strength $mB\sin\theta$ tending to align the moment along the field, where θ is the angle between the directions of the moment and the field. A major attraction of the magnetite detector is that the effect is sufficiently strong to raise it above thermal noise. A very small ($0.1\times0.025\times0.025\mu$m) single domain magnetite crystal that is free to rotate is more than twice as likely to be aligned with its magnetic moment along the direction of a magnetic field of 50μT as against it. There have been several attempts to model biological field detectors based upon magnetite including one involving modified pressure sensors[13], and another involving magnetically switched trans-membrane ion channels[8].

Because the avian compass is an inclination detector[14] which detects the angle between the directions of the field and of gravity, I have attempted to construct[15,16] a model of an avian compass based upon magnetite in which the torque is detected by a hair cell in a macula. The maculae are important acceleration and gravity detectors in vertebrates[17,18]. They are located in the inner ear and consist of roughly hemispherical cavities with hair cells protruding into the cavity normal to its walls. The hair cells are loaded towards their tips with weights (the *statoconia*) consisting of single crystals of calcite. Any acceleration acts upon the statoconia and the torque produced is registered by the hair cells. The hair cells are extremely sensitive[19] and have a large amplification whereby a motion of the tip of a single hair cell of as little as 10-100 nm results in the opening of up to 100 trans-membrane ion channels and the transmission of a train of nerve impulses. The model demonstrates that incorporating single domain magnetite crystals among the statoconia could give rise to a field detection system, based upon a null detector of magnetic torque, with many of the characteristics of the avian compass. Although black magnetic particles (probably magnetite) have been found among the statoconia of the guitarfish[20], the model remains speculative and as yet no firm experimental evidence exists to confirm the operation of a magnetite magnetic field detector in a living creature other than in bacteria.

The first attempt to construct a theoretical model of a magnetic field detector based on an optical detector was by Leask[21]. Whether or not the suggested mechanism of optical pumping is confirmed by experiment, the paper was important in that it directed the attention of experimentalists for the first time to the eye as a magnetic field detector and since that time the Newt[5], the fly[3] and the pigeon[7] have been shown to use optical magnetic field detectors. Optical pumping is an optical/radiofrequency double resonance technique often demonstrated in undergraduate physics teaching laboratories using low pressure metal vapour as specimen. In these circumstances it can detect the direction and the amplitude of the earth's magnetic field to high accuracy. A simplified version of the effect is shown in Figure 4.2a.

Incoming light excites a transition within the photo-receptor molecule between the ground state G and a broadened excited state A. Intermolecular coupling transfers some of this excitation to a pair of triplet states (spin 1) with energy levels C and D. Because of the interaction of the spin components with the earth's magnetic field $\mathbf{B}$, these states differ in energy by $hf = \gamma B$ where γ is a parameter that may depend upon the direction of $\mathbf{B}$ relative to the photo-receptor molecule.

Let us suppose that the transition probability for the transition $D - G$ is much higher than that for $C - G$. The effect of this assumption is that the number of photo-receptors resting in state C is much higher than those in state D. However if the system is irradiated around the radio-frequency f , this will stimulate transitions between C and D and effectively empty state C through transitions to D and hence swiftly to G. The detected output signal could be an increase in the emission of light corresponding to transitions from D to G. The amplitude of the magnetic field may be measured by observing the variation of the output while the radio-frequency f is varied.

It can also be detected at constant f if the strength of B is varied or, because of the dependence of γ on direction, if the direction of $\mathbf{B}$ is varied. The great advantage of the technique is that a radio-frequency photon of low energy triggers the emission of a very much higher frequency optical photon so that the signal-to-noise ratio may be high. The drawbacks of the scheme as a biological mechanism are that it is not easy to visualize a sufficiently stable internal source of the radio-frequency irradiation and that the likely effective widths of the two energy levels C and D in biological tissue are much larger than the energy separation hf so that no distinct resonance may be detectable.

Another suggestion for an optical magnetic field detector is based upon the modulation of the rate of radical pair recombination. Consider a molecule that consists of two portions coupled together by a covalent bond. In its singlet (spin zero) ground state the two electrons participating in the covalent bond have their spins antiparallel. Let us suppose that the molecule is photoexcited to a triplet (spin 1) state which leads to the rupture of the covalent bond. As angular momentum is conserved during rupture, the two radicals will be formed with parallel spins.

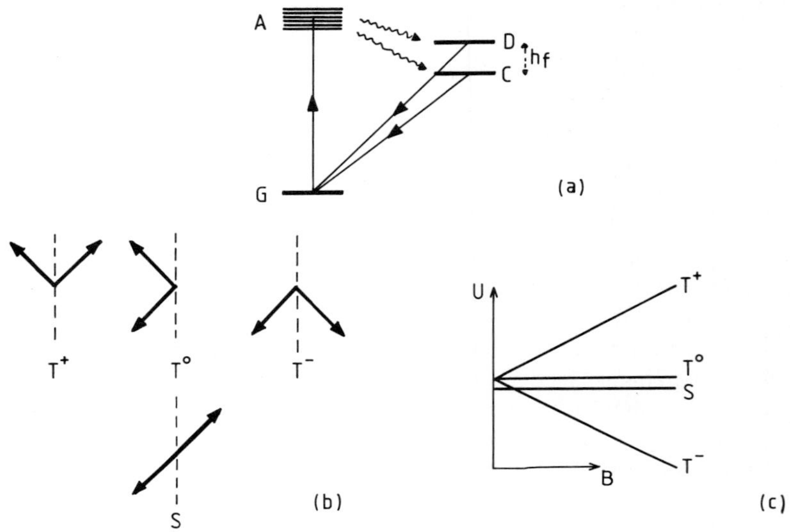

Figure 4.2: *a. An energy level diagram illustrating optical pumping. The incident light excites a transition from the ground state G to a broadened absorption level A. Excitation may then be transferred non-radiatively to the two triplet levels C and D which are separated in energy by their interaction with the applied magnetic field. The transition probability from D to G is assumed to be fast and that from C to G, slow. The population of systems accumulating in C may be rapidly depleted by applying a radio-frequency photon hf which connects C to D. Enhancement of optical emmission from D to G is triggered by the small radio-frequency photon hf.*

b. Semiclassical representation of the the orientation of the two interacting spins in the three triplet states and the one singlet state rotating about the axis of quantization. c. An energy level diagram of the three triplet states and the one singlet state as a function of the applied magnetic field B.

Usually, the immediate recombination into the excited triplet state is ruled out on energetic grounds so that the two radicals can only recombine to form the singlet ground state if the spins become antiparallel. As we shall see, a magnetic field can change the rate of spin orientation and there is a competition between the rate at which the radicals diffuse apart and the rate at which the spins precess to allow recombination in the singlet state. In this manner a magnetic field can change the proportion of radicals that recombine or drift apart to become available for different chemical action which leads to chemistry that depends directly upon the presence and magnitude of a magnetic field.

The technique has been thoroughly reviewed by Steiner and Ulrich[22], its role as a possible explanation of biological effects has been discussed by McLaughlan[23], and a model avian compass based on this effect has been described by Schulten[24]. Here I will attempt a simplified description that illustrates the principles.

The three triplet states of the radical pair may be written as $T^+(a_1a_2), T^-(b_1b_2)$ and $T^0(a_1b_2 + b_1a_2)$ in which the total spin has a positive, negative or zero projection on the axis of quantization. The brackets in each case give the form of the wave function where a represents the up state and b the down state of each electron. The singlet state may be written $S(a_1b_2 - b_1a_2)$. A semi-classical representation of these states in terms of the directions of the individual spins is given in Figure 4.2b.

In zero applied field, the magnetic field acting upon the spin of each radical is due to the magnetic moments of the nuclei within the fragment. The spin of each radical will now precess about this magnetic field at the Zeeman rate for that field. As the direction is arbitrary and different for the two radicals, any re-orientation of the spins is thus possible including the triplet to singlet conversions that would allow recombination.

In an external field large in comparison with the hyperfine fields of the nuclei, both fragments must precess about the same direction and conversion of T^+ or T^- to S is no longer possible, which reduces the possibility of recombination. Thus at applied fields much smaller than the hyperfine fields we would expect no effect but as the field grows in magnitude comparable to the hyperfine field, the rate of recombination should increase and then saturate at a constant higher level when the applied field is much larger than the hyperfine field. In fact, as the applied field increases beyond the range of interest here, other effects start to operate which lead once more to a reduction in recombination.

As this is a complicated mechanism it is worth looking at it again from the slightly different point of view of the energy level diagram as shown in Figure 4.2c.

For an applied magnetic field along the quantization axis this shows the Zeeman splitting of the triplet levels with applied field. The difference in energy between T^0 and S is the exchange energy between the spins of the fragments, here assumed small. Each energy level should be imagined to have a width of about U_{hf} due to the hyperfine interaction between the spins and the nuclear moments. Provided the

(Zeeman) energy splitting in the external field is small in comparison with U_{hf}, all four energy levels overlap and any interconversion is possible. However as the external field increases and the levels diverge it is seen that T^+ and T^- have energies appreciably different from S so that the interconversion of triplet to singlet states is reduced.

The strength of this proposal is that at fields greater than about 1 mT, the mechanism is readily demonstrated and well understood. It also has a direct impact on the chemistry of the system from which biological effects could flow. However as a mechanism for the detection of fields of order 50 μT, some questions remain unanswered. One is whether the relevant photo-excited biological molecules will have a sufficiently small hyperfine interaction that a field effect at 50 μT becomes observable. And a second is whether there will be time for appreciable state conversion under these weak fields to occur before diffusion separates the radical pair. For this second reason particular attention may be focused on biologically sequestered radical pairs with reduced probability of separation by diffusion.

4.4 Effects of Time-varying Magnetic Fields

In this section, I will include mechanisms that require both static and varying magnetic fields, but the generation of electric fields using time-varying magnetic fields as we have discussed above will be dealt with under electric field effects. The best known mechanisms[1] are those of electron spin resonance (ESR) and nuclear magnetic resonance (NMR). Consider a body with angular momentum such as a spinning gyroscope with the axis of spin making an angle θ to the vertical. Because of its mass a static gravitational couple acts upon the gyroscope and this causes its spinning axis to precess about the vertical. Energy must be added to or subtracted from the system in order to change the angle θ. In the case of an atom or a molecule with the angular momentum due to electronic motion and spin, the motion is quantized. When placed in a steady magnetic field, because of its magnetic moment, it is acted upon by a steady magnetic couple which again leads to precession but now the projection of the angular momentum on the direction of the field can only take certain particular values. If a small amplitude oscillating magnetic field of frequency f is applied, such that hf is equal to the difference in energy between two of the quantized energy states, it can cause the system to make a transition between these two states. The resonance is detected either by the change in direction of the spin axis or by the absorption of the applied photon. The technique is of general interest in the present context as it illustrates the enhancement of the signal-to-noise ratio possible by tuning the detector. NMR at 50 MHz is easily detected at room temperature with a signal-to-noise ratio of a few hundred. A 50 MHz photon has an energy hf of about 3.3×10^{-26}J and yet its absorption is easily detected in the presence of room temperature thermal photons of energy kT of about 4.0×10^{-21}J, about

100,000 times bigger than the detected photon. The reason is that the frequency of the resonance is highly stable and the resonance linewidth small so that the detector can be tuned to be sensitive only to a very small range of frequencies around the resonance frequency. By contrast, the background thermal excitation is not tuned and is generally evenly distributed over a very large frequency range. Thus it is only a very small proportion of the total background energy that disturbs the signal detection within its known narrow frequency bandwidth.

A major limitation of conventional magnetic resonance as a technique for measuring very small amplitude magnetic fields is not the signal-to-noise ratio but the fact that in a real system, the energy levels have a finite width. If that width is bigger than the energy level separation due to the field that is to be detected, then no distinct resonance will be observed. A major source of energy level broadening is the background of magnetic fields generated spontaneously within the system under study. Thus, if a molecule is subject to fluctuating magnetic fields due to neighbouring magnetic atoms or nuclei, then no distinct resonance is likely to be observed for applied magnetic fields that are smaller than or comparable to the internally generated fields. This constraint is likely to limit severely ESR or NMR as biological field detectors in our present range of interest below 100 μT.

Another type of magnetic resonance much discussed recently in this context[25] has been cyclotron resonance. Let us consider a free particle of mass m and charge q moving at speed v in the plane perpendicular to the direction of a steady magnetic field $\mathbf{B}_0$. At equilibrium, the particle will move in a circle with the inwardly directed magnetic force qvB_0 balancing the outwardly directed centripetal force. The radius of the circle depends upon the speed v but the frequency at which the particle circulates is independent of that radius and is given simply by

$$f_c = (1/2\pi)(qB_0/m)$$

This synchronism was exploited in the cyclotron particle accelerator. If a small amplitude alternating magnetic field is applied parallel to the steady field at the cyclotron frequency given above, then energy may be transferred from the alternating field to the particle to increase its velocity and hence its radius of motion. In a highly evacuated vessel the cyclotron resonance of electrons and of protons has been detected. It can be shown[1] that a clear resonance is measurable only if the particle may complete on average at least one revolution without a collision to disturb its motion.

However an ion in solution is very different from a free particle in a vacuum in that its motion is highly damped by constant collisions with the solvent molecules. It can be estimated that the mean distance travelled between collisions of a small ion in aqueous solution at room temperature is less than the diameter of the ion whereas the radius of the cyclotron orbit for a bare ion such as K^+ moving with a thermal velocity is of order 1 m. This effectively rules out cyclotron resonance of ions in solution. If, in order to combat this limitation, the ion is constrained to a small orbit mechanically or electrically, then the resonance is again destroyed[26].

Another possible type of ionic resonance is Larmor precession[27]. Here we deal with a tightly bound ion rather than the free ion. Say, we have an ion of mass m and charge q tightly bound to a central point so that it oscillates in space about the point. The Larmor theorem then states that if a steady field B is applied, the motion of the ion will remain unaltered except that the whole system will precess about the direction of the field with a frequency f_l, which is half the cyclotron frequency for the same ion and is given by

$$f_l = (1/2\pi)(qB/2m)$$

A characteristic of the precession is that f_l is independent of the frequency of the ion vibration, it need not be simply harmonic, and may even be driven by stochastic noise[28]. A possible application of this effect is to an ion performing thermally activated vibrations while bound within a protein. Calmodulin[29] is an enzyme which acts as a catalyst in the phosphorylation of myosin. It will only function if it has bound calcium ions. On binding calcium ions, it undergoes a major structural change. If it is assumed[27] that the calmodulin may take up a whole spectrum of marginally different conformations which depend upon the particular position of the calcium ion within the binding cavity, and if it is further assumed that the different conformations have different efficiencies as catalysts, then the extra Larmor motion could have a bearing on the catalysed chemistry. This rather elaborate illustration was chosen because there is experimental evidence that a static magnetic field of magnitude about 50 μT may change the rate of calmodulin catalysed phosphorylation of light chain myosin[30].

In addition to the possible effects of Larmor precession in a static field, it can be shown that resonant Larmor effects are possible at particular frequencies if a small amplitude alternating field is applied parallel or at an angle to the static field. As in magnetic resonance which the effects resemble, a very small amplitude oscillating magnetic field can have a major influence on the motion of the vibrating ion in the static field. These effects have been reviewed elsewhere in some detail recently[27].

4.5 Electric Field Effects

Unlike magnetic fields where the number of mechanisms proposed for biological effects are small, there are numerous possible mechanisms for electric fields. At the microscopic level, many of the control mechanisms in living systems are electrical. Examples of possible sites of action are trans-membrane ion channels and pump proteins, the enzymes and their substrates whose activities can be modulated by charge alteration and changes in electrolytes around the cell. Any one of many possible mechanisms will doubtless lead to biological effects if the internal electric field is sufficiently large. The task here is to assess which mechanism might have a significant effect against the background of thermal agitation when the amplitude of the externally applied field is limited to say 10 kV/m, particularly bearing in mind how effectively electric fields may be screened by ionic fluids.

Because of the effectiveness of mobile ions in cancelling the external electric field in a cell surrounded by a non- conducting membrane described above, it is unlikely that electric fields in our range of interest (static or alternating) will have an appreciable direct effect upon the internal functioning of the cell. For this reason attention has been focused upon possible electrical effects outside the membrane or within the membrane which may influence the internal functioning of the cell indirectly.

If the alternating electric field is generated by applying an alternating magnetic field, then any effect observed could be either magnetic or electric in origin. These alternatives may be readily separated experimentally[31] by exposing the cells in a circular dish which is divided into concentric circular channels. If the dish is placed at the centre of an axially symmetrical coil-assembly and perpendicular to its axis, then, if the diameter of the coils is much greater than that of the dish, the amplitude of the applied field may be taken as constant across the dish. However the induced electric field increases with the radius from the axis as we saw above. Thus by assessing the relative effects in the separate circular channels it can be determined whether the mechanism depends upon the amplitude of the magnetic or electric fields.

One interesting suggestion for an electric field effect has been called electroconformational coupling (see Tsong and Gross, this volume). It has been thoroughly reviewed[32] and has as its central postulate that a protein carrying out some cyclical function, such as a trans-membrane ion pump, undergoes structural changes during the cycle and that these different structures will have different electric dipole moments. Consider one such transition between states A and B in which the electric dipole moment changes by $\mathbf{dP}$. If the transition probabilities between these states are k_{AB} and k_{BA}, then their ratio must[33] obey the relation,

$$k_{AB}/k_{BA} = exp[(U_A - U_B)/kT] \tag{4.10}$$

where U_A and U_B are the energies of the two states. If an electric field -$\mathbf{E}$ acts in the direction of dP then $(U_A - U_B)$ increases by $\mathbf{E.dP}$ and the ratio of the transition rates changes, which, in turn changes the outcome of the cycle. In the case of alternating fields, the possibility of resonant effects arises when the period of the applied field matches that of the enzyme cycle. An attraction of the model is that its effect is cumulative, so that even if the field effect on a single cycle is small it will accumulate over time. The model has been applied[34] to membrane bound Na^+, K^+ATPase which pumps sodium ions out of the cell and potassium ions into the cell using the hydration of ATP to ADP as the power source. As the enzyme is located in the membrane, an amplification of the applied field results as we described above. However it must be remembered that within a $5nm$ thick lipid bilayer surrounding a cell across which is a voltage drop of 100mV, the internal electric field is 2×10^7V/m, which is very large and indeed beyond the electrical breakdown voltage of most materials. Thus the change in the electrical energy of the enzyme conformations brought about by the applied oscillating ELF electric field may represent a very small percentage change in the electrical energy which these same conformations possess in the static trans-membrane field.

Experiments show, however, that oscillating electric fields do change the rate at which rubidium (substituting for potassium) ions are pumped across the membrane of red blood cells[35] or a microsomal preparation from rabbit kidney[36]. The effects are Ouabain-sensitive which does suggest the involvement of Na^+, K^+ATPase. An alternative explanation of the effect[36] has been given based upon changes in the concentrations of ions available for binding to the enzyme at the cell surface and in particular on the relative mobilities of different ions.

We turn now to further considerations of the possible effects of electric fields upon the distribution of particular ions around the cell. If we assume that the cell surface is negatively charged, then close to the surface, there exists the Debye layer within which the voltage created by the surface charges decays to zero very approximately exponentially. This local potential changes the local concentrations of the various ions according to the Boltzmann factor. If at a distance z from the surface, this surface potential is $-V(z)$, then the concentrations of univalent cations of charge q at z is given by $C(+,z)$ where,

$$C(+,z) = C(+,\infty)exp(+qV(z)/kT) \tag{4.11}$$

and $C(+,\infty)$ is the cation concentration in the bulk fluid remote from the cell wall. If $-V(z) = -60$ mV then the concentration enhancement factor is over 11 for univalent ions and 121 for divalent cations, while the concentrations for the equivalent anions are lowered by the same factors. Thus surrounding a negatively charged cell, there exists a sheet of cations at high concentrations trapped close to the surface by the surface potential but free to diffuse along the surface. In an applied electric field, the ions will tend to move tangentially to the surface depleting the cations at one pole of the cell and accumulating them at the other to form an electric dipole until back diffusion limits further motion and a steady state is reached[37]. The effect is called counterion polarization and is well known[38] to lead to very high effective dielectric constants for fluid suspensions of small charged particles. Detailed calculations of this effect show[39] that for a cell of diameter 20 μm in an alternating electric field of frequency 15 Hz, the amplitude of the field at the cell need only be about 4×10^{-3}V/m to produce an effect on counterion redistribution as large as the random effects produced by thermal fluctuations. This field is about 100 times smaller than that required to produce a trans-membrane voltage larger than that caused by thermal fluctuations[40,41] in the absence of tuning or averaging.

However, even at this threshold field the effect is very small and, assuming divalent counterions, it corresponds to a few counterions in every million moving between the hemispheres of the cell under the influence of the applied field. In view of this it is suggested[39] that any biological effect is more likely to be a disturbance of the electrical potential between cells due to the electric dipole moments of the polarized cells than a direct effect of a change in the counterion concentration at the cell surface. These effects provide another example of the extreme granularity of the electrical fluid.

A somewhat similar effect is the electrophoresis of charged membrane-bound molecules in a steady electric field. In an applied field of 400 V/m corresponding to a voltage drop of about 12 mV across a cell, membrane bound Concanavalin A receptors have been observed[42] to migrate towards the negative pole of the cell in a time of the order of 4 hours, in agreement with simplified theory[43]. Similarly, cultured cells taken from *Xenopus laevis* embryos have been observed[44] to elongate perpendicular to an applied field of 500 V/m over an observation period of 90 minutes. Such migration leads to inhomogeneity of different parts of the cell surface and may well play an important role in the differentiation of a developing embryo[45].

Another effect much studied is whether electric fields with amplitudes in our range of interest could change the probability of binding of a substrate to a site when these are oppositely charged. In larger electric fields, this is essentially the Wein 'dissociation field effect' first explained by Onsager in 1934[48]. He showed that the ratio of the transition probabilities operating between a state in which two oppositely charged ions are closely associated (an ion pair) and a state in which the two ions are apart (separate ions) is changed by the presence of an applied electric field such as to favour the separated state. The mechanism is simply that the electrical forces acting on the two ions are opposite in direction no matter what the direction of the field. A recent thorough theoretical investigation of ligand receptor binding using the classical Langevin equation[47] concludes that for applied electric and magnetic fields in our range of interest any effects will be totally obscured by thermal fluctuation unless the ligand can be confined within a cavity or held by strong endogenous electric fields such as to markedly increase the ligand escape time.

Finally, when calculating signal-to-noise ratios to assess whether a particular mechanism for interaction with an electric field is realistic, is is usual take a spherical cell about 10μm in radius. It has recently been pointed out that much higher signal-to-noise ratios are possible if the cell is elongated in the direction of the applied field[40], or if a chain of cells is interconnected say by gap junctions along this direction[48]. In these cases, a bigger voltage drop is encountered for the same applied field. Also, living systems are capable of adaptation so that if a given signal must be detected to ensure survival, the system may change[49] to enhance its ability to detect that particular signal. In electronic design, it is well established that if the amplitude and time characteristics of the signal to be detected are well-known, then a special 'matched' filter may be designed which will ensure an optimally high signal-to-noise ratio for the detection of this particular signal.

4.6 Conclusion

I have attempted to describe some of the theoretical mechanisms that have been suggested for the interaction of low-frequency low-amplitude electric and magnetic fields with living cells. The choice of mechanisms is somewhat arbitrary and determined by personal opinion as to which are physically the more probable. With few exceptions, such as the oriented motion of magnetotactic bacteria, none of the mechanisms has been established as being involved in particular experimentally observed phenomena. One of the most serious criticisms of this subject has been the inability of different experimental groups to reproduce supposedly positive results. Good physical reasons may lie behind this apparent randomness such as the recent report[50] that the ability of the fruit fly to detect the earth's magnetic field is totally abolished by switching on a fluorescent light in a neighbouring room with no light path connecting them. The effect has been shown to be due to the radio-frequency signals generated by all fluorescent lamps because surrounding the lamp by a Faraday screen abolished the effect. However, whatever the causes, until reproducible experimental results are published by different experimental groups, the subject will not be taken seriously by the scientific community at large, and theoreticians will not be prepared to expend much energy in attempting to explain experimental results that come and go. On the positive side, the rigour of experimentation is increasing rapidly and agreed experimental protocols (such as careful screening of stray fields) are being adopted so that it should not be too long before we can confirm or reject particular theoretical mechanisms as the explanations for particular experimentally confirmed events.

Bibliography

[1] Bleaney, B.I and Bleaney, B. (1976) In *Electricity and magnetism*, Oxford University Press, Oxford,U.K.

[2] Blakemore, R.P. (1975) Magnetotactic bacteria. *Science* **190** 377-379.

[3] Phillips, J.B. and Sayeed, O. (1993) Wavelength-dependent effects of light on magnetic compass orientation in *Drosophila melanogaster. J. Comp. Physiol A* **172** 303-308.

[4] Kirschvink, J.L. and Kirschvink, A.K. (1991) Is geomagnetic sensitivity real? Replication of the Walker-Bitterman magnetic conditioning experiment in honey bees. *Amer. Zool.* **31** 169-185.

[5] Phillips, J.B. and Borland,S.C. (1992) Behavioural evidence for use of a light-dependent magnetoreception mechanism by a vertebrate. *Nature, Lond.* **359** 412-144.

[6] Wiltscho, W. and Wiltscho, R.(1990) Magnetic orientation and celestial cues in migratory orientation. *Experientia* **46** 342-352.

[7] Semm, P. and Beason, R.C. (1990) Sensory basis of bird orientation. *Experientia* **46** 372-378.

[8] Kirschvink, J.L. and Gould, J.L. (1981) Biogenic magnetite as a basis for magnetic field detection in animals. *Biosystems* **13** 181-201.

[9] Mann, S., Moench, T.T. and Williams, R.J.P. (1984) A high resolution electron microscope investigation of bacterial magnetite. Implications for cyrstal growth. *Proc. R. Soc. Lond.* **B221** 385-393.

[10] Walcott, C. Gould, J.L. and Kirschvink, J.L. (1975) Pigeons have magnets. *Science* **205** 1027-1029.

[11] Beason, R.C. and Nichols, J.E. (1984) Magnetic orientation and magnetically sensitive material in a trans-equatorial migratory bird. *Nature Lond.* **309** 151-153.

[12] Mann, S., Sparks, N.H.C., Walker, M.M. and Kirschvink, J.L. (1988) Ultrastructure, morphology, and organization of biogenic magnetite from sockeye salmon *Oncorhynchus nerka*: implications for magnetoreception. *J. Exp. Biol.* **140** 35-49.

[13] Semm, P. and Beason, R.C. (1990) Responses to small magnetic variations by the trigeminal system of the Bobolink. *Brain Res. Bull.* **25** 735-740.

[14] Wiltscho, W. and Wiltscho, R. (1988) Magnetic orientation in birds. *Curr. Ornithol.* **5** 67-121.

[15] Edmonds, D.T. (1992) A magnetite null detector as the migrating bird's compass. *Proc. R. Soc.* **B249** 27-31.

[16] Edmonds, D.T. (1993) A model avian compass based upon a magnetite null detector. *J. Nav.* **46** 359-363.

[17] Wilson, V.J. and Jones, G.M. (1979) in *Mammalian vestibular physiology*, Chapter 3C, New York: Plenum Press.

[18] Smith, C.A. (1985) The middle ear. In *Form and function in Birds* (eds. A.S. King and J. McLelland) pp273-310, London:Academic Press.

[19] Ashmore, J.F. (1991) The electrophysiology of hair cells. *Ann. Rev. Physiol.* **53** 465-476.

[20] O'Leary, D.P.O., Vilches-Troya, J., Dunn, R.F. and Campos-Munoz, A. (1981) Magnets in guitar fish vestibular receptors. *Experientia* **37** 86-88.

[21] Leask, M.J.M. (1977) A physiochemical mechanism for magnetic field detection by migratory birds and homing pigeons. *Nature, Lond.* **267** 144-145.

[22] Steiner, U.E. and Ulrich, T. (1989) Magnetic effects in chemical kinetics and related phenomena. *Chem. Rev.* **89** 51-147.

[23] McLaughlan, K. (1992) Are environmental magnetic fields dangerous? *Physics World* January 1992.

[24] Schulten, K. (1982) Magnetic fields in chemistry and biology. *Adv. Solid State Physics* **22** 61-83.

[25] Smith, S.D., McLeod, B.R., Liboff, A.R. and Cooksey, K. (1987) Calcium cyclotron resonance and diatom mobility. *Bioelectromagnetics* **8** 215-227.

[26] Halle, B. (1988) On the cyclotron resonance mechanism for magnetic field effects on transmembrane ion conductivity. *Bioelectromagnetics* **9** 381-385.

[27] Edmonds, D.T. (1993) Larmor precession as a mechanism for the detection of static and alternating magnetic fields. *Bioelectrochemistry and Bioenergetics* **30** 3-12.

[28] Edmonds, D.T. (1993) Larmor precession as a mechanism for the biological sensing of small magnetic fields. In *Electricity and Magnetism in Biology and Medicine* (ed. M. Blank) pp.553-555, San Francisco, San Francisco Press.

[29] Babu, Y.S., Bugg, C.E. and Cook, W.J. (1988) Structure of Calmodulin refined at 2.2 Å resolution. *J. Mol. Biol.* **204** 191-204.

[30] Markov, M.S., Wang, S. and Pilla, A.A. Effects of low frequency sinusoidal and dc magnetic field effects on myosin phosphorylation in a cell-free preparation. *Bioelectrochemistry and Bioenergetics* **30** 119-125.

[31] Misakian, M. and Kaune, W.T. (1990) Optimal experimental design for it in vitro studies with ELF fields. *Bioelectromagnetics* **11** 251-255.

[32] Tsong, T.Y. and Astumian,R.D. (1987) Electroconformational coupling and membrane protein function. *Prog. Biophys. Mol. Biol.* **50** 1-66.

[33] Klein, J.K. and Meijer, P.H.E. (1954) Principle of minimum entropy production. *Phys. Rev.* **96** 250-255.

[34] Tsong, T.Y. and Astumian, R.D. (1986) Absorption and conversion of electric field energy by membrane bound ATPases *Biolectrochemistry and Bioenergetics* **15** 457-476.

[35] Serpersu, E.H. and Tsong, T.T. (1983) Stimulation of a Ouabain-sensitive Rb+ uptake in human erthrocytes with an external electric field. *J. Mem. Biol.* **74** 191-201.

[36] Blank, M. and Soo, L. (1990) Ion activation of the Na,K-ATPase in alternating currents. *Bioelectrochemistry and Bioenergetics* **24** 51-61.

[37] Schwarz, G. (1962) A theory of the low-frequency dielectric dispersion of colloidal particles in electrolyte solution. *J. Phys. Chem.* **66** 2636-2642.

[38] Pethig, R. (1979) *Dielectric and Electronic Properties of Biological Materials*, New York. John Wiley and Sons.

[39] Polk, C. (1992) Counter-ion polarization and low frequency, low electric field intensity biological effects. *Bioelectrochemistry and Bioenergetics* **28** 279-289.

[40] Weaver, J.C. and Astumian, R.D. (1990) The response of living cells to very weak electric fields: The thermal noise limit. *Science* **247** 459-462.

[41] Adair, K.A. (1991) Constraints on biological effects of weak extremely-low-frequency electromagnetic fields. *Phys Rev.* **43** 1039-1048.

[42] Poo, M. and Robinson, K.R. (1977) Electrophoresis of concanavalin A receptors along embryonic muscle cell membrane. *Nature Lond.* **265** 602-605.

[43] Jaffe, L.F. (1977) Electrophoresis along cell membranes. *Nature, Lond.* **265** 600-602.

[44] Luther, P.W., Peng, H.B. and Lin, J.J. (1983) Changes in cell shape and actin distribution induced by constant electric fields. *Nature, Lond.* **303** 61-64.

[45] Jaffe, L.F. and Nuccitelli, R. (1977) Electrical controls of development. *Ann. Rev. Biophys. Bioeng.* **6** 445-476.

[46] Onsager, L. (1934) Deviations from Ohm's law in weak electrolytes. *J. Chem. Phys.* **2** 599-634.

[47] Chiabrera, A., Bianco, B., Kaufman, J.J. and Pilla, A.A. (1992) Bioelectromagnetic resonance interactions:endogenous field and noise. In *Interaction Mechanisms of Low-level Electromagnetic Fields in Living Systems* (eds. B. Norden and C.Ramel), Oxford University Press.

[48] Pilla, A.A., Nasser, P.R. and Kaufman, J.J. (1993) On the sensitivity of cells and tissues to therapeutic and environmental electromagnetic fields. *Bioelectrochemistry and Bioenergetics* **30** 161-169.

[49] Barnes, F.S. (1993) Possible mechanisms by which biological systems can separate coherent signals from noise.In *Electricity and Magnetism in Biology and Medicine*, (ed. M. Blank), San Francisco, San Francisco Press.

[50] Sayeed, O. and Phillips, J.B. (1993) Radio frequency radiation disrupts magnetic orientation by *Drosophila melanogaster*. 1993 Conference of Royal Institute of Navigation on *Orientation and Navigation. Birds, Humans and other Animals*, Oxford U.K.

Chapter 5

The Language of Cells – Molecular Processing of Electric Signals by Cell Membranes

Tian Y. Tsong and Carol J. Gross

5.1 Introduction

There are many potential mechanisms whereby a cell can sense electromagnetic signals originating endogenously or from an external source. To understand these mechanisms, we need to recognize certain unique properties of molecules in cell membranes which are crucial for sensing.

5.1.1 Anisotropic Reactions in A Cell Membrane

A transmembrane protein does not have as many degrees of freedom as a molecule in a homogeneous solution. Because a transmembrane protein is embedded in the lipid bilayer whose local viscosity is high, it has a relatively slow translational mobility (1/100 or smaller than that in an aqueous phase). It may be allowed to rotate along the axis normal to the membrane surface but rotation along other axes is not permissible. Flip-flop of molecules across the lipid bilayer is also too slow to be useful. In addition to the limited mobility, a transmembrane protein usually exhibits different affinities for ligands and reactivities to other molecules on the two sides of a membrane. For example, Na,K-ATPase of erythrocytes exhibits higher affinity for K^+ on the outer, than the inner surface of the plasma membrane whereas affinity for Na^+ is the converse: higher on the inside than outside. Similarly, the binding site for ouabain is external and that for ATP is internal. Hormone receptors have even more complex asymmetries across the cell membrane. On account of the compartmentalization of chemical reactions in the cell by membranes, these reactions cannot be dealt with by conventional equilibrium and kinetic analyses. Energy coupling between an energy producing reaction, e.g. the hydrolysis of ATP, and an energy consuming reaction, e.g. pumping of an ion up its concentration gradient, cannot be treated by simple mass action. As we shall see below, the membrane reaction can be adequately treated as a vector-to-vector interaction, and the reactivity of molecules considered separately along each space coordinate. A membrane is recognized to be an anisotropic medium, distinguishable from an isotropic medium such as an homogeneous solution, or a gas phase, in which reactants can approach each other from any direction, and their reactivities are indistinguishable in all directions[1,2]. A reactant in an anisotropic medium may become anisotropic and react only with other molecules approaching in the right orientation. Figure 5.1 uses simple pictures to explain these concepts.

The distinction between an isotropic reactant and an anisotropic reactant is sometimes arbitrary, depending not only on the mobility of the molecule but also on the frequency of an oscillatory driving force (see below). For example, a molecule tumbling with a relaxation time of nanoseconds may be considered isotropic if it reacts with another molecule at a slow rate, say, in millisecond. However, when it interacts with polarized light of a much higher frequency than the rotational frequency of the molecule, it becomes anisotropic. In other words, when it is exposed to a short pulse of polarized light, only those molecules with their transition moments falling in the plane of polarization and parallel to the electric vector of the light beam will be excited and those molecules with their transition moments perpendicular to the plane of polarization will not be affected. This phenomenon is the basis of nanosecond fluorescence depolarization spectroscopy. As we shall discussed below for a biological system, the oscillatory driving force is often of low frequency. In this case, a fast tumbling molecule is isotropic, while oriented molecules, molecules in a microstructure, or molecules in a highly viscous medium will be anisotropic.

5.1.2 Vectorial Driving Force

Similarly, a driving force can be directional. The chemical potential of a proton, sodium, potassium, or chloride gradient is vectorial because only a channel or an enzyme which allows the passage of an ion in the direction of its gradient can use this potential energy. Channels or enzymes in the wrong orientation will not be able to transduce the energy of the gradient.

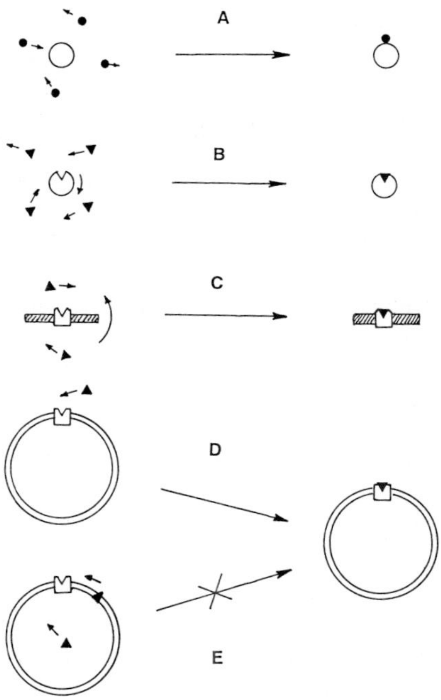

Figure 5.1: *Isotropic and anisotropic chemical reactions. Ligand (filled circles or triangles); macromolecule (large open circle, large open circle with a cavity, large open squares with a cavity).*

Case A, isotropic: The ligand can bind to the macromolecule at any location, and both ligand and macromolecule can diffuse in solution with no apparent directionality. Case B: The macromolecule has a specific binding site for the ligand. If its rotational relaxation is much faster than its diffusive collision with the ligand, the reaction is isotropic. If the rotational relaxation is slow compared to the collision rate of ligand, the ligand binding reaction becomes anisotropic. Case C: The macromolecule is embedded in a piece of membrane. This situation is similar to Case B, although, in most cases, the reaction would be anisotropic. Cases D and E: Compartmentalization of reaction. The ligand can only approach the macromolecule, immobilized in the cell membrane, from the extracellular space. These reactions are anisotropic.

This concept has been implicit in most current treatments of membrane transport. However, many of these discussions are limited in scope and have not explicitly recognized a more important aspect of the vector- to-vector interaction, namely, *the interaction of an anisotropic molecule with a vectorial driving force which oscillates between two opposite polarities*, such as the interaction of a transmembrane enzyme and an alternating electric field (AC field). A symmetrical, bipolar AC field has a time-average field intensity of zero. However, effects produced by such interactions are generally non-linear, and effects due to the positive phase and those due to the negative phase of an AC field will not cancel each other[3,4]. In addition, there are special kinetic effects which transfer the energy in the applied field to specific conformational states of a membrane protein, so that the energy stored transiently in these conformational states can be coupled to a ligand binding reaction to perform chemical work[1–4]. An oscillating potential of defined frequency and amplitude possesses all the characteristics of a signal, and if such an oscillating potential can be recognized by a molecule, that molecule functions, in effect, as a signal transducer. This is the basic premise we will pursue in order to understand the language of cells[5]. When the process is reversed - and most reactions under physiological conditions are reversible - the cell should also be able to generate an electric signal. Figure 5.2 shows a molecule with two conformational states of different electric charge separations, or dipole moments, interacting with electric fields of either positive polarity, negative polarity or oscillating polarity.

The cell can process electric fields of different magnitudes. We shall arbitrarily define the magnitudes, expressed as transmembrane potentials, to be high if it is 200 mV or greater, medium if it is between 1 and 200 mV, and low if it is below 1 mV. We now have a fairly good understanding of the effects of the high and medium electric fields on ATPases of cell membranes, but we know very little as to how a cell, or molecule, can transduce a low level electrical signal. Most of the experimental observations on the effects of low level electric fields on cell function are not very easily reproduced experimentally, in part because the ambient electrical noise is often of the same orders of magnitude, or much higher than those of the applied electric fields. However, if one begins to examine possible mechanisms, based on established principles of physics and chemistry, one might be able to design better experiments and obtain more reliable results to enable us to draw useful conclusions.

5.1.3 Signal and Language

An oscillating electric field that can be recognized or transmitted by a cell for communication, either to coordinate or regulate reactions inside the cell, or to establish a link with other cells, may be called a language[5]. Typical oscillatory electric fields that exhibit characteristics of a language are shown in Figure 5.3. These wave forms have intensity factors and frequency factors. The former is determined by the interaction energy and the latter reflects the rate of a chemical process being affected

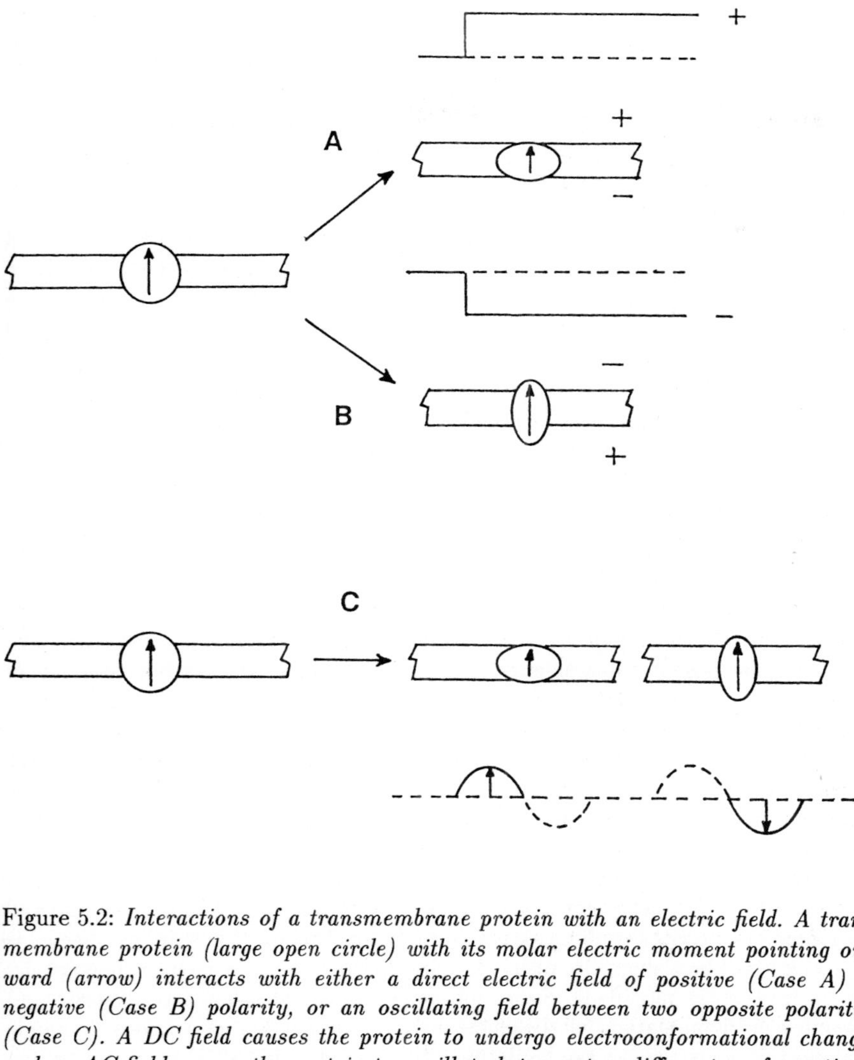

Figure 5.2: *Interactions of a transmembrane protein with an electric field. A transmembrane protein (large open circle) with its molar electric moment pointing outward (arrow) interacts with either a direct electric field of positive (Case A) or negative (Case B) polarity, or an oscillating field between two opposite polarities (Case C). A DC field causes the protein to undergo electroconformational changes and an AC field causes the protein to oscillate between two different conformations.*

by the field[1,3]. Mechanisms for the transduction of different levels of electric signals will be discussed below.

As mentioned earlier, the electric signal in a cell is mostly processed by the membranes. The plasma membrane obviously handles intercellular communication and should be able to relay a signal from the outside to internal organelles. Because such a function involves processing electric signals in a broad range of intensities and waveforms, the plasma membrane is enriched with various types of ion channels, ATPases, and proteins which perform redox reactions[4]. Mitochondrial membranes are the sites for the synthesis of ATP. ATP synthesis requires a considerable amount of free energy (10 - 15 kcal/mol under normal physiological conditions) and transducing this magnitude of free energy demands a large electric field (a transmembrane potential of 200 mV or greater). The inner membrane is able to support such high fields. The nuclear membrane is known to be porous, which would make it less suitable for sustaining an electric potential. However, recent experiments show that there are ions channels of large conductivities in the nuclear membrane which may modulate gene expression. Membranes of organelles obviously can also dispense electric signals[6–8]. These membranes are easily influenced by pH and ions.

The relevance of the electric activation of an enzyme to the *in vivo* regulation of cellular function has often been questioned by critics. Experiments show that an oscillatory electric field is more efficient for energy coupling than a direct current (DC) electric field in the case of Na,K-ATPase. However, most cell physiologists believe that the transmembrane potential of a cell or an organelle is constant, except during the passage of an action potential. For example, the resting potential of a neuron is -70 mV and that of the mitochondrial inner membrane, 200 mV. A constant electric potential across a cell membrane will hardly serve as a signal, which, by definition, should contain specific information. These supposedly constant potential values are really averages measured over a long time interval and at a low spatial resolution. Molecules in the cell membranes are much more dynamic, and their function depends on the electric potential in their immediate vicinity. Only the short range electrostatic interactions play a role in electric signal transduction. Because of the very heterogeneous arrangement of molecules and charges in a cell membrane, its electrical state is also far from homogeneous. When an ion channel is open, there will be a short circuit in its vicinity, where the transmembrane potential will be transiently modified. The membrane thus provides both steady-state and dynamic electric potentials for the efficient control of membrane reactions. Dynamic local electric fields are not randomly fluctuating electric fields[6], they contain many messages and energy packets essential for the regulation and the coordination of intra- and inter-cellular activities. They are generated and sustained by metabolic energy and are different from equilibrium noise. In order to decipher the language of cells, one must study how these transmembrane proteins can respond to electric stimulation, and how applied AC fields of specific frequency and amplitude can affect specific cellular functions.

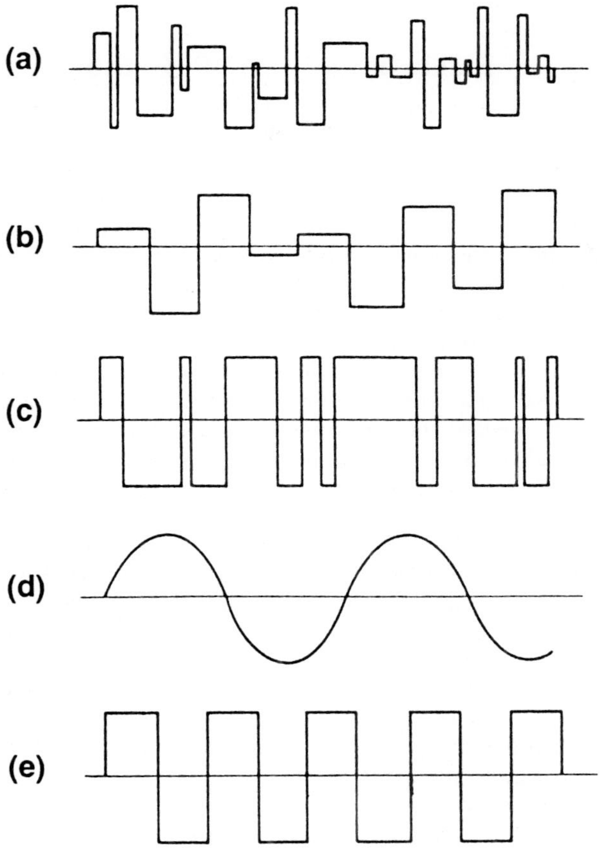

Figure 5.3: *Various Waveforms of Electric Fields Characteristics of A Signal Waveforms (a), (b) and (c) are fluctuating electric fields and (d) and (e) are regularly oscillating electric fields. In (a) both amplitude and lifetime change according to some error functions; in (b) only amplitude changes; and in (c) only the lifetime changes. (d) is a cosine wave and (e) is a square wave or meander function signal. An oscillating or fluctuating electric field is considered a signal if it is energy-sustained. An equilibrium noise is not a signal.*

5.2 Electric Activation of Membrane ATPases

5.2.1 Electric Field Induced ATP Synthesis in Energy Transducing Membranes

According to Mitchell's chemiosmotic hypothesis, the synthesis of ATP from ADP and Pi derives its energy from the electropotential energy of protons. The electrochemical potential energy consists of ΔpH the proton gradient and $\Delta\Psi_{membr}$ the membrane potential and either term can be a driving force. Most experiments designed to measure ΔpH across the mitochondrial membranes, e.g. by phosphorous NMR of lipids, have failed to detect any significant value. It would appear that the membrane potential is the dominant term *in vivo* for ATP synthesis. Experiments measuring equilibrium partition of ions or charged dyes have confirmed that $\Delta\Psi_{membr}$ is close to 200 mV for activated mitochondrial membranes. Witt and coworkers in 1976 reported that by exposing thylakoids to pulsed electric fields of about 1 kV/cm and milliseconds duration, ATP was formed in the absence of light (see reviews[1−4]). However, the ATP synthesized was less than one per enzyme per electric pulse. And it was later shown that an electric pulse simply propitiated the release of nucleotides tightly bound to the enzyme. So, enzyme catalysis due to the pulsed electric field has not been demonstrated in these experiments. The low yield of ATP in these experiments was not unexpected, as the authors used electric fields that were not of sufficient intensity to supply adequate free energy for ATP synthesis. ATP synthesis in these membranes requires approximately 10 to 15 kcal/mol of energy. For the transmembrane potential to play a role, a pulsed electric field must generate an E_{membr} sufficiently large so that with a reasonable value of ΔM, the molar electric moment of the enzyme, the interaction energy, $\Delta M E_{membr}$, will be of the same order of magnitude as 15 kcal/mol. In subsequent experiments with mitochondrial membranes, we used much higher electric fields of 10 - 30 kV/cm which could generate $\Delta\Psi_{membr}$ of 50 - 150 mV across the submitochondrial particles. Because the high fields also produced enormous heating, pulses of microseconds duration were used. ATP yield was still low, less than one per enzyme per pulse. However, when we added dithiothreitol (DTT) to the medium to prevent lipid peroxidation which might occur during electric treatment, ATP yield increased to 5 - 10 ATP per enzyme per electric pulse. Enzyme catalysis had occurred. This proved to our satisfaction that electric energy alone can fuel *de novo* ATP synthesis. In our experiments, the suspension medium usually contained rotenone or cyanide, or both, to inhibit oxidative phosphorylation and electron transport. Therefore, the free energy for ATP synthesis could not have come from the normal oxidative phosphorylation pathway.

In other words, the cell can respond to an electric field and capture energy from it to drive an energetically unfavorable reaction. The level of electric field to accomplish this task is high, and can hardly be considered a language. Nevertheless, these experiments give us some insight into how an electric field may interact with a cell at the molecular level. The electric field interacts directly with the ATPase such that the effects of electric fields are completely abolished with inhibitors of the ATPase. We also showed that the integrity of cell membranes was essential: membranes permeailized either by ionophores, uncouplers, or electroporation also abolished the observed effects. Permeabilized membranes are short- circuited by ion leakage and hence cannot maintain a transmembrane electric potential. Consequnetly, no electroconformational coupling can occur and no ATP is synthesized.

The effects of the sulfhydrial reagent, DTT, to promote enzyme catalysis is less clear at the moment. A tentative explanation is that when an ATP is released, there is oxidation of certain SH groups, which would have to be reduced again, either by the electron transport reaction *in vivo*, or by DTT *in vitro*, before the active enzyme is regenerated.

Obviously, the amount of free energy for driving the synthesis of ATP from ADP and Pi depends on the concentrations of ATP, ADP and Pi in the medium. The phosphorylation potential is a term used to express the level of free energy required for overcoming the uphill reaction. In experiments where the phosphorylation potential was low, the level of electric fields required for ATP synthesis was also low, as expected. We have used low level AC fields to stimulate ATP synthesis in a submitochondrial suspension containing hexokinase to convert the newly formed ATP quickly into glucose-6-phosphate. This ATP trap significantly reduced the phosphorylation potential of the mixture, and AC fields as low as 60 V/cm were able to induce *de novo* ATP synthesis. The rate of ATP synthesis was low, but after 10 - 30 minutes of continuous stimulation, more than 10 ATP per enzyme was obtained. With AC fields, there appeared to be a window around 10 Hz for effective energy coupling (to be published).

5.2.2 AC Induced Cation Pumping by Na,K- ATPase

The AC stimulation of cation pumping by the Na,K-ATPase of human erythrocytes presents a more complete set of data for analysis than studies with other enzymes. Human erythrocytes in suspension were stimulated by AC fields of intensities up to 30 V/cm and of frequencies from 1 Hz to 10 MHz. Ouabain sensitive pumping of K^+, Rb^+ and Na^+ was monitored with radioactive tracers. The AC stimulation experiment was performed initially at 4°C to avoid ATP-dependent pumping. This made the background activity low, and it was possible to measure with confidence the field-dependent part of the cation transport activity. The crucial point was to show that the AC fields in the range used in these experiments did not cause nonspecific leakage of these ions. We did not detect ouabain-insensitive transport of these ions under the influence of the applied fields. All electric-field stimulated activity was inhibited by ouabain, indicating that Na,K-ATPase was the prime target. The salient features of these experiments[9–11] are summarized below.

The AC stimulated only the Na^+ and the K^+ pumping modes of the enzyme. Other modes, e.g. the Na^+/K^+ exchange mode was not affected. This is shown clearly in the data presented in Table 5.1. The net AC stimulated pumping (the quantity shown in the column of S - OS (difference between AC-stimulated, and AC-stimulated in the presence of ouabain) or S - NS (difference between AC-stimulated, and non-stimulated)), in this experiment, was 11amol/RBC/h for Rb^+ and 16 amol/RBC/h for Na^+. In terms of ions per molecule, the highest net stimulated activities we have obtained so far are approximately 20 Rb^+ and 35 Na^+ per enzyme per second, at 4° C. At higher temperatures, pumping of cations was increased, but the ATP-dependent pumping activity also increased, and at a much faster rate. At 37°C, we could no longer measure net AC stimulated activity. This could simply mean that when Na, K-ATPase is functioning at its maximal capacity, the applied electric field can no longer further stimulate its activity, suggesting that both ATP-dependent and AC-stimulated activities are by the same mechanism.

TABLE I

Electric field-stimulated pump transport of Rb^+ and Na^+ at 3.5 °C

For measurements of ion concentration by flame photometry and ion movement using radioactive tracers see "Experimental Procedures." Each value is the mean of 3–5 measurements. Standard deviation is given in parentheses. 1 amol = 1 attomole = 1×10^{-18} mol. 1 amol/RBC/h = 0.0108 mmol/liter cells/h. Values varied for erythrocyte samples from different individuals. Data given in this table were obtained from blood samples of a single individual. With samples from different individuals, Rb^+ influx values were in the range 10–20 amol/RBC/h and Na^+ efflux values were in the range 15–30 amol/RBC/h.

	Cellular ion conc			Medium ion conc				Measured ion movement						
	Na	K	Rb	Na	K	Rb	Mg	NS	S	ONS	OS	NS-ONS	S-OS	S-NS
		mM			*mM*						*amol/RBC/h*			
20 V/cm a.c., 1.0 kHz														
Rb influx	6	75	27	2.5	0	12.5	2	13.0 (0.3)	23.5 (1.2)	10.1 (0.6)	11.1 (0.15)	2.9 (0.6)	12.4 (1.2)	10.5 (1.2)
Rb efflux	6	65	15	2.5	0	12.5	2	42.1 (1.7)	43.4 (1.1)	41.7 (1.5)	41.6 (1.5)	0.4 (1.7)	1.8 (1.5)	1.3 (1.7)
Na influx[a]	6	75	0	150	5	0	2	3.2 (<0.1)	3.54 (0.2)	4.0 (0.1)	6.2 (0.3)	−0.8 (0.1)	−2.7 (0.2)	0.4 (0.2)
Na efflux[a]	6	75	0	150	5	0	2	4.3 (2.0)	6.2 (0.6)	1.7 (0.1)	1.9 (0.8)	−1.9 (2.0)	4.3 (0.6)	1.9 (2.0)
20 V/cm a.c., 1.0 MHz														
Rb influx	6	75	27	2.5	0	12.5	2	10.6 (3.8)	10.4 (3.5)	8.8 (1.8)	8.9 (1.6)	2.1 (3.8)	1.5 (3.5)	−0.5 (3.5)
Rb efflux	6	65	15	2.5	0	12.5	2	38.3 (2.0)	37.7 (1.0)	40.4 (0.3)	39.5 (1.1)	−2.1 (2.0)	−1.8 (1.0)	−0.6 (2.0)
Na influx[a]	6	75	0	150	5	0	2	6.1 (0.6)	6.9 (0.9)	6.6 (0.3)	6.9 (<0.1)	−0.5 (0.6)	0.0 (0.9)	0.8 (0.9)
Na efflux[a]	6	75	0	150	5	0	2	4.0 (2.7)	20.8 (3.2)	2.0 (0.1)	5.3 (1.8)	2.0 (2.7)	15.5 (3.2)	16.8 (3.2)

[a] In Na^+ influx and efflux experiments, Rb^+ was not added because our intention was to demonstrate the active pumping of Na^+. K^+ was present on both sides of the membrane.

Another important observation is that ATP was apparently not required for the AC stimulation of the cation pumping. Erythrocyte in which ATP was depleted to less than 10 μM (compared to the normal concentration of approximately 1 mM) responded to AC stimulation with equal efficiency. This result suggests that the AC fields did not stimulate ion pumping by ATP hydrolysis. As the experimental conditions were such that any pumping of a cation would require input of energy, the logical conclusion is that the applied AC fields fueled the ion pumps.

A systematic study revealed several important features of the AC stimulation experiment. First, at an optimum frequency of an AC field (i.e., 1 kHz for Rb^+ pumping and 1 MHz for Na^+ pumping), the stimulated ion fluxes reach maximal values at a field strength of 20 V/cm (amplitude). Conversely, with a 20 V/cm AC field, the maximal Na^+ efflux was observed at 1 MHz, and maximal Rb^+ influx was observed at 1 kHz. This means that the two pumps of the enzyme can function independent of each other, i.e. they could be uncoupled. At 1 kHz, the potassium pump is activated but not the sodium pump. The converse is true when a 1 MHz AC field is used. These observations of amplitude and frequency windows are crucial for formulating our enzyme models based on the concept of electroconformational coupling. By determining the position of windows, thermodynamic and kinetic parameters of the enzyme catalytic reaction can be determined. Figure 5.4 gives a complete set of data for the AC stimulated cation pumping by Na, K-ATPase.

Our experiment also showed that although the potassium and the sodium pumping modes can function independently of each other, both ions are required for the enzyme to respond to electrical stimulation, and the K_m's for these cations are identical to those of the ATP-dependent pumping activities.

5.3 Electroconformational Coupling Model

5.3.1 Electric Field Interaction with A Transmembrane Protein

The experiments described above have unequivocally shown that a membrane ATPase can capture energy from an electric field to synthesize ATP or to pump an ion up its concentration gradient. These results can be explained by the concept of the electroconformational coupling[1–3]. Although it is possible that a pulsed electric field could simply provide an electrophoretic driving force to stimulate the translocation of protons through the ATPases, no evidence has been found that proton translocation actually took place during the exposure of these membranes to electric fields. As the electric pulses used were short, mostly in microseconds, and proton translocation in these membranes is slow and needs much longer times (milliseconds), it is unlikely that the pulsed electric field induced ATP synthesis was by field driven translocation of protons to provide the chemiosmotic energy for ATP synthesis. The ECC

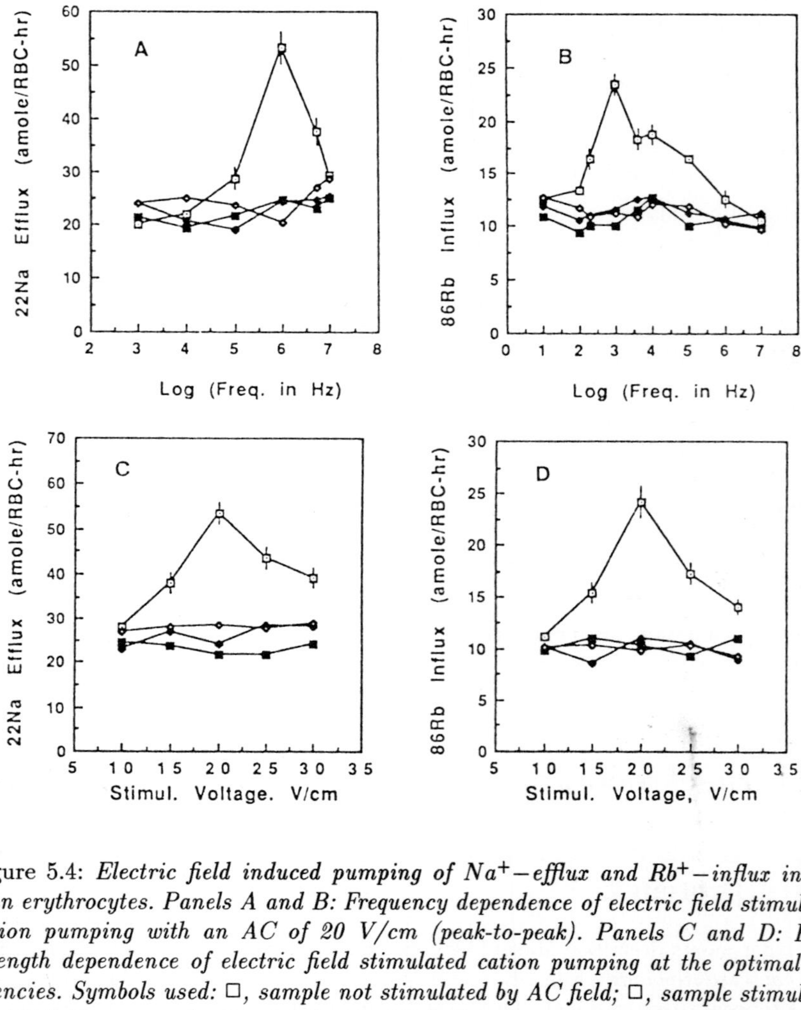

Figure 5.4: *Electric field induced pumping of Na^+−efflux and Rb^+−influx in human erythrocytes. Panels A and B: Frequency dependence of electric field stimulated cation pumping with an AC of 20 V/cm (peak-to-peak). Panels C and D: Field strength dependence of electric field stimulated cation pumping at the optimal frequencies. Symbols used: □, sample not stimulated by AC field; □, sample stimulated by AC field; ○, sample stimulated with AC field in the presence of 0.2 mM ouabain; ●, sample not stimulated with AC field in the presence of 0.2 mM ouabain. The temperature was 4°C. Stimulation was continued for 1 hour. Taken from Liu et al.*[11].

model postulates that the free energy required for ATP synthesis is derived from the applied field and transiently stored as conformational energy of the ATPase[1,2]. The same concept can be applied to the Na,K-ATPase of human erythrocytes. The use of an AC field is to enforce conformational oscillation, thus facilitating the turnover (regeneration) of the enzyme in its catalytic cycle[1,2]. In the simple conformational change,

$$A \underset{k_{BA}}{\overset{k_{AB}}{\Longleftrightarrow}} B$$

where A and B are two conformation states at equilibrium and k_{AB} and k_{BA} are the rate constants, the basic thermodynamic relationship for electroconformational coupling is,

$$\left[\frac{\partial \ln K}{\partial E_{membr}}\right]_{P,V,T} = \frac{\Delta M}{RT} \tag{5.1}$$

Equation (5.1) simply states that an effective transmembrane electric field can shift the conformational equilibrium of Scheme I towards B if ΔM, the difference in the molar electric moment of the B state and the A state, is positive, and towards A if it is negative. The effective field strength of a transmembrane electric field is,

$$E_{membr} = 1.5 R_{cell} E^0 / d_{membr} \tag{5.2}$$

where R_{cell}, E^0, and d_{membr} are the radius of the cell, the amplitude of the applied AC field, and the thickness of the lipid bilayer, respectively. Molar electric moment is the sum of the field interactions with charges and electric dipoles of a molecule, in molar quantities.

$$\Delta M = N_0 \sum_{i or j} (z_i q_i \delta d_i + \delta \mu_j) \tag{5.3}$$

where N_0 is the Avogadro number, $z_i q_i$ is the number of charges in the i-th moiety, and δd_i and $\delta \mu_j$ are the change in the distance of the separation of two charges between B and A states, and change in the j-th dipole moment, respectively. The equilibrium constant K has the usual meaning of $K = [B]/[A] = k_{AB}/k_{BA}$. Integration of this differential equation gives,

$$K_E = K_0 exp[-\Delta M E_{membr}/RT] \tag{5.4}$$

The term $\Delta M E_{membr}$ is free energy of interaction between the electric field and the molar electric moment of a molecule. Because E_{membr} is oscillatory and K_0 constant, K_E is oscillatory. In the case, where both k_{AB} and k_{BA} are much faster than $1/f_{AC}$(f_{AC}, being the frequency of the AC field), K_E will oscillate synchronously with E_{AC}, or E_{membr}. [A] and [B] will oscillate with 180° separation in phase, i.e. when [A] reaches a maximum, [B] will approach a minimum, and vice versa. This phenomenon is due to simple electroconformational changes. If E_{membr} or ΔM is large, or both are large, the amplitude of the oscillation can be a nearly all-to-none oscillation. This means that the applied AC field will drive the protein to assume almost exclusively either the B form or the A form, depending on the polarity of the field. When the field is turned off, the zero-field equilibrium will be restored by dissipating its captured free energy into the solution. The maximal amount of energy transferred from the AC field to the solution is $\Delta M E_{membr}$. To utilize this captured energy, the enzyme must prevent conversion of energy into non-usable forms, e.g. heat loss in solution, by coupling its conformational change to a chemical reaction. This coupling enables the system to harness the electric energy to drive an endergonic reaction. The theory of ECC will be demonstrated using a simple membrane transport model.

5.3.2 Coupling of An Electric Potential to A Chemical Potential

The four-state cyclic membrane transport model can couple the conformational energy of a transporter to the chemical potential energy of a neutral ligand or the electrochemical potential energy of an ion. This property of the ECC does not depend on direct interaction between an electric field and a ligand[3,4]. An electric field interacts only with the transporter, P, which can assume one of the two forms, either P_{left}, with its substrate binding site facing the left of a membrane, or P_{right}, with its binding site facing the right. The ligand, L, is also distinguishable in two different compartments, either in the left or the right side of the membrane.

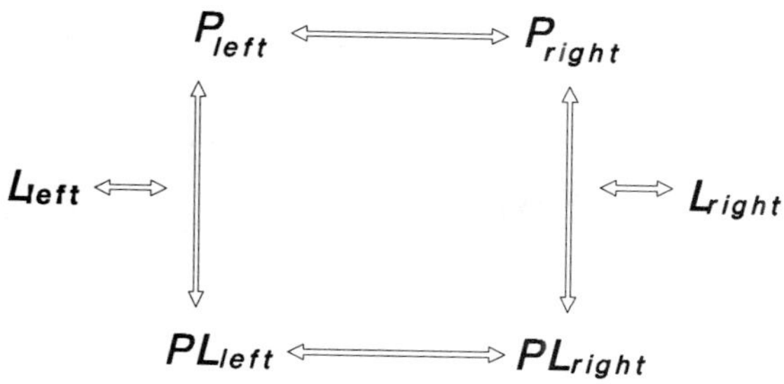

I

Here the transitions between P_{left} and P_{right} and between PL_{left} and PL_{right} involve a ΔM and are influenced by the applied field, according to Equation (5.1). In our initial analysis, a neutral L will be considered[3,4,12,13]. In such a case, the two ligand binding steps do not have ΔM, and are not affected by an electric field. If this transport system has some asymmetry, for example, the affinity of P_{left} for L_{left} is different from that of P_{right} for L_{right}, then an oscillating electric field will cause the catalytic wheel to turn. And as a result, there will be a net pumping of L from one side to the other. The direction of the wheel spin will depend on the affinities of the two binding reactions. If the affinity is higher on the left hand side, the wheel will turn counterclockwise, and if it is higher on the right hand side, the wheel will turn clockwise. Accompanying the rotation of the wheel, the ligand will accumulate at the side with the lower affinity. The four state transport model has become a molecular motor to pump a ligand up its concentration gradient. It uses electric energy instead of chemical bond energy of ATP. There are many conditions to satisfy before the pump will work. In addition to a ΔM for P_{left} and P_{right} and a difference in ligand affinity for the two conformational states, another important requirement is that the turnover rate of the enzyme must match the frequency of the AC field. This last property of the system implies that a cell can recognize an electric signal by virtue of the kinetic characteristics of the transport system.

5.3.3 Windows for Effective Coupling

There are at least three windows we are now aware of which are associated with coupling of electric energy to the chemical potential energy of a concentration gradient or the γ-phosphodiester bond energy of ATP[1–4,13]. These are the frequency window, the amplitude window, and the ligand concentration window. All three have been theoretically derived and experimentally observed. In theory, some assumptions are made to simplify the mathematics. The relationships among several rates are such that the conformational changes of the transporter are much greater than $1/f_{AC}$ and the ligand association/dissociation rates are much smaller than $1/f_{AC}$. The first assumption ensures that the transfer of electric energy to the conformational energy of the protein is close to the equilibrium at any time. This condition permits the maximal efficiency of energy transfer. The second assumption ensures that a ligand can always be furnished with sufficient energy to overcome its gradient, thus preventing back flux or slippage in transport. Under these conditions, it was shown that the efficiency of energy transduction can approach 100% if the ligand is neutral or if it is a cation of the same sign as the gating charges of the transporter. The frequency window and the window for the ligand concentration are interrelated. Changing AC intensities will cause the optimal field intensities to change, and vice versa. A three-dimensional plot is devised to indicate their interrelationship, as shown in Figure 5.5.

The window for the AC amplitude can be modeled in two ways. First, the ΔM as defined in Equation (5.3) lacks the induced dipole term, $\alpha N_0 E_{membr}{}^2$ (α being the polarizability of molecule) which, being squared, does not depend on the sign of E_{membr}. When E_{membr} becomes large, it will become the dominant term for ΔM, making the sign of E_{membr} irrelevant, the enzyme conformational state will be locked into the B form, and will not oscillate effectively. The result is a decrease in the efficiency of energy transduction. This interpretation was suggested in our first publication of the ECC model[3] . However, Markin *et al*[14] have investigated a "Channel-Enzyme" model, in which a transmembrane enzyme is given certain properties of a channel. A channel enzyme can open its ion binding site either to the left side or to the right side of membrane but not to both sides at the same time (see Fig. 5.6 for a model).

This latter property distinguishes a channel-enzyme from a conventional membrane channel. A cation can also interacts with an AC to change its affinity. When these two properties are taken into account, the field induced cation pumping shows a window of AC amplitude for optimal energy coupling. If this interpretation is correct, the Na, K-ATPase will have some characteristics of a channel-enzyme. This would not be surprising. Our previous experiment on electroporation of human erythrocytes suggested that the K^+-pump was perforated by a pulsed electric field; a strong electric pulse was able to puncture a pore through that potassium pump, thus, inducing a membrane conductance that could be blocked by ouabain. This result suggests that K^+-pump may have a channel-like structure. It is not known if the Na^+-pump also has a channel-like structure.

The above results indicate that optimal frequency, amplitude, and ligand concentration for the transduction of electric signal are not fixed for a transmembrane protein. Depending on the state and the environment of the protein, these windows can change. For example, the optimal frequencies shown in Fig. 5.4 for the Na^+- and the Rb^+-pumps would change when the experimental conditions are varied. This ability of a transmembrane protein confers the protein ability to detect or generate a variety of electric signal. A cell can use this property of the ECC mechanisms to adapt to different needs by modifying the state of a protein and regulate the concentration of a ligand. Some preliminary data are available which show that

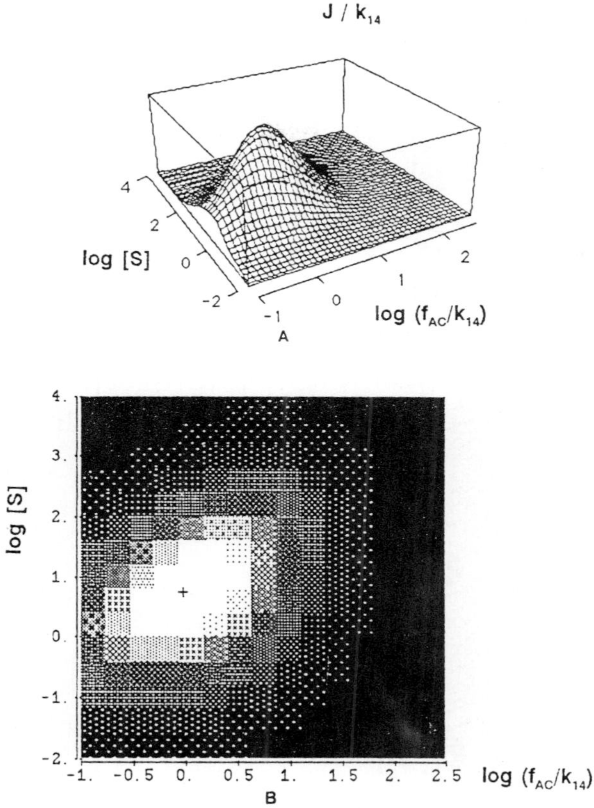

Figure 5.5: *Windows for the transduction of electric energy by the electroconformational coupling model. A: Flux of substrate, J/k_{14}, is plotted versus substrate concentration, [S], and frequency of the AC field, f_{AC}/k_{14} for Scheme I. At a constant substrate concentration, there is a window for AC frequency; and at an AC frequency, there is a window for substrate concentration. B: The pump activity in different regions is expressed in different degrees of gray-shades to mimic the transparency of windows. Taken from Markin and Tsong*[13].

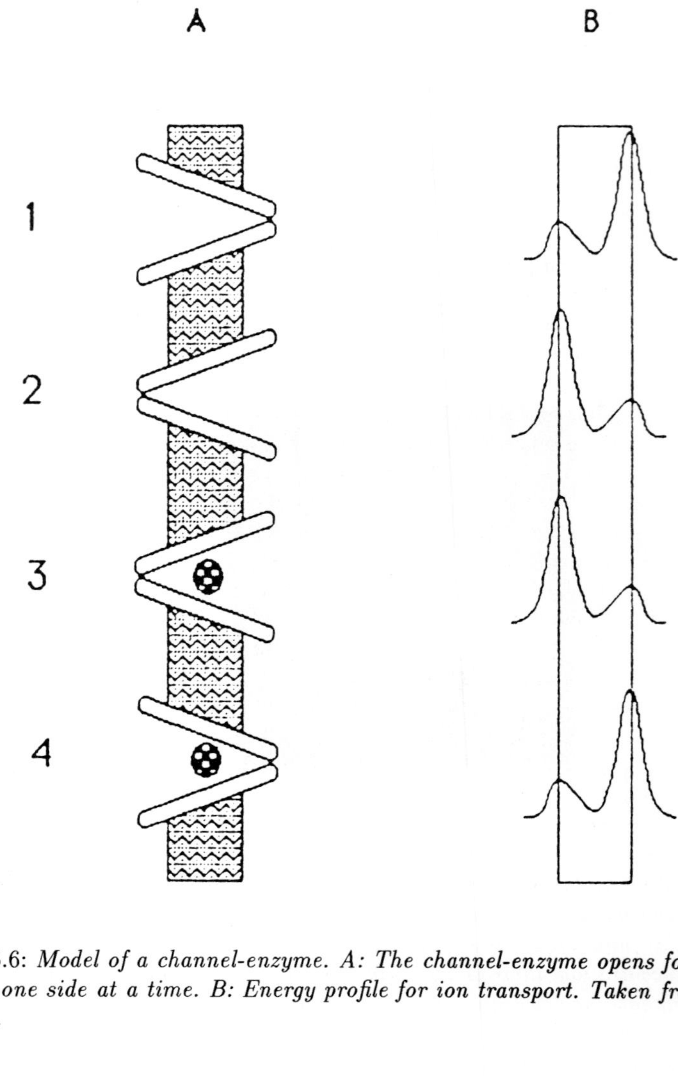

Figure 5.6: *Model of a channel-enzyme. A: The channel-enzyme opens for substrate passage one side at a time. B: Energy profile for ion transport. Taken from Markin et al.* [14].

the optimal frequency for the hydrolysis of ATP by an Ecto-ATPase changes values with changing temperature. Other mechanisms for regulating the electric response of a transmembrane protein are discussed in Section 5.

5.4 Transduction of Low Level Electric Signals

A low intensity electric field is insufficient to drive a conformational change of a protein or other molecule unless the ΔM associated with the conformational change is unusually large. In that case, where strong cooperativity exists among highly structured molecules, $n_H\Delta M$ should be substituted for ΔM in Equations (5.1) and (5.4), where n_H is the number of molecules which undergo conformational transition concertedly, i.e., the Hill coefficient of the system. Because n_H is unlikely to be very large, a low level electric field will not be able to influence the activity of a membrane protein by the ECC mechanisms. Many physical models have been proposed. Weaver and Astumian have considered thermal electric noise as the limit of sensitivity[15]. Barnes has proposed several models, one of which is the neural network model, with which one can train a signal transducer to recognize a signal even though the signal-to-noise ratio in this case is much less than unity[16]. However, most of these models emphasize physical principles of electromagnetics to explain how a weak electromagnetic signal can be detected, and rarely discuss molecular mechanisms for the recognition of a signal by a cell. We have proposed an "Oscillatory Activation Barrier (OAB)" model for the transduction of very weak electric signals[17] based on the following premises. First, a weak electric signal contains little energy for fueling a chemical reaction, so a weak electric field cannot influence the reaction by shifting its chemical equilibrium. Second, in a spontaneous chemical reaction, the free energy of reaction is the driving force, so a low level electric field can only exert its influence by altering the rate of the reaction and no energy input from the electric field is required. Thus, a low field can interact with a spontaneous chemical reaction by changing its kinetics rather than its energetics. Third, there are many ways the rate of a chemical reaction can be altered. A familiar example is a catalyst mediated reaction. A catalyst enhances the rate of a reaction by changing the structure of the transition state to reduce the activation barrier of the reaction.

In the OAB model, a low level electric field will induce a resonance between the field and a vibrational mode of the transition state of the enzyme/substrate complex. An enzyme/substrate complex involves electrostatic interactions. The product dissociation is often the rate limiting step. The vibrational mode responsible for product dissociation may include movement of charges and the enzyme/substrate complex will then be sensitive to an oscillatory electric field. In other words, the activation barrier of the rate limiting step in an enzyme catalysis is oscillatory, with a characteristic frequency. Thus, this barrier may interact with a low level electric field to produce a resonance between the field and the barrier. When this happens, the rate of product formation will be affected because of the non-linear dependence of rate on the barrier height. The basic concept of the OAB model is explained in Figure 5.7.

Markin *et al*[17] have investigated the properties of a Michaelis-Menten enzyme where the activation barrier behaves like a simple harmonic oscillator. The complete catalytic cycle of an enzyme is represented by a two step mechanisms:

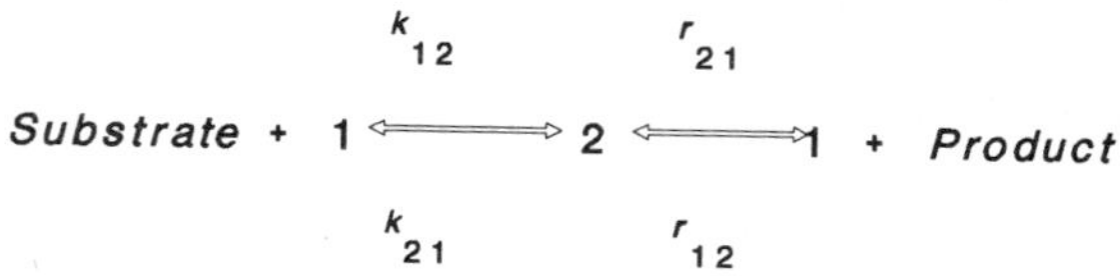

II

The first step is arbitrarily assigned the rate-limiting step, and all other faster reactions are combined and represented by the second step. The model assumes that only k_{12} and k_{21} are affected by an electric field. For an AC, the mean values of the two rate constants are,

$$< k_{12} >_{AC} = k_{12}F(E^0_{AC}, \omega_{AC}) \quad (5.5)$$

$$< k_{21} >_{AC} = k_{21}F(E^0_{AC}, \omega_{AC}) \quad (5.6)$$

where $F(E^0_{AC}, \omega_{AC})$ is a function which may be derived once an enzyme model is decided. E^0_{AC} is the amplitude and ω_{AC} the angular frequency of the AC field. $F(E^0_{AC}, \omega_{AC})$ was derived for Scheme II and the overall rate was compared for the zero-field and the in-field conditions. For an activation barrier behaving like a simple harmonic oscillator, with a characteristic frequency of $\omega^0/2\pi$, the amplitude of the net AC- induced barrier oscillation can be determined by numerically fitting experimental data on the AC-stimulated enzyme activity. Our experimental results on the AC-stimulated ATP hydrolysis activity of Ecto-ATPase from chicken oviduct were analyzed with the OAB model. It was shown that a 30 percent enhanced ATP hydrolysis could translate into an increased barrier oscillation of 2.5 kT, or 1.5 kcal/mole. The amplitude of the AC potential that the enzyme was exposed to, in the detergent solubilized form, was approximately $10\mu V$. This is a low level AC field, too weak to produce an interaction energy comparable to the barrier oscillation estimated from the data, of 2.5 kT. Nevertheless, as we emphasized above, the 2.5 kT change in barrier can be derived from the free energy of the ATP hydrolysis instead of from the interaction energy of the AC and the enzyme. The free energy of ATP hydrolysis under these experimental conditions would be approximately 15 kT. The

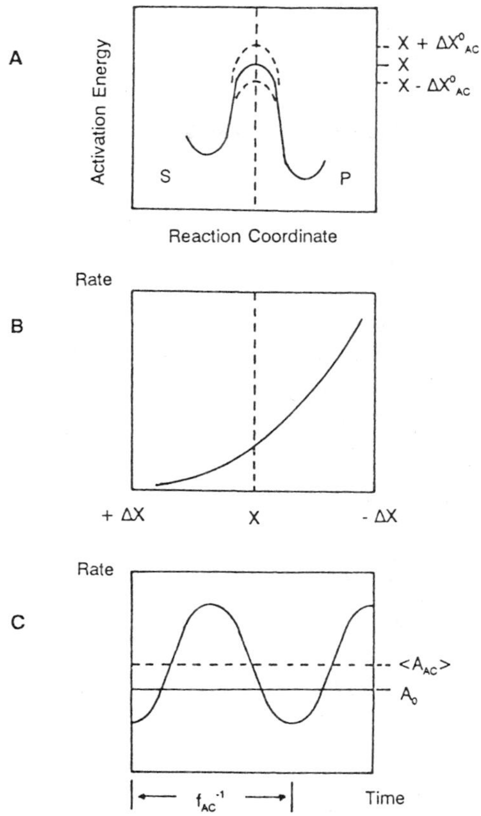

Figure 5.7: *Oscillatory activation barrier model for electric stimulation of the rate of a spontaneous reaction. Panel A: Activation barrier of an enzyme catalyzed reaction, X, is induced to oscillate by an AC field, with the amplitude of oscillation ΔX^{o}_{AC}. Panel B: The non-linear dependence of rate on the activation energy (Arrhenius activation). Panel C: The overall-rate oscillates as the barrier oscillates. The mean overall rate, $\langle A_{AC} \rangle$, in the presence of an AC is greater than the rate without the barrier oscillation, A_o, because of the non-linearity of rate on the activation energy. S and P denote substrate and product, respectively. Taken from Tsong*[2].

OAB model showed similar properties to the two features of the AC stimulated ATP hydrolysis activity of the Ecto enzyme, namely, a window for the AC frequency and the saturation of field effects with increasing AC amplitude.

5.5 Generation of Electric Signals in Cells

Organisms can generate electricity and electric signal[7,8]. Electric eels use electric shock to fend off enemies. Electric fishes use very weak electric organ discharge to warn enemies or to track prey. Rays, skates and sharks can perceive very low level electric signals. In organs or tissues, electric signals are constantly deployed to regulate or coordinate activities with other parts of the body. Neurons transmit and decipher electric impulses continually. How are the electric activities of a cell related to the biochemical reactions? We know that stepwise electric potentials (DC fields) of proper amplitudes can open membrane ionic channels. The experiments discussed in this chapter show also that AC fields of defined frequencies and amplitudes can activate membrane transporters or transmembrane enzymes. However, mechanisms for a membrane channel or an enzyme to generate an electric signal have received little attention. According to the ECC model, an electric field can transmit its energy to drive an enzyme reaction. A reverse chemical process should generate an electric field of a defined frequency and amplitude from the same enzyme reaction. Up to this point, we have not considered the reversible process in the ECC model. To give a more complete account of the ECC model, we shall consider all plausible combinations of charges in a transporter and a ligand. Different behaviors arising from these charge combinations will be compared, including the ability for reverse energy transduction, i.e. the conversion of a gradient energy into an electric potential energy.

In total, there can be only six different types of charge combinations between a transporter and a ligand[18]:

System 0: no gating charges for transporter and no net charges for ligand.

System 1: gating charges of either sign for transporter and neutral ligand.

System 2: no gating charges for transporter and charged ligand of either sign.

System 3: gating charges of either sign for transporter, but neutralized by charges of ligand, i.e. no gating charges for the transporter/ligand complex.

System 4: gating charges of either sign, but having opposite sign for gating charges of the transporter/ligand complex.

System 5: gating charges of either sign for transporter; the transporter/ligand complex having gating charges of the same sign.

Table 5.2: Comparison of five electrically active membrane transport systems

System mechanism	zL	zP	zPL	maximal grad → elec	efficiency elec → grad	(%)
1	0	1	1	100	100	ECC
2	1	0	1	8.7	0	ECC/Rect
3	1	-1	0	8.7	0	ECC/Rect
4	2	-1	1	50	0	Rect
5	1	1	2	100	100	ECC

Notes: zL, zP, zPL, ECC, and Rect denote, respectively, the net charge of ligand, the gating charge of protein, the gating charge of protein/ligand complex, energy conversion by the ECC mechanisms, and energy conversion by the electric rectification of ion. See Markin and Tsong[18] for details.

Obviously System 0 will not interact with an electric field. It is electrically silent, and the ECC model does not apply to this type of systems. System 1 has been discussed earlier, because here, an electric field interacts directly and only with the transporter but not with the ligand. This allows us to exploit the main premise of the model, which is, that an electric field can induce conformational change in the protein. It turns out that System 1 is a very efficient energy transducing system. The system can also perform the reverse energy transduction to generate an electric signal using the chemical potential energy of a concentration gradient. Table 5.2 summarizes some important properties of the five electrically active systems and their differences. These features are: 1) An electric field can transfer its energy to a chemical reaction, in this case the transport of a ligand across a cell membrane, by two mechanisms, namely the ECC and the electric rectification of ions. Energy transduction in Systems 1 and 5 are purely by the ECC mechanisms, and in System 4 is purely by electric rectification mechanisms. In contrast, Systems 2 and 3 can use both mechanisms. 2) Energy coupling is the most efficient in Systems 1 and 5; the theoretical maximum being 100 percent. The theoretical maximum for energy coupling purely by rectification is 50 percent, as is the case for System 4. Systems 2 and 3 are less efficient, the maximum being around 8.7 percent. 3) Reverse energy transduction can only be done with Systems 1 and 5. In other words, the ECC mechanisms can use the gradient energy of a ligand to generate an electric signal, but the ion rectification mechanisms cannot.

This last finding is an important revelation and points to a fact which has hitherto received little attention from investigators, namely that an ion channel, functioning by passive open/close, switch-like mechanisms, cannot perform reverse energy transduction, and its role in signal transduction is severely limited. If, on the other hand, the opening and closing of a channel is actively controlled by other energy consuming chemical processes, the channel would be able to perform reverse energy transduction. In that case, the opening-closing of a channel is an enforced conformational change typical of an ECC system.

Another important revelation of these analyses is that by modifying the gating charges of a transporter, the transporter can be changed from one type to another. For example, phosphorylation or dephosphorylation of a transporter can change the gating charges of the transporter and the transporter/ligand complex, thereby converting the transport system from one type to another. Thus, charge modification of proteins by different means can play a role in the regulation of membrane enzymes, receptors and transporters via the ECC mechanisms.

Because our analyses of the ECC and the OAB models have been largely based on the sinusoidal waveform, one may ask whether in a cell membrane, a transmembrane electric potential can exhibit such a waveform, or a waveform with similar regularity. If not, then would the deviation from such regularity invalidate the conclusion made above? This question has two different aspects. First, one asks if the transmembrane electric field of a cell is constant, regularly oscillating, or fluctuating, and second, if the transmembrane electric field is fluctuating, can the ECC and the OAB models still function? To answer the first point, we note that most textbooks list constant values for the transmembrane potential of different cells, e.g. -70 mV for a neuron, 200 mV for a mitochondrion, -240 mV for a yeast cell. The ECC model does not function with a constant electric potential unless the potential is modulated to become oscillatory[1–4]. One should realize that the potential values reported in the literature do not have time and spatial resolution. As discussed above, it is the electric field near a transporter protein that is relevant for the ECC model because the coulombic interaction depends on the inverse square of the distance between two interacting bodies and the interaction energy falls rapidly over distance. Thus, it is the local electric field rather than the mean electric field which is relevant to the ECC model. The electric potential in the vicinity of a transport protein, when resolved to a millisecond or a smaller interval, is unlikely to be constant because of the movement of ions or redox reactions occurring in its vicinity. The local electric potential is most likely to be fluctuating. If, on the other hand, these ion movements or redox reactions are well regulated or controlled, the field fluctuation may be far from random. Although the waveform may not appear to be regularly oscillatory, it may follow a certain distribution depending on the reactions taking place in the vicinity. In Figure 5.3, we listed several waveforms, but there may be other wave distributions. Are these seemingly irregular waveforms capable of driving an endergonic reaction by the mechanisms of ECC discussed above? Astumian *et al* have investigated this question by computer analysis of a simple four state ECC model, such as one shown in Scheme I[19]. They have investigated several cases, one with constant amplitude but varying pulse width according to an error function centering around a frequency, a second with constant pulse width but varying amplitude according to the error function centering around an amplitude, and the third, with both amplitude and pulse width varying. In all three cases, after a short latent period, they observed the absorption of electric energy by the transporter to pump a substrate up its concentration gradient. This was surprising because, according to the second law of thermodynamics, no random noise should be able to perform chemical work. It was then shown that chemical work was done only if the apparently randomly fluctuating electric field was actually sustained by an energy source, i.e., the waveform or the distribution of an electric field was not altered by reciprocal interactions with the transporter. If there were reciprocal interactions, then there was no energy transduction. Thus, these results did not violate the second law of thermodynamics.

Experimental testing of these results is presently being done. Marszalek and Xie in my laboratory have generated waveforms according to certain distributions of amplitude and pulse width by a computer programme which controls a waveform generator. The computer generated waveforms are then used to stimulate cation pumping as one does with the sinusoidal waveform. These experiments should provide information on the efficiency of energy transfer between a membrane protein and an energy sustained apparently fluctuating electric field. This last point is emphasized because it is often misunderstood that energy can be absorbed from a purely random fluctuating field or from membrane electric noise. Equilibrium noise that contains no energy cannot drive a chemical reaction away from its chemical equilibrium. Only a sustained energy source can drive such a reaction. In our computer simulation, an energy sustained waveform does not change its shape upon interacting with a transporter. Thus, the waveform is autonomous. If there is reciprocal interaction between the field and the electric dipoles of the transporter, no energy can be extracted from the field.

5.6 Perspective

In this chapter, we have shown that oriented molecules or molecules in a microstructure may exhibit reactivities which are unusual for molecules in a homogeneous solution. The anisotropic reactivity of these molecules allows them to interact with an oscillatory driving force to produce many chemical effects. Among these remarkable effects is the ability of molecules to convert energy from one form to another. This ability does not require extraordinary features. Any simple membrane transporter or membrane integrated enzyme (Michaelis-Menten type) can perform the task provided that it satisfies certain conditions or specific requirements. Two such requirements are that a transporter or an enzyme exhibit unequal affinities for a ligand or substrate on the two sides of the membrane, and that the protein can respond to a periodic potential, e.g. an oscillating electric field, an acoustic or sonic wave, or other types of oscillating driving forces. Because the transduction of energy or signal depends on kinetic characteristics of the system, there are windows for the amplitude and frequency of the periodic field, and the ligand concentration. We propose that these characteristics satisfy the concept of the language of cells. Our future tasks are to search for experimental evidence of: 1) electroconformational changes of membrane proteins, 2) oscillations of local transmembrane electric potential of a cell, 3) using such potential to trigger or stimulate specific functions of a cell, a tissue or an organisms, and 4) communicating with cells using electric fields of defined frequency and amplitude.

The concept discussed here should provide some strategies for dealing with some important issues confronting biomedical research. Can one develop therapeutic procedures using electromagnetic signals, in the so called "electromagnetic medicine"? Where irregularities of cellular signaling are detected, can one correct these irregularities by appropriate intervention? Furthermore, if one can decipher the electromagnetic language of cells, would we be able to communicate directly with cells by these signals to command or to influence cellular transports or biochemical reactions? Much work leading up to these possibilities have been in progress in medicine. Is there scientific basis for such practices? Can we neutralize the harmful effects of enviromental electromagnetic fields by an interference signal which can cancel or abolish the harmful fields, as is now done in acoustic noise reduction devices? These will be the challenges of the future for biochemical research.

Acknowledgments

T.Y. Tsong is indebted to his current and former colleagues, and collaborators, Drs. J. Tessie, E. H. Serpersu, B. E. Knox, D. S. Liu, F. Chauvin, V. S. Markin, P. Marszalek, T. D.Xie, R. D. Astumian, H. Westerhoff, and Y. D. Chen for their contribution to this project. Support of this work by the United States National Science Foundation and Office of Naval Research is acknowledged.

Bibliography

[1] T.Y. Tsong, Electrical modulation of membrane proteins: Enforced conformational oscillations and biological energy and signal transductions. *Annu. Rev. Biophys. Biophys. Chem.* **19**, (1990) 83-106.

[2] T.Y. Tsong, Molecular recognition and processing of periodic signals in cells: Study of activation of membrane ATPases by alternating electric fields. *Biochim. Biophys. Acta* **1113**, (1992) 53-70.

[3] T.Y. Tsong, and R.D. Astumian, Absorption and conversion of electric energy by membrane ATPases. *Bioelectrochem. Bioenerg.* **15**, (1986) 457-476.

[4] T.Y. Tsong and R.D. Astumian, Electroconformational coupling and membrane protein function. *Prog. Biophys. Mole. Biol.* **50**, (1988) 1-45.

[5] T.Y. Tsong, Deciphering the language of cells. *Trends in Biochemical Sciences (TIBS)* **14**, (1989) 89-92.

[6] L.J. DeFelice, *Introduction to Membrane Noise*, Plenum Press, New York.(1981)

[7] T.H. Bullock, R. Orkand, and A. Grinnell, *Introduction to Nervous System.* W.H. Freeman, San Francisco.(1977)

[8] T.H. Bullock and W. Heiligenberg, eds. *Electroreception.* Wiley & Sons, New York.(1986)

[9] E.H. Serpersu and T.Y. Tsong, Stimulation of a ouabain-sensitive Rb^+ uptake in human erythrocytes with an external electric field. *J. Membrane Biol.* **74**, (1983) 191-201.

[10] E.H. Serpersu and T.Y. Tsong, Activation of electrogenic Rb^+ transport of Na,K-ATPase by an electric field. *J. Biol. Chem.* **259**, (1984) 4757-4763.

[11] D.S. Liu, R.D. Astumian, and T.Y.Tsong, Activation of the Na^+−pumping and the Rb^+−pumping modes of Na,K-ATPase by oscillating electric field. *J. Biol. Chem.* **265**, (1990) 7260-7267.

[12] V.S. Markin, T.Y. Tsong, R.D. Astumian and B. Robertson, Energy transduction between a concentration gradient and an alternating electric field. *J. Chem. Phys.* **93**, (1990) 5062-5066.

[13] V.S. Markin and T.Y. Tsong, Frequency and concentration windows for the electric activation of a membrane active transport system. *Biophys. J.* **59**, (1991) 1308-1316.

[14] V.S. Markin, D.S. Liu, J. Gimsa, R. Strobel, M.D. Rosenberg and T.Y. Tsong, Ion channel-enzyme in an oscillating electric field. *J. Membrane Biol.* **16**, (1992) 137-145.

[15] J. Weaver and R.D. Astumian, The response of living cells to very weak electric fields: The thermal noise limit. *Science* **247**, (1990) 459-461.

[16] F.S. Barnes, Some engineering models for interactions of electric and magnetic fields with biological systems. *IEEE Transaction on Education.* In press.

[17] V.S. Markin, D.S. Liu, M.D. Rosenberg and T.Y. Tsong,Resonance transduction of low level periodic signals by an enzyme: An oscillatory activation barrier model. *Biophys. J.* **61**,(1992) 1045-1049.

[18] V.S. Markin and T.Y. Tsong, Electroconformational coupling for ion transport in an oscillating electric field: Rectification versus active pumping. *Bioelectrochem. Bioenerg.* **26**, (1991) 251- 276.

[19] R.D. Astumian, P.B. Chock, T.Y. Tsong, Y.D. Chen and H.V. Westerhoff, Can free energy be transduced from electric noise? *Proc. Natl. Acad. Sci. USA* **84**, (1989) 434-438.

Chapter 6

Electromagnetic Fields and Biomembranes

Robert P. Liburdy

6.1 Signal Transduction and the Cell Membrane

The plasma membrane is the primary channel through which the cell communicates with its environment. This membrane has specialized receptors on the surface for binding ligands, such as hormones, peptides, and proteins, which subsequently trigger a cascade of biochemical reactions — the signal transduction cascade — that culminates in cell growth and proliferation. Also present in the cell membrane are specialized ion-channels regulating cation and anion fluxes which are crucial for the proper functioning of the cell. Both receptor-site binding and ion-channel gating are accompanied by structural changes in the local organization of the lipid bilayer that trigger diverse biochemical pathways to transduce signals to intracellular sites[1,2]. Thus, it is important for us to understand how the cell membrane can interact with electromagnetic fields. The signal transduction cascade is an amplification system that enables a receptor-binding event to cause large changes in the cell. This means that when the cell surface is influenced by an imposed electromagnetic field, the disturbance can propagate down the signal transduction cascade to influence a variety of cell functions including gene expression, cell growth and proliferation.

A number of laboratories, including ours, have recently provided experimental evidence that signal transduction is influenced by electromagnetic fields[3–9]. In this chapter, we shall examine the experimental evidence that electromagnetic fields in the microwave frequency range interact with the cell membrane. I shall concentrate on research from our laboratory on two natural cell membrane systems, the erythrocyte and the lymphocyte, and a synthetic system, the liposome bilayer vesicles,in an attempt to lay out a foundation for understanding how electromagnetic fields can have biological effects via influences on the cell membrane.

6.2 Studies on Erythrocytes

These studies are based on the underlying hypothesis that membrane structure and conformation are important for coupling to external electromagnetic fields[10–21]. Electromagnetic fields interact with biological systems according to how well the energy is transferred to components of the target membrane. For microwave fields, membrane components that are dipolar, such as polar amino acid side chains and cell-surface associated bound water, will undergo field orientation at microwave frequencies. This can be thought of as rotational motion with an associated time constant similar to the wavenumber of the impressed oscillating electromagnetic field. At sufficiently high levels of absorbed power, such induced motion will result in heating; in general, specific absorption rates (SAR) of less than approximately 1 - 10 mW/gm for a localized or partial body exposure are not considered to have thermal effects[22].

6.2.1 Membrane Permeability and Structural Phase Transition Studies.

In studies on rabbit erythrocytes, we have shown that sodium cation permeability is enhanced during exposure to 2,450 MHz fields (0 - 100 mW/kg, SAR) in a striking, temperature-dependent manner[12,13,15]. In these studies microwave-enhanced sodium transport was observed only at 17 - 19.5°C which coincides with the presence of a membrane structural phase transition known to exist in the erythrocyte bilayer (Figure 6.1). We tested the hypothesis that the structural phase transition is critical for microwave interaction by modifying the erythrocyte membrane with cholesterol to eliminate T_c. This biochemical modification rendered the bilayers completely unresponsive to microwaves (Figure 6.1). The increase in permeability at T_c was shown to be (a) linearly dependent on the electric field strength (0 - 600 V/m), (b) reversible, (c) enhanced by oxygen, and (d) reduced by the presence of the antioxidants ascorbic acid and β-mercaptoethanol[15]. The later findings raise the possibility that free radical species may be operative in this interaction. This is an important consideration and will be raised again in studies on the simpler liposome vesicle systems. Five other laboratories have independently reported a permeability increase at T_c temperatures in the erythrocyte during microwave exposures[23–27]. These findings represent the only biological effect of microwave bioeffect that has been observed in six different, independent laboratories.

A theoretical framework for understanding how low-level electromagnetic fields may interact with biological membrane at structural phase transition has been provided

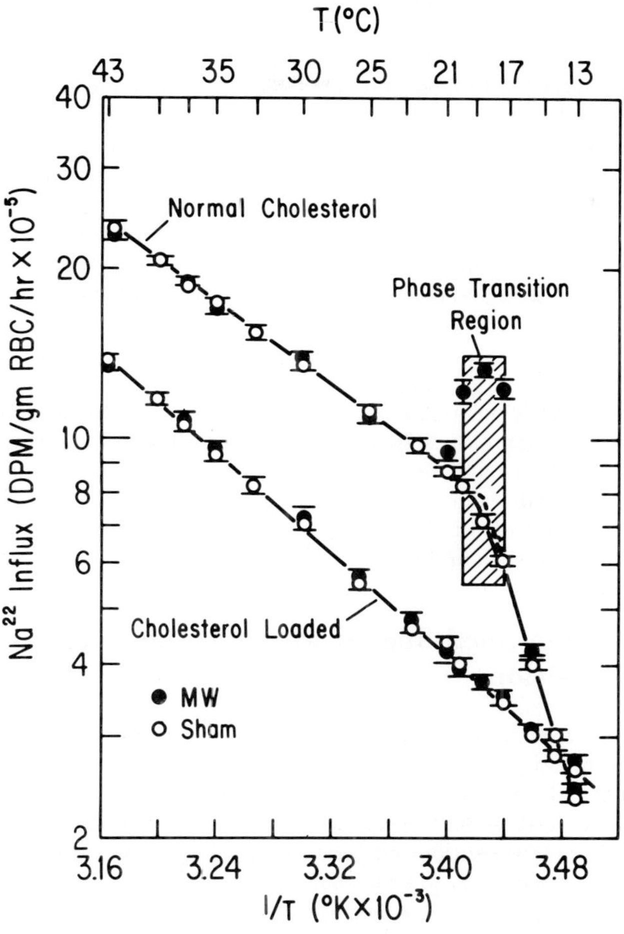

Figure 6.1: *Arrhenius plot of sodium influx for microwave-exposed normal and cholesterol-enriched rabbit erythrocytes. Microwave exposures (2450 MHz) at 100 mW/gm, 30 min, atmospheric pO_2. Sham exposures at isothermal temperatures. Cells at* 10^{10}*/ml, Ringer's buffer, pH 7.4. Mean ± S.D.*

by Bond and Wyeth[8,29]. At the phase transition temperature, T_c, the bilayer approaches a critical state characterized by extreme sensitivity to external perturbations. For example, in liposome systems constructed of very pure phospholipid components, an increase in membrane permeability is seen only near T_c as temperature is increased[30]. In more complex biological membrane systems, such as the erythrocyte and the lymphocyte (see below), a structural phase transition is more usually observed as a change in slope of the activity curve in an Arrhenius plot, as shown in Figure 6.1. At the transition temperature the membrane components physically co-exist in two or more phases, giving the break in the Arrhenius plot. It is at T_c that the system can be perturbed by the external low-level electromagnetic fields. Bond and Wyeth show that the isothermal electric susceptibility is given by an expression that goes to infinity as the temperature of the system approaches T_c,

$$\chi_T = (1/\varepsilon_o)(\delta P/\delta E)_T \sim 1/[T - Tc] \tag{6.1}$$

where P is the electric polarization, E is the magnitude of the electric field, and T is temperature. As T approaches T_c the electric susceptibility would therefore become infinite and the effect on a molecule of a low-level electromagnetic field would be greatly enhanced. For example, the electric dipole moment of the phosphocholine head group has been reported to be relatively large at approximately 19 D and, is, thus a candidate for such an interaction[31].

6.2.2 Protein-Shedding Studies at the Structural Phase Transition.

In addition to investigating the movement of ions across the membrane bilayer, we have also investigated how microwave fields influence the binding of protein to the cell surface. This relates to the first step in the signal transduction cascade discussed above. In these studies we asked the question whether microwave fields at the structural phase transition temperature would result in the release of proteins from the cell surface. This phenomenon is termed "protein-shedding" and since erythrocytes are not lysed during field exposures, the protein species involved in this response are not integral membrane proteins associated with the lipids that span the bilayer, but more likely, represent the class of peripheral proteins associated with the bilayer through cationic electrostatic and van der Waal coupling.

We first reported that microwave fields can remove protein from cell surface using the sensitive technique of polyacrylamide gel electrophoresis (PAGE)[12,13,18]. Following exposure to microwaves (2450 MHz, 60 mW/gm, 17° C, 30 minutes) or to isothermal sham treatment, cells were immediately spun in an Eppendorf microfuge for 10 seconds (12,000 rpm) and the cell-free supernatants were collected for analysis by PAGE. Protein bands were visualized by silver staining. It is important to reemphasize that erythrocytes are not lysed by the microwave fields and that the proteins released from the cell membrane are, thus, not integral proteins associated with lipid that span the bilayer, but are loosely bound peripheral proteins, as mentioned above. Peripheral proteins also play a critical role in signal transduction[12,13,18].

Figure 6.2 depicts the PAGE analyses for cells maintained at different oxygen tensions: 5mmHg; 760mm Hg; 150mm Hg. These values were chosen to represent hypoxic, hyperoxic and atmospheric (250mm Hg) values for pO_2. Oxygen is of interest because haemoglobin is the major protein present in the erythrocyte and oxygen transport is the biological function of red cells *in vivo.* In addition, the autoxidation of haemoglobin to methaemoglobin produces the superoxide radical and hydrogen peroxide, and this process is enhanced by oxygen[18]. Therefore, we were interested in determining if microwave fields influenced protein-shedding in an oxygen-dependent manner.

Figure 6.2 shows that sham treatments (S) result in the release of proteins from the erythrocyte which is a normal process. Comparison of band intensities as a function of oxygen tension, however, indicates that atmospheric oxygen leads to bands that are less intense than for hypoxic and hyperoxic conditions. Both lack of oxygen and oxygen saturation cause the release of more proteins, thus, protein-shedding is oxygen dependent. Although greater protein-shedding under hypoxic conditions is not understood, it is reasonable to assume that hyperoxic conditions may enhance protein-shedding through oxygen's ability to form free-radicals, as discussed above.

There are signicant differences in protein-shedding between microwave and sham treatments. Microwave treatments at T_c results in at least eleven protein species ($< 31,000Da$) being shed from the erythrocyte. Shedding of $26,000$ and $24,000Da$ proteins is unique to microwave treatments, with enhanced release of $28,000$ and $< 15,000Da$ species. Interestingly, microwave-induced protein-shedding displays an oxygen dependence similar to that seen for sham treatment, discussed above. This same pattern, with greater response to microwaves under hypoxic and hyperoxic conditions, was also observed when $^{22}Na^+$ influx was followed in erythrocytes exposed to the same microwave field; antioxidants blocked this microwave effect[15]. In the studies shown in Figure 6.2, microwave treatment is associated with enhanced protein-shedding compared to its isothermal sham, at all three oxygen tension conditions. The enhancement of protein-shedding at 760 mm Hg, which corresponds to oxygen saturation of haemoglobin, suggests a greater involvement of free-radicals in the presence of the microwave field. This response is most prominent for 28,000 - 31,000 and $< 15,000Da$ species. These protein-shedding results and the $^{22}Na^+$ influx results[15] discussed above represent the first evidence suggesting that free radicals may be involved in the biological effects of microwaves.

Protein-shedding was further investigated in two ways[18]. First, we performed similar experiments at non-T_c temperatures and obtained no evidence for protein-shedding in microwave fields. This links the protein-shedding effect to T_c. Second, we performed calcium chelating studies in which EDTA was used to remove calcium in the buffer. The hypothesis was that divalent cations, which act to stabilize the pe-

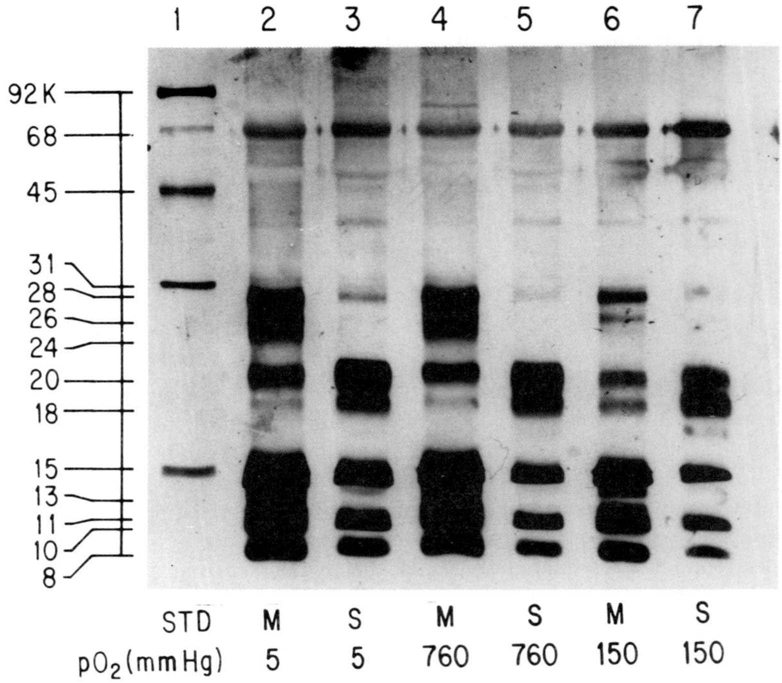

Figure 6.2: *Protein-shedding for rabbit erythrocytes at Tc. SDS-PAGE electrophoretogram of cell-free supernatants following 2450 MHz microwave (M, 60 mW/gm) or sham (S) exposure for 30 min at hypoxic (5mm Hg), hyperoxic (760 mm Hg), or atmospheric (150 mm Hg) oxygen tension. Cells as in Figure 6.1 plus 0.01% BSA. Molecular weight standards were phosphorlyase B (91 kDa), BSA (68 kDa), OVA (45 kDa), carbonic anhydrase (31 kDa), and lysozyme (14 kDa).*

ripheral proteins to the cell surface, if removed, would result in the same pattern of protein-shedding as seen with the microwaves. This was indeed the case, suggesting that cationic salt bridge formation is destabilized by microwaves at T_c.

The above studies indicate that protein-shedding is influenced by microwaves, and that this interaction can be mediated by several factors. Temperature (T_c), oxygen, and the calcium cation are capable of modulating the microwave effect. A temperature and calcium dependence for microwave protein-shedding can be understood through structural considerations of the cell membrane, as discussed above. Oxygen, under hyperoxic conditions in the presence of microwaves, most likely acts indirectly on the cell membrane through enhanced free radical formation.

Protein-shedding has been confirmed by us using a second technique, quantitative high performance liquid chromatography (HPLC)[19]. In these studies human erythrocytes were treated with microwaves at T_c, as above, and the cell-free supernatants analyzed for the presence of protein using four different types of interactive HPLC columns; gel permeation, anion exchange, hydrophobic interaction, and reverse phase. The quantitative nature of HPLC enabled us to determine that following microwave treatment, the erythrocyte experiences a two-fold increase in total protein-shedding. Protein-shedding due to microwave treatment involves the release of approximately 450 fgm of protein per erythrocyte, and this corresponds to ~1% of the total protein mass of the cell (cytosol plus membrane), but to greater than 50% of the protein mass of the cell membrane or stroma. Naturally the erythrocyte does not renew these proteins identified above.

Figures 6.3 and 6.4 give examples of HPLC chromatographs employing a reverse phase column. For this column proteins are bound to a relatively nonpolar support matrix, C-8 bonded silica, if they possess domains that are relatively nonpolar. These protein species are subsequently eluted from the column using a gradient of increasing nonpolarity, e.g. replacing water with acetonitrile. The elution profile for sham treatment shows a single major peak at 6.45 minutes, corresponding to a relatively polar species, with no major, prominent peaks identified during the gradient elution. The microwave treatment profile, however, shows three prominent peaks eluting well into the acetonitrile gradient at 21.15, 22.00, and 22.72 minutes; the last two peaks were attenuated (AT) to permit on-scale tracing. This cluster of relatively non polar protein species are not seen in the sham-treated group; and constitute 51% of the total protein mass detected. Thus, microwave treatment resulted in a significant release of specific, nonpolar protein species.

Apparently, charge and polarity are involved in destabilizing these peripheral proteins during microwave treatment. The loss of these proteins from the cell surface represents a significant structural change to the cell surface, and, as discussed above, such a modification to the cell surface can have significant effects on the transduction of signals across the cell membrane.

6.3 Studies on Lymphocytes

The above studies on erythrocyte raise the important question of whether the phenomena of increased ion transport and protein-shedding are generalizable to other cell types. The lymphocyte is significantly different from the erythrocyte as it is nucleated and its cell membrane is responsible for a great variety of transport and specialized receptor binding functions. For example, we have previously shown that nonthermal radio frequency radiation fields increase the rotational motion of immunoglobulin (Ig) in solution and also result in the release of Ig bound to antibody receptor sites on the surface of B-lymphocytes[11]. Thus, it seemed to us that this cell type would be an important subject for further studies on ion transport and

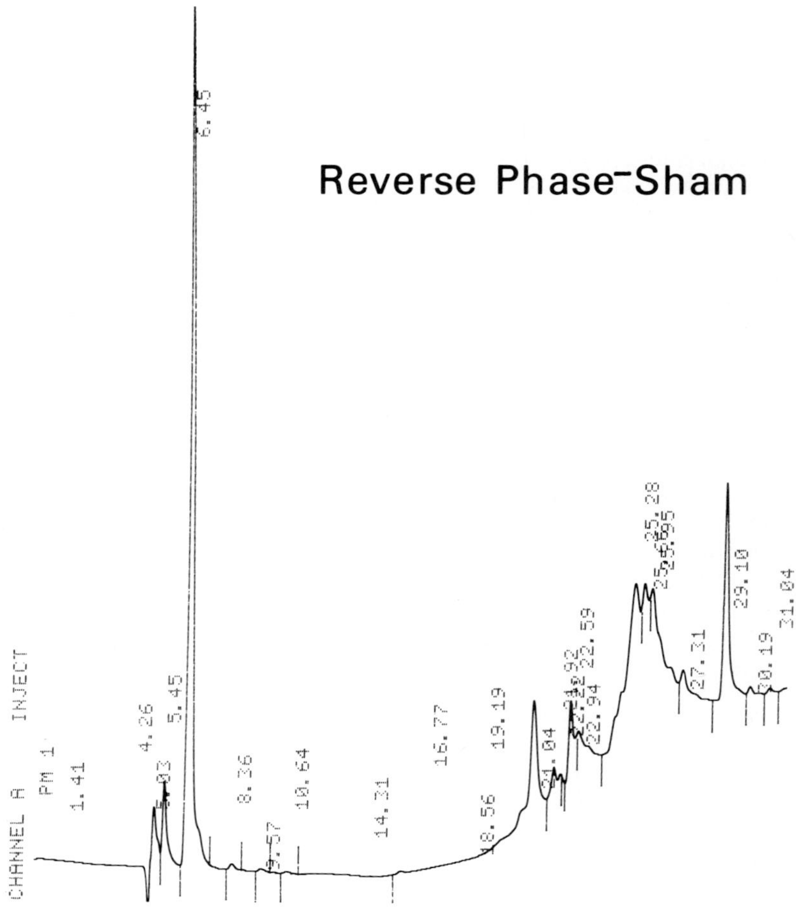

Figure 6.3: *Reverse phase HPLC of cell-free supernatants from sham-treated human erythrocytes. Exposures as in Figure 6.2. Linear gradient elution over 20 min using water plus 1% TFA and an increasing concentration of acetonitrile (5-95%). A C-8 silica gel support was employed at 0.5 ml/min. Abs at 280 nm.*

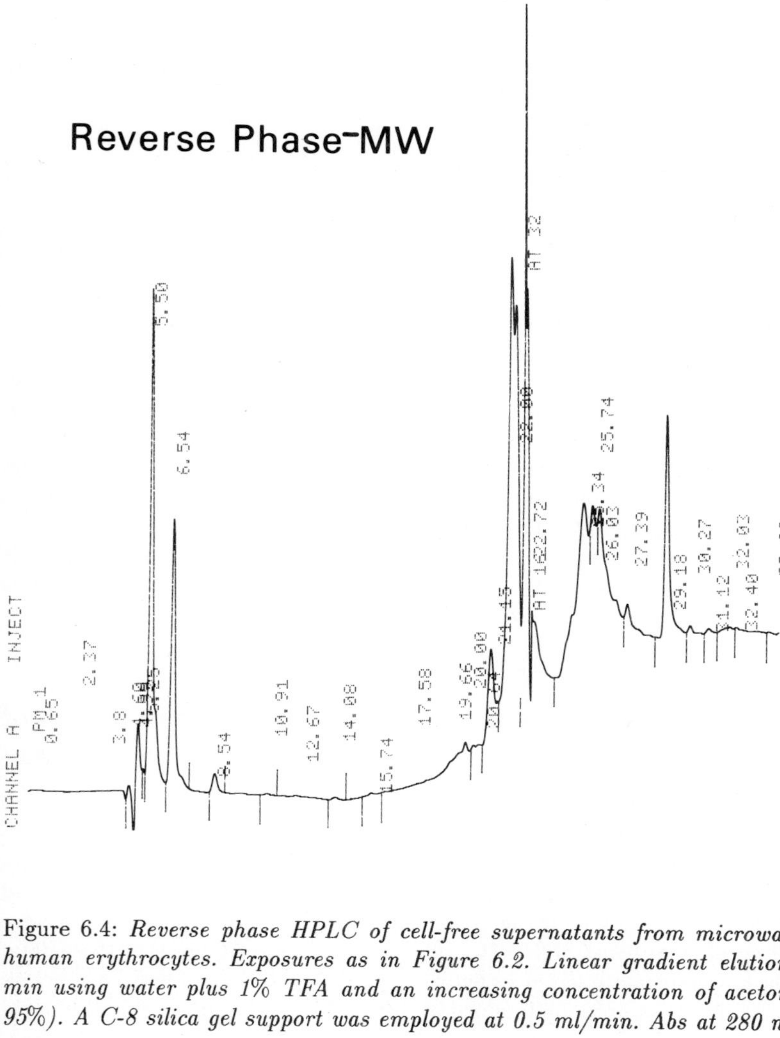

Figure 6.4: *Reverse phase HPLC of cell-free supernatants from microwave-treated human erythrocytes. Exposures as in Figure 6.2. Linear gradient elution over 20 min using water plus 1% TFA and an increasing concentration of acetonitrile (5-95%). A C-8 silica gel support was employed at 0.5 ml/min. Abs at 280 nm.*

protein-shedding in microwave fields.

6.3.1 Arrhenius Analysis of Active and Passive Transport of Sodium in Rat Spleen Lymphocytes.

The working hypothesis for these studies was that microwave effects would be most prominent at the structural phase transition for the lymphocyte. This parallels work performed on the erythrocyte. To determine if lymphocytes display a membrane structural phase transition, an Arrhenius plot was constructed for both passive and active sodium transport. Identification of T_c is associated with a break in the slope of the Arrhenius plot and this temperature would be a candidate for microwave sensitivity experiments. Figure 6.5 displays data from these experiments and we observe that rat spleen lymphocytes exhibit a maximum value for passive accumulation of $^{22}Na^+$ (Ouabain-treatment) in the lymphocyte at 90 minutes at a temperature of 37°C. Total sodium accumulated (untreated cells) reaches a minimum at 37°C. Thus, the difference, which corresponds to the effect of active sodium transport, achieves a maximum at 37°C, as shown. Since the pump *extrudes* sodium, this means pump activity is minimized at 37°C. This is interesting and may reflect an adaptive advantage associated with the physiological temperature of 37°C. Taken together, the data indicate that rat spleen lymphocytes exhibit a break at 37°C in the Arrhenius plots for both passive and active sodium transport.

6.3.2 Microwave Exposures Facilitate the Accumulation of Sodium in the Lymphocyte During Passive and Active Transport at T_c.

The Arrhenius analysis presented in Figure 6.5 demonstrates that spleen lymphocytes exhibit a T_c at 37°C. To determine if microwave fields influence active and passive transport of sodium in the lymphocyte at T_c, microwave exposures were performed for 90 minutes at 37°C, 4°C and at 40°C. Figure 6.6 presents these data, which show no significant differences between lymphocytes treated with microwaves at 40°C and control samples maintained at 40°C. When exposures were conducted at 37°C, however, an approximately two-fold increase in sodium accumulation during passive transport was observed in response to microwave treatment. Sodium accumulation in the lymphocyte during active sodium transport was also increased approximately 1.6 fold. Passive transport data indicate that sodium enters the cell in greater quantities in the presence of the microwave field; this is interpreted as an effect on permeability. Active transport data indicate that sodium accumulates in the cell due to inhibition of an pump action.

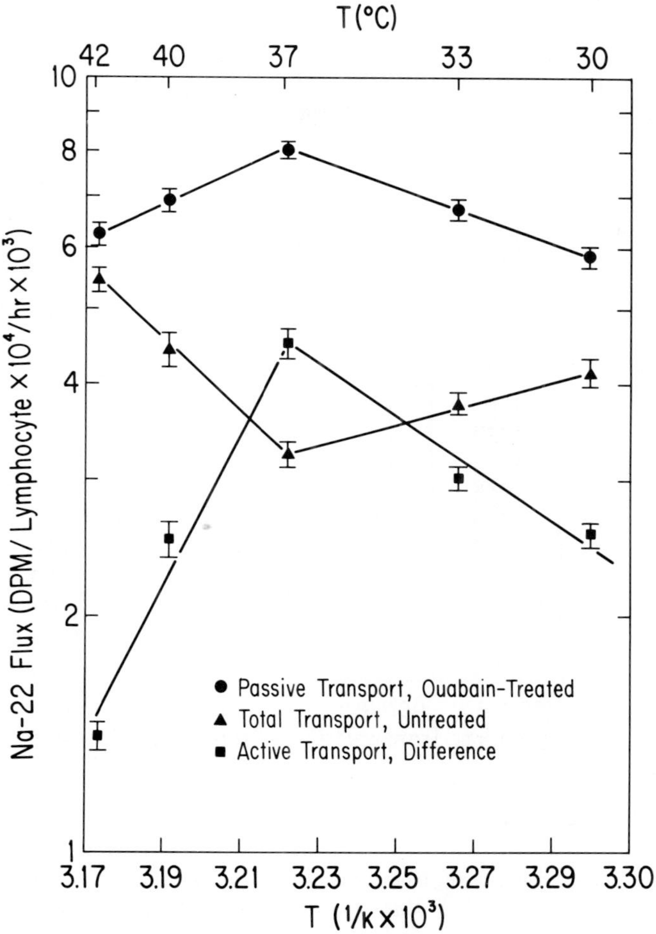

Figure 6.5: *Temperature dependence of sodium transport in the rat spleen lymphocyte. An apparent structural transition is present at 37°C.* $^{22}Na^+$ *flux represents accumulation of sodium in the cell after 90 minutes. Lymphocytes at* $5 \times 10^6/ml$*. Mean* $\pm$*S.D.*

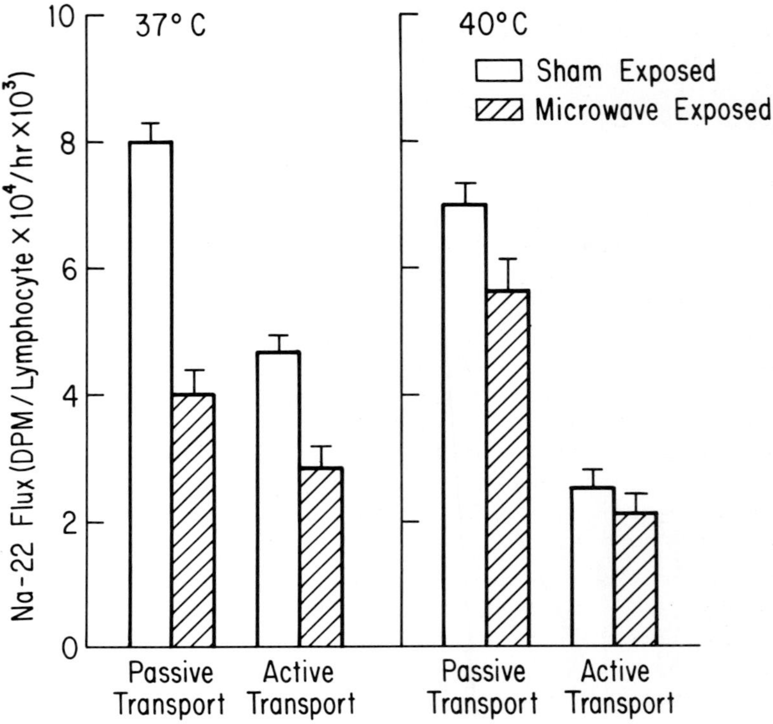

Figure 6.6: *Effect of microwave fields on sodium transport in the rat spleen lymphocyte for exposure at T_c (37°C). $^{22}Na^+$ flux represents accumulation of sodium in the cell after 90 minutes. Passive and active transport as in Figure 6.4. Microwave exposures at 2450 MHz, SAR = 7 mW/gm, 90 minutes. Sham exposures in unenergized waveguide, 90 minutes. Mean ± S.D.*

6.3.3 Protein-Shedding Studies at the Structural Phase Transition.

These studies were performed in a parallel manner as that for erythrocytes described above. The hypothesis was that the lymphocyte which displays a T_c at 37°C (Figure 6.4) also responds to microwave fields at this temperature by undergoing protein-shedding. The rat spleen cell populations were incubated at 37°C for 90 minutes in the presence of a microwave field, as in the transport studies, or at 37°C in an unenergized microwave waveguide. Cell-free supernatants were harvested as for the erythrocytes and analyzed by PAGE.

Figure 6.7 depicts a typical PAGE electrophoretogram with sham-treated samples in lanes 1 - 4 (20, 15, 10, and >50 μg/well) and microwave-treated samples in lanes 6 - 9 (all 10 μg/well). Both microwave and sham treatments yielded a variety of proteins spanning the molecular weight range of $91,000$ to $14,000Da$ that were resolved by PAGE. Two noticeable differences, however, were apparent between the microwave and sham-treated groups. First, protein species shed at molecular weight $< 15,000Da$ are more pronounced in the microwave-treated samples. Second, there is a unique triplet band at $44,000Da$ that is apparent only in the microwave-treated samples. These results indicate that a relatively short exposure to microwave fields at T_c results in protein-shedding from the lymphocyte and that the microwave-induced pattern is different from that due to an isothermal sham treatment.

The results of the lymphocyte studies extend observations presented above on the erythrocyte that microwave fields alter membrane transport and protein-shedding at T_c. The research described here is significant for several reasons. First, the lymphocyte is critical to the immune system. The lymphocyte exhibits a T_c at 37°C and that is where microwave alterations in both active and passive cation transport are also observed. In addition, microwave-induced protein-shedding for the lymphocyte occurs at the same temperature.

The lymphocyte studies raise several interesting questions. The observation that microwave fields influence both passive and active sodium transport in the lymphocyte at 37°C indicates that a relatively short term exposure of 90 minutes can alter lymphocyte membrane function. Previous reports have demonstrated that sodium transport in the lymphocyte is important for key intracellular activities[32–36]. For example, Na/K transport is involved in lymphocyte cell volume homeostasis, intracellular enzyme function, membrane transport and signal transduction. The transport of sodium, potassium, and calcium are early biochemical events in lymphocyte proliferation, B-cell maturation, and antibody production. During signal transduction, one of the earliest measurable changes in the lymphocyte immediately after antigen stimulation is an abrupt increase in the activity of the sodium/potassium pump.

As cation transport is an early biochemical event linked to cellular proliferation, we speculate that microwave alterations of cation transport have the potential to lead to subsequent changes in nuclear processes of DNA and protein synthesis. Recently,

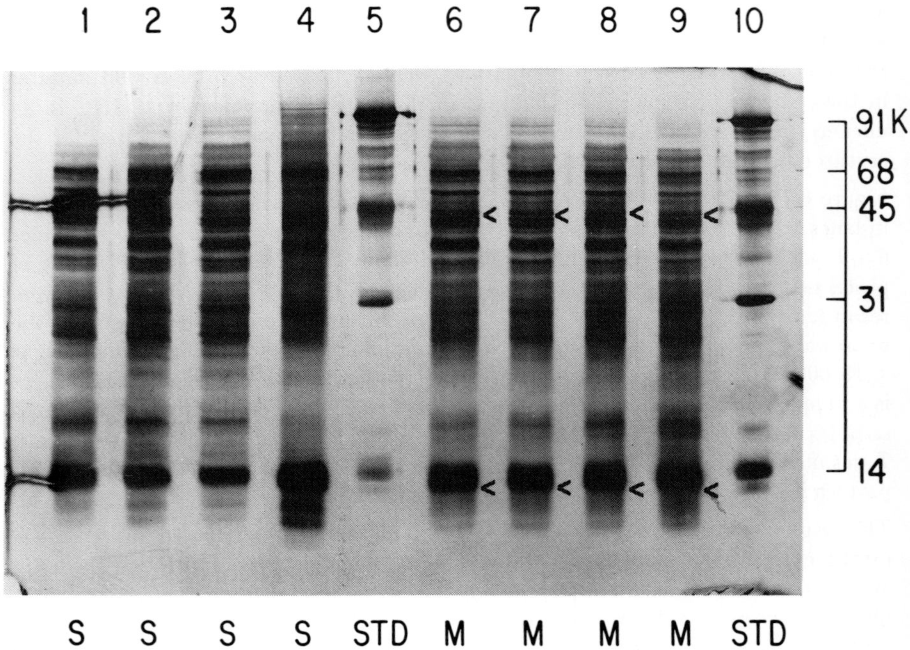

Figure 6.7: *Protein-shedding in rat spleen lymphocyte at 37°C (T_c). Microwave (M) and Sham (S) treatments as in Figure 6.6, and cells were rapidly centrifuged and the cell-free supernatants were analyzed by PAGE. Lanes 1 - 4: Sham at 20, 15, 10, >50 ug/well. Lanes 6-9: 10 ug/well. Standards as in Figure 6.2.*

Cleary and colleagues[37] reported that a two hour exposure to 2,450 MHz fields (SAR < 50 mW/kg, 37°C) resulted in increased DNA synthesis in human peripheral blood lymphocytes. Based on our observations using rat lymphocytes exposed to 2,450 MHz fields at an identical temperature, we consider it likely that these workers have also triggered an increase in cation transport, during their exposures. Thus, a cellular interaction mechanism as described here, involving an early increase in cation transport, may be involved in enhanced DNA synthesis.

In our lymphocyte experiments an increase in the passive accumulation of sodium was observed during microwave treatment at T_c. This effect by itself would lead to an increase in intracellular sodium. When active transport was examined during microwave treatment, an elevation in sodium accumulation in the lymphocyte was detected. This means that the Na/K pump, which extrudes sodium, was inhibited. Interestingly, Allis and Sinha-Robinson[26] reported in a careful study that 2,450 MHz microwaves significantly inhibits by 35% the Na/K pump of human erythrocytes when treated at T_c. This raises the possibility that the sodium pump in other cell types, in addition to the erythrocyte and lymphocyte, may be inhibited by microwave fields. The observation of microwave-induced protein-shedding for the lymphocyte is of interest with regard to signal transduction. Unlike the erythrocyte, the lymphocyte has the ability to regenerate and recycle plasma membrane proteins[38]. Our studies demonstrate that both microwave and sham-treated cells display a wide range of molecular weight proteins that are shed during incubation at T_c for 90 minutes (Figure 6.6). Microwave treatment, however, leads to the release of a unique protein band at $44,000 Da$ and to the additional release of unique bands at $< 15,000 Da$. At present, the significance of the loss of these protein species is not known. Little research is available on the characterization of protein-shedding in the lymphocyte or any nucleated cell. Limited literature is available on blood storage alterations but such studies deal with alterations associated with erythrocytes maintained at 4°C. One reference for human lymphocytes does report that overnight storage (18 hours) at 4°C leads to a marked decrease in the percentage of cells staining positive with anti-Leu2a (supressor cells), which results in a significant depression of the measured ratio of helper to supressor T lymphocytes[39]. This effect was not observed if storage was at 22°C; storage effects at 37°C were not studied. A $44,000 Da$ calcium-binding protein was isolated from the lymphocyte cytoskeleton[40]. Such calcium-binding proteins have been further studied by Geisow[41,42]. The $44,000 Da$ protein that is released from the lymphocyte by microwaves at T_c might be related to the $44,000 Da$ calcium-binding protein mentioned above. It is interesting to note that calcium played an important role in protein-shedding in the erythrocyte: we observed that removal of calcium from the buffer induced the shedding of proteins similar to that induced by treatment with microwave fields[18]. Calcium stabilizes proteins at the cell surface through cation bridges that are apparently disrupted by microwave fields.

Our observation that microwave fields lead to the release of endogenous protein from the lymphocyte is complementary to an earlier study from our laboratory[11], which demonstrated that nonthermal levels of a microwave field (2,500 MHz, 0.117 mW/gm) removed cell-surface bound polyclonal anti-IgG from Ig receptor sites present on murine B-lymphocytes. Taken together with the present findings we see that endogenous as well as non-endogenous protein may be released from the lymphocyte cell surface by microwave fields.

6.4 Liposome Vesicle Studies

The erythrocyte and lymphocyte studies suggest that cell membrane can interact with microwave fields via components such as proteins, carbohydrates, glycoproteins, glycolipids, and phospholipids. To address the question of whether a simple bilayer system comprised of only pure phospholipids interact with microwave fields, a series of experiments were conducted using unilamellar liposome vesicles[14,16,20,21]. The results described below reveal that phospholipid bilayers experience a dramatic permeability increase at T_c in microwave fields. This response to microwave fields enables liposomes loaded with a chemical agent to "microinject" substances into target cells such as lymphocytes by using the microwaves to initiate membrane fusion between the liposome and the target cell. Finally liposomes comprised of phospholipids in the fluid phase, far above T_c, are shown to experience microwave permeability effects. Since our initial report in 1985, two other research groups have reported similar increases in liposome permeability during microwave treatment[43,44].

6.4.1 Liposome Bilayer Organization and the Structural Phase Transition.

Liposome vesicles are comprised of synthetic phospholipid components that form a bilayer structure based on the amphoteric nature of the phospholipid molecule. The hydrophobic effect spontaneously drives fully hydrated, relatively hydrophobic, long-tailed hydrocarbon chains to associate with the tail regions of other phospholipids, and, in doing so, the relatively polar headgroup become exposed to the aqueous environment. This process can be experimentally manipulated to result in the formation of spherical vesicles of different sizes, and having different numbers of concentric bilayer membranes. For example, ultrasound can be used to facilitate the rapid formation of vesicle bilayers. Figure 6.7 illustrates the structural organization of a unilamellar bilayer[45]. There are two structural forms, a crystalline-solid state or gel state, and a fluid form known as the liquid-crystalline state. In the solid or gel state the fatty acid chains are fully extended in the all-trans positions and the phospholipid molecules can pack together tightly in a hexagonal array. For phosphatidylcholine the acyl chains pack at a tilt angle, as shown. In this configuration the bilayer is thickest, the surface area associated with each phospholipid head group is minimized, and the entrapped volume of the vesicle is minimized[46].

A remarkable feature of liposome vesicles is the structural transition they undergo when the phospholipid components in the gel phase are converted to the liquid-crystalline or fluid state. By increasing the temperature, this conversion can be easily achieves. However, for liposomes comprised of charged phospholipids, other physical factors have been shown to trigger this process such as pH, ionic strength, and divalent cations. The temperature at which this conversion takes place is termed the structural phase transition temperature, or T_c.

During this phase conversion the phospholipid molecules undergo fast lateral and rotational diffusion. This is illustrated in Figure 6.8 as the formation of rotational isomers about the C-C acyl chains in the all-trans state when a *gauche* isomer forms, producing a 120° rotation about the C-C bond. This results in a kink that leads to irregularities in packing within the bilayer, and as temperature is increased an increasing number of kinks are produced. The result is a global change in the structure of the bilayer such that the bilayer becomes slightly thinner with an increase in the surface area associated with each phospholipid head group and in the volume of the vesicle. This change in phase is continuous and occurs over a narrow temperature region of 1-2°C, and is believed to be of first order as there are latent heat and volume changes, as mentioned above.

The most striking functional change in the liposome bilayer associated with the structural phase transition is the marked permeability increase that occurs at T_c[30]. The physical change believed to be responsible for the permeability changes at the transition is depicted in Figure 6.9. Both the liquid crystal and gel phase are shown as being relatively closely packed so that the bilayer is structurally impermeant to solute at temperatures above and below T_c, respectively. At T_c, however, both phase states coexist and this leads to packing irregularities at the interfacial boundaries delineated by the two different phases. As temperature is increased or decreased the liquid-crystalline or the gel phase will predominate, respectively. As T_c is approached, the initial step of formation of a defect and melting of one phase state is very rapidly followed by the rest of the bilayer in a highly cooperative manner by fast lateral diffusion of lipid species within the bilayer. It is these packing irregularities in the boundary layers between interfacial lipids that create 'pores' or space for solutes to cross the bilayer.

6.4.2 Microwave - Triggered Liposome Permeability: Phase-Transition Liposomes and the Role of Cholesterol.

In these studies we investigated whether pure phospholipid vesicles would experience a permeability change in a microwave field identical to that employed in the

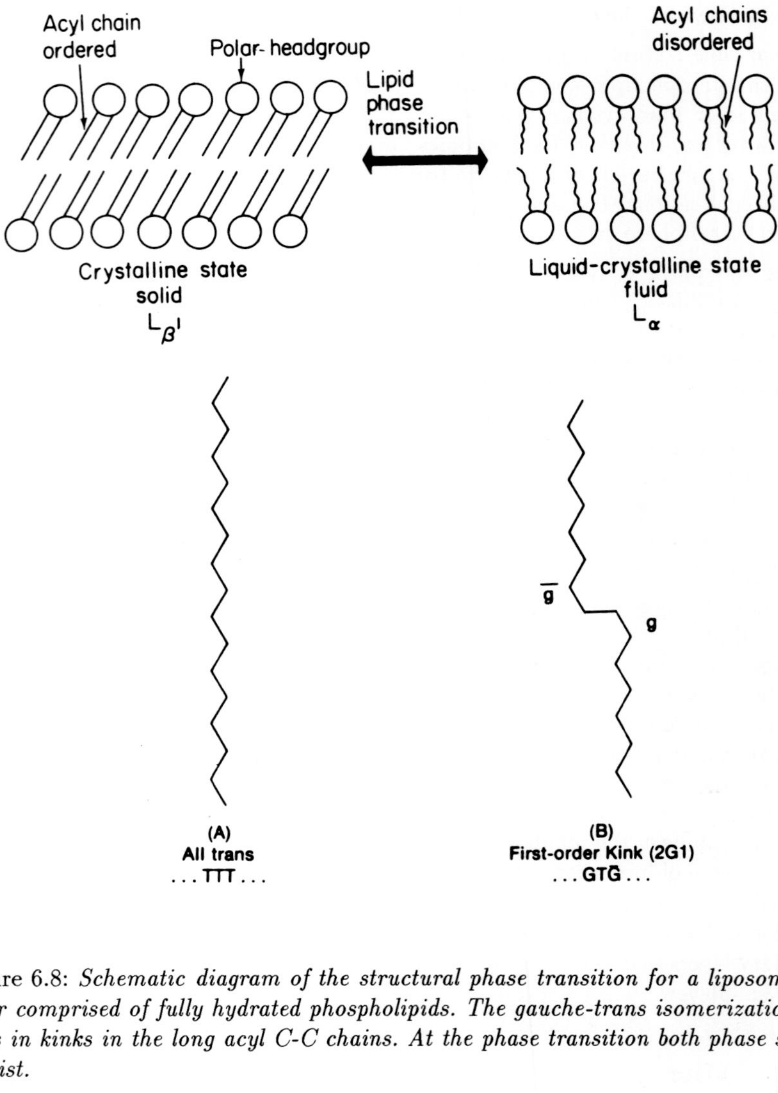

Figure 6.8: *Schematic diagram of the structural phase transition for a liposome bilayer comprised of fully hydrated phospholipids. The gauche-trans isomerization results in kinks in the long acyl C-C chains. At the phase transition both phase states coexist.*

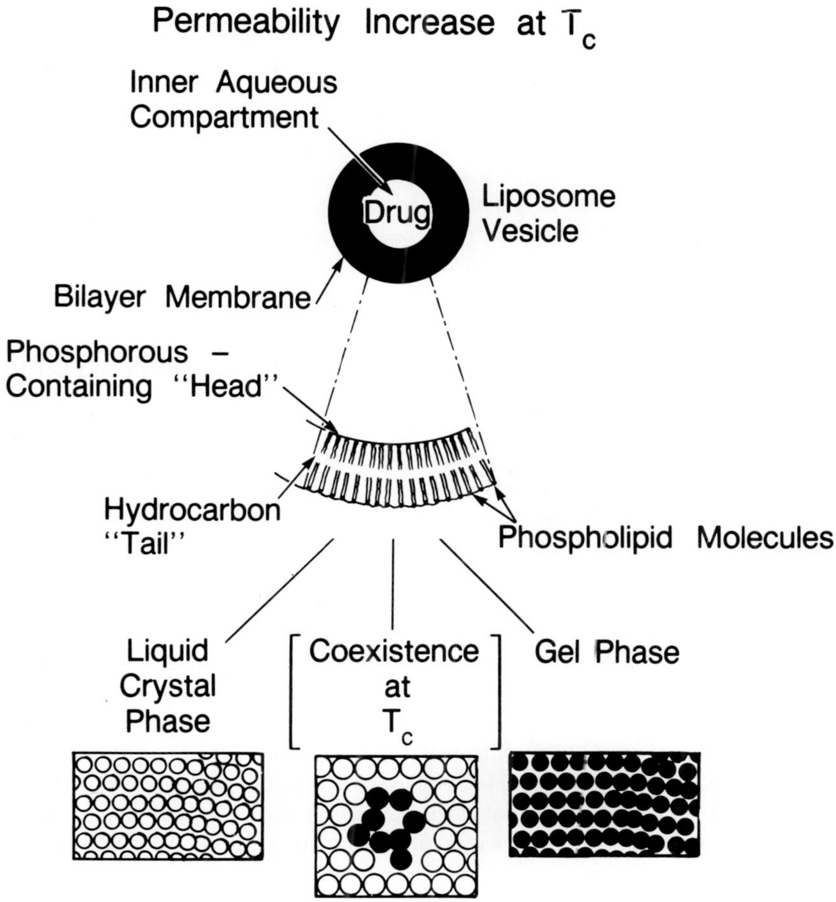

Figure 6.9: *Representation of the packing discontinuity between phase states at T_c for a liposome vesicle. Interfacial boundary discontinuities are believed to result in permeability increases in the bilayer.*

erythrocyte and lymphocyte studies, described above. Unilamellar liposome vesicles of 200 nm diameter were formed from dipalmitylphosphatidylcholine (DPPC) and dipalmitylphosphatidylglycerol (DPPG) in a 4:1 (w/w) ratio with tritiated cytosine arabinofuranoside (^{3}H-ARA-C; MW 243) loaded inside the vesicle as a permeability marker. Both DPPC and DPPG have acyl chains comprised of 16 carbon atoms, and display a T_c of approximately 39.5°C. The head groups of DPPC carry a net neutral charge while those of DPPG carry a net negative charge, so that the vesicles are slightly negatively charged to overcome tendencies of neutral vesicles to associate which compromises stability of the preparation. In addition, the bilayer membrane of these liposomes were modified during their formation to include trace amounts of ^{14}C-DPPC in the bilayer so that degradation of phospholipid in the bilayer could be followed by monitoring ^{14}C-DPPC indicating lysolecithin release.

Microwaves were observed to significantly increase the permeability of DPPC:DPPG liposomes at temperatures well below T_c. Importantly, evidence for membrane rupture was not evident. Figure 6.10 shows the drug release profiles for these liposomes as a function of temperature. The y-axis is scaled in this and the following figures so that data is presented as a percentage of drug released by the detergent Triton X - 100, and this is defined as 100% release. In the absence of microwave (isothermal sham treatment) the liposomes showed a nominal T_c at approximately 39.5°C, with no evidence of release of phospholipid from the bilayer. In contrast, when microwaves were present drug release was observed at temperatures as low as 31°C, which is eight degrees below T_c. In these experiments bulk sample temperature was monitored using an electromagnetic transparent probe and sample temperature was uniform to ±0.08°C during microwave exposures. Thus bulk temperature gradients do not explain the permeability increase. It is possible that micro-hot spots might be formed in the bilayer if the phospholipids themselves, or if bound water associated with the surface of the vesicles, absorbs microwaves preferentially[20]. What is striking about these findings is that microwaves shift the structural phase transition to lower temperatures, and microwaves spread out the temperature range associated with the phase transition.

The interaction mechanism, whatever its specific molecular nature, appears to involve a permeability change operationally identical to that associated with the structural phase transition. Specific physical changes that occur during the phase transition may be crucial to how microwaves interact with the phospholipid bilayer. During phase transition, the rotational motion of the end segments of the acyl carbon chains undergo a *gauche-trans* isomerization, discussed above, and, importantly, they exhibit rotational relaxation frequencies in the range 2-12 MHz[14]. The polar head groups exhibit a much lower rotational frequency of approximately 4 MHz; significantly, these polar head groups become markedly hydrated during phase transition. Thus at the molecular level, we have speculated that it is possible for the 2.45 GHz microwaves employed here.to excite a *gauche-trans* chain, and through the excitation of water, which is a strong microwave absorber, to facilitate polar head

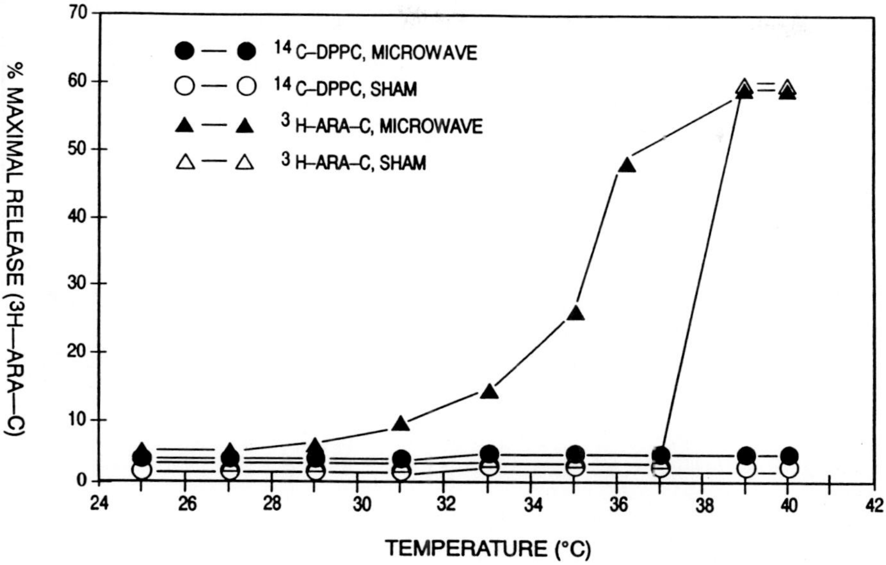

Figure 6.10: *Microwave-stimulated drug release from phase transition (T_c) liposomes. Drug release profiles for liposomes treated with microwaves (60 mW/gm, 10 min) or by an isothermal waterbath (sham, 10 min). The phase transition for DPPC:DPPG liposomes occurs at 39.5° C. The aqueous phase marker inside the liposomes was tritiated cytosine arabinofuranoside ^{3}H-ARA-C, and the membrane marker was ^{14}C-labeled dipalmitylphosphatidylcholine ^{14}C-DPPC. Maximal release was achieved with 1% Triton X-100.*

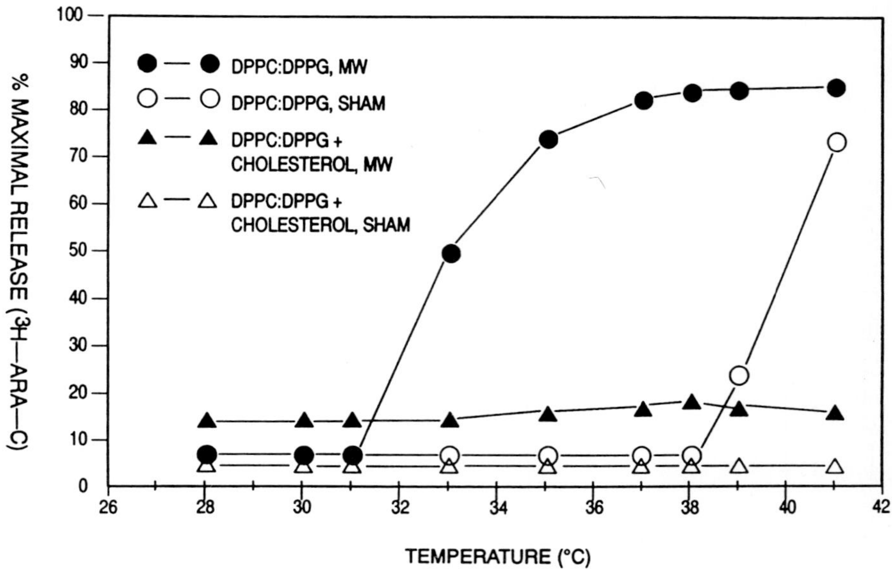

Figure 6.11: *Effect of microwave fields on drug release from phase transition (T_c) liposomes modified by cholesterol enrichment. Conditions as in Figure 6.10, except that DPPC:DPPG liposomes were modified with 30 mole % cholesterol.*

group hydration. These events are candidates for the necessary physical changes to induce permeability increases operationally identical to that associated with a phase transition.

In a parallel approach to that for the erythrocyte experiments, liposomes were modified so that the bilayer incorporated 30 mole % cholesterol to eliminate the phase transition. This enabled us to test the hypothesis that T_c is critical for the permeability change. Figure 6.11 shows drug release profiles as a function of temperature for DPPC:DPPG liposomes that have no cholesterol or 30 mole % cholesterol in the bilayer. Sham isothermal treated liposomes (no cholesterol) show a nominal T_c of 39.5°C, and microwaves are seen to trigger of drug release at temperatures as low as 33°C. This is consistent with the data presented in Figure 6.10. When cholesterol is present, these liposomes do not exhibit a permeability increase at T_c during isothermal treatment at temperatures from 28-41°C. This is expected and indicates that the cholesterol incorporated in the bilayer blocks the endothermic cooperative interactions between adjacent phospholipid molecules[47]. Cholesterol inserts into the bilayer with its hydroxyl group oriented at the polar headgroup region and its hydrocarbon tail in the hydrophobic, apolar core of the bilayer. The physical presence of the rigid sterol nucleus significantly interferes with ordered packing and phase-correlated motion of the acyl chains of the phospholipids.

When cholesterol-modified liposomes are treated with microwaves there is an increase in membrane permeability seen across the temperature range studied but this increase is modest and approaches 20% of maximal release. Interestingly, there is no temperature-dependence detected in this effect. Thus, cholesterol incorporation is seen to block a dramatic increase in permeability observed at lower temperatures for microwave exposure of normal liposomes, discussed above. It is interesting that microwaves induce a small but significant increase in permeability in the cholesterol-liposomes. Indicating that microwaves are capable of disordering bilayer organization sufficiently to allow drug molecules to traverse the bilayer. The molecular mechanism could be similar to the microwave excitation of *gauche-trans* isomerization, discussed above, with cholesterol acting as a steric block to reduce the magnitude of the effect.

6.4.3 Microwave-Triggered Liposome Permeability: Non-Phase-Transition Liposomes and Net Surface Charge.

The interesting finding that cholesterol-modified liposomes which do not exhibit a phase transition, were still able to respond to microwaves, led to a series of experiments using non-phase transition liposomes with no cholesterol. Furthermore, as our original observation that microwaves release the drug cytosine arabinofuranoside from T_c liposomes was made with liposome vesicles carrying a net negative surface charge[14], we have investigated liposomes that carry a net positive surface charge.

Figure 6.12 depicts the solute release profiles for the three model liposomes employed plotted as a function of temperature. The aqueous phase solute marker in each of these liposome systems is 6-carboxyfluorescein (6-CF). Solute release was determined at each temperature after a 10 minute incubation by measuring 6-CF fluorescence and comparing it to maximal release induced by Triton X-100. DPPC:DPPG liposomes (net negative surface charge) display a T_c at approximately 39.5°C shown in the figure. DPPC:lysinyl- phosphatidylethanolamine (95:5 mole ratio) liposomes carrying a net positive charge and also exhibit a nominal T_c at 39.5°C (DPPC:PE). Notice, however, that the release of 6-CF begins at 36°C since the addition of lysinyl-PE, an impurity, slightly broadens the phase transition region. Finally, phosphatidylcholine-lysinyl-PE (PC:PE;95:5 mole ratio) also carries a net positive charge but, importantly, this liposome is in the fluid phase and does not display a phase transition over this temperature range. This is indicated by a baseline release of approximately 7 - 9% detected at all temperatures studied.

When these three liposome systems are monitored for 6-CF release in the presence

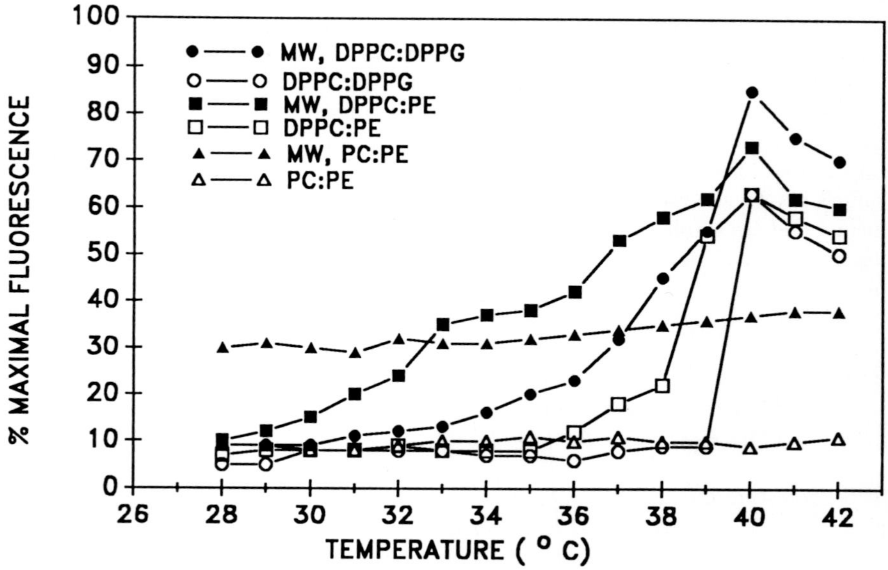

Figure 6.12: *Response of phase transition (T_c) and non-phase transition (non-T_c) liposomes to microwave fields. Microwave treatment (MW, 6 mW/gm, 10 min) or isothermal sham treatment. Liposomes were loaded with the aqueous phase dye marker 6-carboxyfluorescein (6-CF) inside of the liposomes at self- quenching concentrations (100 mM). T_c liposomes: DPPC:DPPG (negative surface charge) and DPPC:PE (positive surface charge). Non-T_c liposomes: PC:PE (positive surface charge). Maximal release was achieved with 1% Triton X-100.*

of a microwave field, as above, we found that the permeability of each liposome was influenced, but in different ways. The two T_c liposomes (MW, DPPC:DPPG and MW, DPPC:PE) respond to the microwave field in a temperature dependent manner with enhanced release at Tc and a broadening of the solute release profile. For example, 6-CF release from DPPC:DPPG was detected at 37°C which is 3~°C below T_c where release is normally first detected. The positively charged vesicles (MW,DPPC:PE) displayed an even greater broadening with solute release detected at temperatures as low as 31°C which is ~ 9°C below the nominal T_c. In contrast, the non-T_c vesicles (PC:PE) responded in a different manner with enhanced solute release at all temperatures studied and in a temperature-independent manner. This response is characterized by a release of solute at levels of 30 - 40% of maximal release at all temperatures. It is striking that this behavior is similar to that observed for the cholesterol-DPPC:DPPG liposomes, which were converted to non-T_c liposomes by chemical modification (Figure 6.10).

6.4.4 Microwave-Triggered Liposome Permeability: Role of Free Radicals.

In erythrocyte experiments oxygen was observed to enhance microwave interaction with the membrane and result in protein-shedding (Figure 6.2). This observation first raised the possibility of free radical involvement in microwave bioeffects, and studies were performed with liposomes to determine if a simple phospholipid bilayer system would also display such an oxygen dependence.

Figure 6.13 shows drug release profiles for DPPC:DPPG liposomes loaded with ^{3}H-ARA-C that also contain trace amounts of ^{14}C-DPPC. In the absence of microwaves, and for buffers that contained saturated oxygen (O_2), saturated nitrogen (N_2), or atmospheric oxygen (ATM), the liposomes showed a normal T_c at approximately 39.5°C . These liposomes also displayed no release of ^{14}C-DPPC over the temperature range studied. Therefore the presence of saturating levels of oxygen had no effect on normal phase transition drug release for this liposome vesicle. In contrast, microwaves are seen to trigger drug release at temperatures as low as 25°C in a saturated oxygen environment (pO_2 = 760 mm Hg) while no release of ^{14}C-DPPC is evident. Thus oxygen significantly enhanced drug release during microwave treatment without disruption of the membrane. When microwave exposures were conducted for liposomes maintained at atmospheric oxygen (ATM, pO_2 = 150 mm Hg) or in a saturated nitrogen environment (pO_2 = 1-2 mm Hg) drug release was observed at temperatures of 33°C, which is six degrees below T_c. Release of phospholipid from the bilayer was not evident.

The liposomes used in these studies were constructed from saturated phospholipids, so that they possessed no oxidizable acyl carbons. Unsaturated phospholipid chains are susceptible to oxidation which can lead to permeability changes. The oxygen

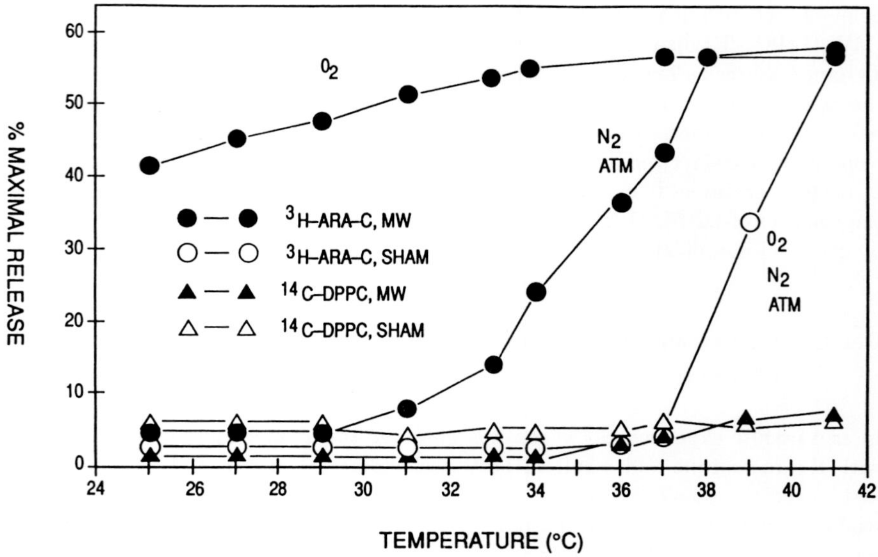

Figure 6.13: *Effect of oxygen on microwave-stimulated drug release. Liposomes were exposed as described in Figure 6.10, with the exception that prefiltered oxygen, nitrogen, or atmospheric air was used to saturate the buffer sample.*

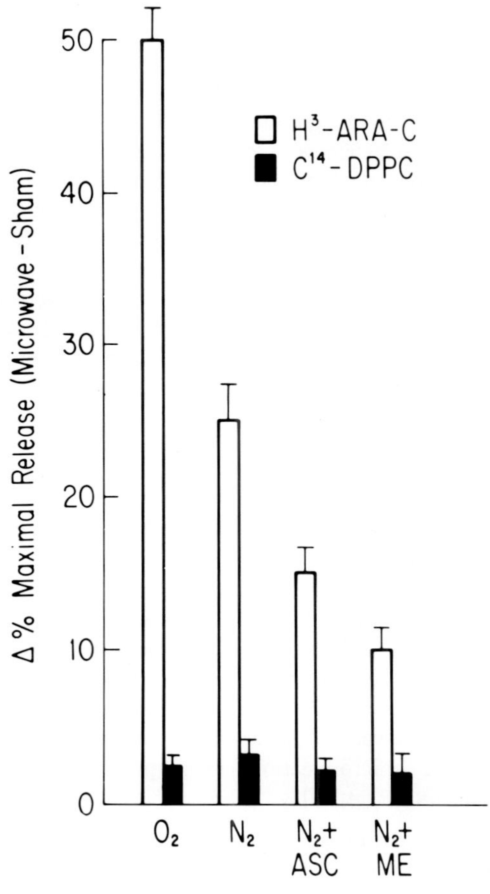

Figure 6.14: *Effect of antioxidants on microwave- stimulated drug release. Exposures as in Figure 6.10, except that ascorbic acid (ASC, vitamin C) or β-mercaptoethanol (βb-ME) was present at a 5x and a 1x mole ratio of antioxidant to lipid, respectively. Mean ± S.D.*

effect described above cannot be interpreted as a result of simple oxidation. This effect might be due to the action of a reactive oxygen free radical in the bilayer. To investigate whether a free radical species may be involved in the oxygen effect, liposomes were suspended in saline buffer and exposed to microwaves in the presence of ascorbic acid (10 mM)(ASC) or β-mercaptoethanol (2 mM)(ME), each a potent free radical scavenger. Figure 6.13 presents results from microwave and sham exposures at 37°C. Microwave exposure of liposomes at 37°C for O_2 and N_2 saturated buffer resulted in a net increase of ~50% and 25% maximal release over sham values, respectively. When exposures were conducted using N_2 saturation in the presence of ascorbic acid or β-mercaptoethanol the net difference between microwave and sham was reduced to ~17 and ~10% maximal release, respectively. No significant changes were found in ^{14}C-DPPC, and this was corroborated by thin-layer chromatography. The results indicate that two different free radical scavengers can act to reduce, but not eliminate, microwave-stimulated drug release, and that the observed permeability effect is not associated with lipid degradation. These findings support the oxygen-dependence observed in the erythrocyte studies, discussed above, and provide evidence that free radicals may be involved in the interaction. This argues for a chemical basis, at least in part, for the interaction.

6.4.5 Microwave-Triggered Liposome Permeability: Membrane Fusion.

The demonstration we have made that microwaves can act to release an aqueous marker from both T_c and non-T_c liposome vesicles, holds promise for triggering membrane fusion events. In this regard, it is of interest to determine whether liposome vesicles bound to a target cell surface can still respond to microwave fields and undergo enhanced solute release. If such an interaction does occur it could form the basis of a controlled "microinjection" technique in which liposomal solute is transferred directly into a target cell.

In these experiments, purified rat spleen lymphocytes were complexed *in vitro* with the three liposome systems described in Figure 6.12. Each liposome vesicle type was found to bind electrostatically to the lymphocyte surface in significant quantities so that fluorescent measurements of 6-CF transfer into the lymphocyte were possible once the dye was released from the interior of the liposome where it was quenched at 100 mM. Lymphocyte-liposome complexes thus formed were thoroughly washed and subsequently treated with a microwave field or sham exposed, as in Figure 6.12, at either 24° C for 15 minutes or at 34°C for 30 minutes. These temperatures were chosen since they were well below the nominal T_c of 39.5°C for the T_c liposomes (Figure 6.12). The complexes were then extensively washed and the amount of 6-CF transferred into the lymphocyte target cell was assayed spectrofluorometrically. The microphotograph shown in Figure 6.15 is a light transmission microphotograph that depicts lymphocytes coated with DPPC:DPPG liposome vesicles containing 6-CF

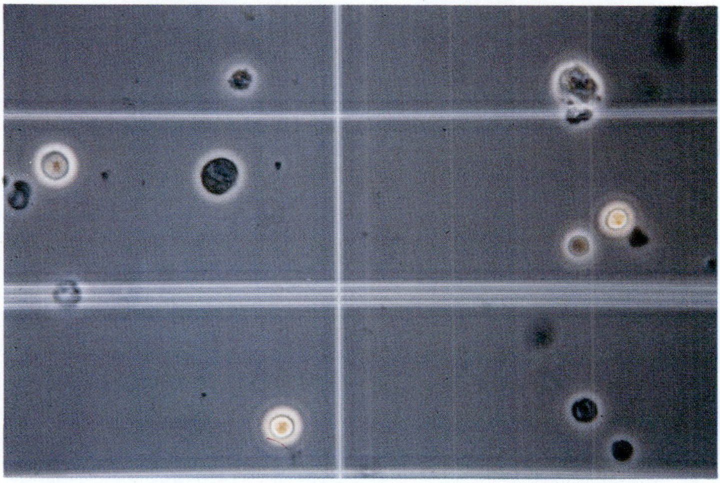

Figure 6.15: *Phase contrast, transmitted light photomicrograph of human lymphocytes coated with liposomes loaded with 6-CF after microwave exposure. Size heterogeneity is apparent in the population of peripheral blood lymphocytes, and cells have a normal appearance. Microwave treatment as in Figure 6.12. Compare to Figure 6.16 which shows 6-CF dye transfer to lymphocytes.*

after microwave treatment. This preparation of human peripheral lymphocytes is heterogeneous in size and the cells range in diameter from approximately 5-7 microns; the cells appear normal and are not lysed. Lymphocytes are the larger dark cells, while the bright smaller cells are erythrocytes. Figure 6.16 shows the field of cells illuminated using fluorescence excitation light for 6-CF. As can be seen microwave treatment resulted in significant transfer of 6-CF dye to the interior of the lymphocytes, as indicated by the green fluorescence visualized inside the lymphocytes. The red erythrocytes contain hemoglobin that quenches 6-CF fluorescence. This result can be interpreted as evidence for 'microinjection' of 6-CF into the lymphocytes, and can be explained most simply by postulating that the dye was released from the liposomes during microwave treatment and the free dye (a) passively diffused down a concentration gradient near the cell surface into the cell, or (b) the liposome vesicle fused with the lymphocyte membrane and directly transferred dye into the target cell.

In order to quantify this phenomenon for the three liposome systems mentioned above, fluorescence measurements were made of the target cell populations after microwave or sham treatments. In Figure 6.17 is depicted the results of these experiments. Each of the vesicle systems, both T_c and non-T_c, were influenced by the

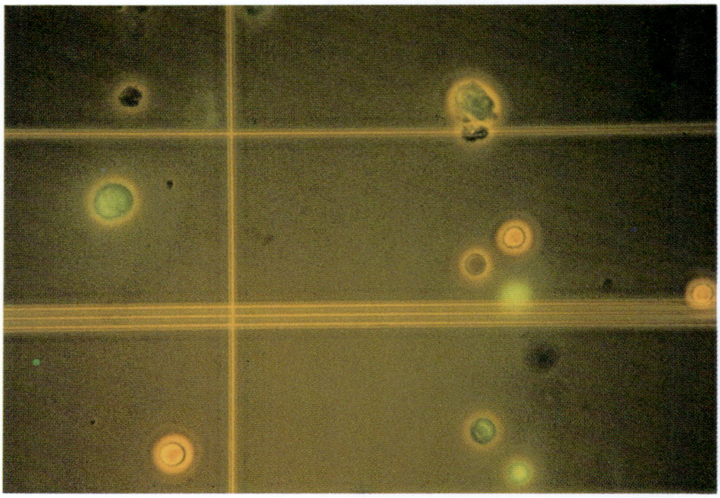

Figure 6.16: *Flourescence photomicrographof the same human lymphocytes as in Figure 6.15. Microwave treatment has transferred the dye 6-CF from liposome vesicles on the surface of the lymphocytes to the interior of the lymphocyte and is visualized as uniform green fluorescence inside of the cells.*

microwave field. DPPC:DPPG vesicles exhibited the greatest increase in 6-CF transfer to the target lymphocyte with a 8.6 fold and 8.2 fold increase of 6-CF transfer at 25°C and 34°C, respectively. The least responsive vesicles were the non-T_c liposomes, PC:PE, with a 1.41 fold increase in 6-CF transfer at 34°C.

In the liposome experiments described above, the idea that phospholipids bilayers can interact with microwave fields in the absence of membrane proteins was demonstrated using several different liposome systems. Phase transition liposomes carrying either a net negative or a net positive charge underwent microwave-induced permeability increases at pre-Tc temperatures, where they are not normally leaky. It is not known why positively charged liposomes responded to the microwave fields with greater release at temperatures further removed from T_c.

An interesting finding is that non-T_c vesicles are responsive to microwave fields. We observe that two non-T_c systems respond to microwaves by increased membrane permeability: PC:PE of Figure 6.12 and cholesterol-loaded DPPC:DPPG of Figure 6.11. Significantly, as the bilayers of both of these liposomes are fluid at the temperatures employed during microwave exposures, the results represent direct evidence that relatively fluid membranes can be permeablized by microwave fields. Since our observations were first made[48], another laboratory group has recently demonstrated, in careful studies, qualitatively similar findings employing a non-Tc fluid phase li-

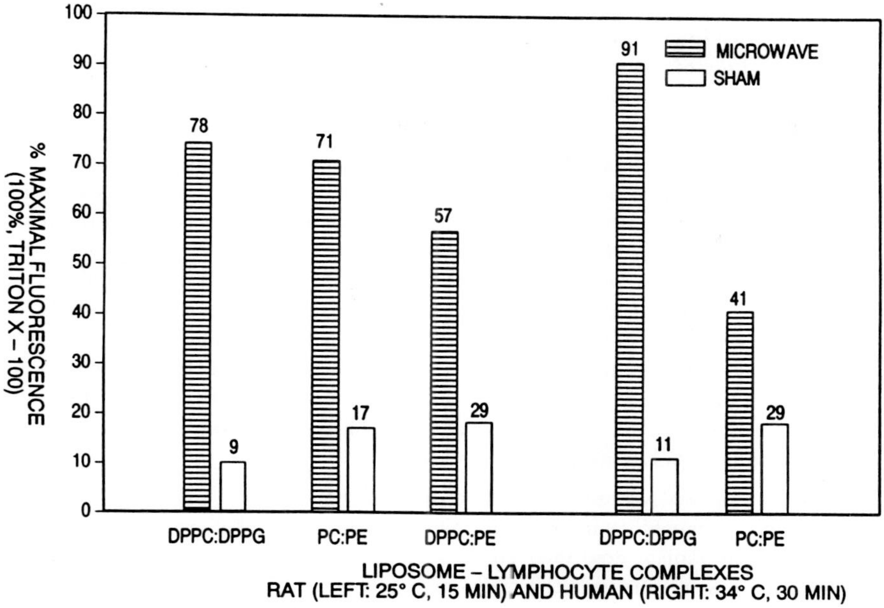

Figure 6.17: *Quantitative measurement of microwave- triggered "microinjection" of 6-carboxyfluorescein from liposomes into lymphocyte target cells. Influence of T_c and Non-T_c Liposome Systems. Microwave exposures of liposome-coated lymphocytes were as in Figure 6.12. Sham exposures were at isothermal temperatures. T_c Liposomes: DPPC:DPPG, DPPC:PE. Non-T_c Liposome: PC:PE.*

posome formed from phosphatidylcholine and exposed to 2,450 MHz fields at 38 mW/gm for 10 minutes at 38°C[43]. Both findings taken together support the phenomenon of microwave-triggered permeability in fluid phase liposome vesicles.

The exact molecular mechanism for the bilayer permeability effect is not clear at this time. Importantly, however, we have observed that the erythrocyte bilayer is rendered completely insensitive to microwave fields when cholesterol is added to 30 mole %[15], an amount that completely obliterates long-range order necessary for a phase transition. This finding suggests to us that microwave coupling in the liposome phospholipid bilayer may also be dependent on the ability of membrane components to exhibit unimpeded long-range order, as is discussed below. It is also of interest to compare the response in Figure 6.12 of T_c liposomes to microwave fields at temperatures where they are relatively solid ($< T_c$) to that where they are relatively fluid ($> T_c$). We see that the most pronounced increases in permeability due to microwaves occur at temperatures below T_c where the bilayer is solid. At T_c both solid and fluid phases coexist, by definition. Above T_c there is increased drug release associated with microwave field treatment, but these increases are less pronounced. Thus, regardless of magnitude, we see that the bilayer can respond to microwave fields in the solid or in the fluid state.

What is the explanation for this? Our original interpretation[14] of liposome vesicle interactions with microwave fields was that the permeability effect depends on the ability of the bilayer to be able to undergo a phase transition at T_c. In those studies we were working with liposomes in the solid state at temperatures just below but near T_c, and we knew that the addition of cholesterol to the erythrocyte membrane, which abolishes long- range order required for a phase transition, abolished microwave-sensitivity. Based on these findings, as discussed above, we hypothesized that microwave fields induced rotational motion in the phospholipid acyl chains, which exhibit rotational frequencies in the 2-12 GHz range, and this caused packing irregularities and pore formation. Such induced motion would be similar, therefore, to that experienced at T_c when relatively solid phase phospholipid acyl chains are converted to relatively fluid phase moieties.

However, we presently observe that two different non-T_c, fluid phase liposome systems exhibit permeability increases in microwave fields. We believe that the general interaction mechanism described above represents, perhaps, the simplest coupling mechanism. This means that the microwave field leads to permeability increases in fluid phase lipids by triggering additional rotational motion in the phospholipid acyl chains that results in packing irregularities and an increase in bilayer permeability. This is consistent with our original interpretation of a microwave permeability change mediated by enhanced or increased rotational motion of the phospholipid acyl chains.

An interesting application of microwave-triggered liposomal drug release is the 'microinjection' technique described above that employs a fluid-phase non-Tc liposome. This technology should complement existing approaches for drug or chemical delivery to cells employing membrane fusion[49]. The transfer of chemical agents, sequestered in a liposome bound to the surface of a target cell, across the membrane bilayer of the target cell to its interior space might have potential as a basis for an effective "microinjection" technique for chemical and drug delivery.

Acknowledgments

This work was supported by the Department of Energy, Office of Health and Environmental Research, under Contract DE- AC03-76SF00098.

Bibliography

[1] N.G. Noel, *Cell Signalling* (The Guilford Press, New York, 1989).

[2] P. Gardner,. *Cell*: **59** (1989) 15-20.

[3] R.P. Liburdy, in *Interaction Mechanisms of Low-Level-Electromagnetic Fields in Living Systems* (Eds. B. Norden and C. Ramel, Oxford University Press, New York, 1992), pp 217-239.

[4] R.P. Liburdy, in *Interaction Mechanisms of Low-Level Electromagnetic Fields in Living Systems* (Eds. B. Norden and C. Ramel, Oxford University Press, New York, 1992), pp 259- 279.

[5] R.P. Liburdy, *Annals of the New York Academy of Sciences* (Eds. R. Magin, R.P. Liburdy and B. R. Persson) **649** (1992) 74-95.

[6] R.P. Liburdy, *FEBS Letters* **301** (1992) 53-59.

[7] M.G. Yost and R.P. Liburdy, *FEBS Letters* **296** (1992) 117-122.

[8] R.A. Luben, *Health Physics* **61** (1991) 15-28.

[9] W.R. Adey, In *Interaction Mechanisms of Low-Level Electromagnetic Fields in Living Systems* (Eds. B. Norden and C. Ramel, Oxford University Press, New York, 1992), pp 47-77.

[10] R.P. Liburdy, *Radiation Research* **77** (1979) 34-46.

[11] R.P. Liburdy and A. Wyant, *Int. Journal. of Radiation Biol.* **46** (1984) 67-81.

[12] R.P. Liburdy and A. Penn, *Bioelectromagnetics* **5** (1984) 283-291.

[13] R.P. Liburdy and A. Penn, *Annals of the New York Academy of Sciences* **435** (1984) 302-305.

[14] R.P. Liburdy and R.L. Magin, *Radiation Research* **103** (1985) 266-275.

[15] R.P. Liburdy and P.F. Vanek, Jr,*Radiation Research* **102** (1985) 190-215.

[16] R.P. Liburdy and T.S. Tenforde, In *Biophysical Effects of Steady Magnetic Fields* (Eds. G. Maret, J. Kiepenheuer, and N. Boccara, Springer-Verlag, Heidelberg, F.R.G., 1986) pp.44-51.

[17] R.P. Liburdy, T.S. Tenforde and R.L. Magin, *Radiation Research* **108** (1986) 102-111.

[18] R.P. Liburdy and P.F. Vanek, Jr., *Radiation Research* **109** (1987) 382-395.

[19] R.P. Liburdy, A.W. Rowe and P.F. Vanek, Jr., *Radiation Research* **114** (1988) 500-514.

[20] M.A. Stuchly, S.S. Stuchly, R.P. Liburdy and D. A.Rousseau *Phys. Med. Biol.* **33** (1988) 1309-1324.

[21] T.S. Tenforde and R.P. Liburdy *J.Theoret. Biol.* **133** (1988) 385-396.

[22] American National Standards Institute, *Safety Level of Electromagnetic Radiation with Respect to Personnel* C95.1-1991 (The Institute of Electronics and Electrical Engineers, NY. NY., 1991).

[23] R.B. Olcerst, S. Belman, M. Eisenbud, W.W. Mumford and J. Rabinowitz, *Radiation Research* **82** (1980) 244-256.

[24] P.D. Fisher, M.J. Pozansky and W.A.G. Voss, *Radiation Research* **92** (1982) 411-422.

[25] S.F. Cleary, F. Garber and L.M. Liu, *Bioelectromagnetics* **3** (1982) 453-466.

[26] J.W. Allis and B.L. Sinha-Robinson, *Bioelectromagnetics* **8** (1987) 203-212.

[27] W.G. Lotz and J.L. Saxton, *Membrane Permeability in Erythrocytes Exposed to High Peak Pulsed Microwaves.* Eleventh Annual Bioelectromagnetics Society Meetings (Tucson,AR, USA, June 18-22, 1989) Abstract P-2-15.

[28] J.D. Bond and N.C. Wyeth, in *Mechanistic Approaches to Interactions of Electric and Electromagnetic Fields with Living Systems* (Eds. M. Blank and E. Findl, Plenum Press, New York, 1987) pp.291-300.

[29] J.D. Bond and N.C. Wyeth,*J. Chem. Phys.* **85** (1986) 7377-7379.

[30] D. Papahajopoulos, K. Jacobson, S. Nir and T. Isac, *Biochim. Biophys. Acta.* **311** (1973) 330-348.

[31] P. Bernardi and G. D'Inzeo, In *Electromagnetic Biointeraction: Mechanisms, Safety Standards, Protection Guides* (Eds. G. Franceschetti, O.P. Ghandi, and M. Grandolfo, Plenum Press,New York, 1989) pp. 27-57.

[32] M.R. Quastel and J.G. Kaplan, *Experimental Cell Research* **63** (1970) 230-233.

[33] R. Hesketh, In *Biochemistry of Cell Walls and Membranes II.* Vol. **19** (Ed. J.C. Metcalfe, University Park Press, Baltimore, MD, USA, 1978) pp 63-91.

[34] P.M. Rosoff and L.C. Cantley *Proc. Nat. Acad. Sci. (USA)* **80** (1983) 7547-7550.

[35] G.B. Segel, G. Kovach and M.A. Lichtman, *J. Cell. Physiol.* **100** (1979) 109-117.

[36] G.B. Segel, W. Simon and M. A. Lichtman, *J. Clin. Investigation* **64** (1979) 834-841.

[37] S.F. Cleary, Li-Ming Liu, and R. E. Merchant, *Bioelectromagnetics* **11** (1990) 47-56.

[38] B. Pernis, *Immunology Today* **6** (1985) 45-49.

[39] W.H. Dzik, *New England Journal of Medicine* **309** (1983) 435-436.

[40] R.J. Owens and M.J. Crumpton, *Biochemical J.* **219** (1984) 309-316.

[41] M.J. Geisow and J.H. Walker, *Trends in Biochemical Sciences* **11** (1986) 420-423.

[42] M.J. Geisow, J.H. Walker, C. Boustead and W. Taylor,*Biochemical Society Transactions* **15** (1987) 800-802.

[43] E. Saalman, B. Norden, L. Arvidsson, Y. Hamnerius, P. Hojevik, K.E. Connell and T. Kurucsev, *Biochim. Biophys. Acta* **1064** (1991) 124-130.

[44] A.R. Orlando, G. Mossa, M.C. Annesini, A. Zevano and G.D'Inzeo, *First World Congress for Electricity and Magnetism in Biology and Medicine, Lake Buena Vista, FL, USA*, June 14 - 19 (1992).Abstract K-3.

[45] G. Legaly, A. Weiss and E. Stuke,*Biochim. Biophys. Acta.* **470** (1977) 331-341.

[46] J.F. Nagle, *Ann. Rev. Phys. Chem.* **31** (1980) 157-195.

[47] B.D. Ladbrooke, R.M. Williams, D. Chapman, *Biochim. Biophys. Acta.* **150** (1968) 333-340.

[48] R.P. Liburdy and B. Fingado,*Bioelectromagnetic Society Meeting, San Antonio, TX,* June 10 - 14 (1990). Abstract B- 4-5.

[49] D. Papahadjopoulos D. and A. Gabizon,*Annals of the New York Academy of Sciences* **507** (1987) 64-74.

Chapter 7

Can Weak Magnetic Fields (or Potentials) Affect Pattern Formation?

Mae-Wan Ho, Adrian French, Julian Haffegee, and Peter T. Saunders

7.1 The Physical versus the Chemical Approach to Morphogenesis

We know a great deal, nowadays, about the molecular and genetic constituents of organisms, but almost nothing of the organism itself. The problem of how body pattern is laid down in development remains practically untouched since Hans Driesch and others introduced the concept of an 'embryonic field', drawing attention to the coherent, regulative nature of the processes involved in determining body pattern. The embryonic field is a domain in which pattern is roughed out very early in development, to undergo more and more detailed differentiation later on. It is approximately 1 mm in linear dimensions in all animals and plants, regardless of the adult size of the species. This domain contains some 10^{20} molecules of many different kinds. How does macroscopic pattern arise in such a diverse molecular system?

Alan Turing's 1952 trend-setting paper on chemical morphogenesis[1] shows how a simple system of chemical reaction and diffusion could lead to patterns of chemical 'morphogens' starting from homogeneity. This has given rise to a whole genre of reaction-diffusion models whereby 'pre-patterns' of chemical morphogens are formed, which are then 'interpreted' into body pattern by something else, perhaps, special genes. What is missing from these models is an account of the forces that *mobilize* molecules of different kinds into a pattern. The only plausible 'chemical' identified, thus far, that fits the description of a morphogen is calcium ions. Calcium ions interact electrically with all kinds of proteins, in particular, cytoskeletal proteins, and can trigger electromechanical forces that have patterning effects. Several 'viscoelastic' models of morphogenesis[2,3] are based on calcium-cytoskeletal protein interactions, and emphasize the *physical* forces involved in patterning, much in the tradition of D'Arcy Thompson[4] and others before him.

7.1.1 The Electrodynamical Morphogenetic Field

Also in the same tradition are Burr and others (see Bischof, this volume), who suggested that the morphogenetic field is electrodynamical. Electric fields have indeed been detected in all developing and adult organisms, and by successive generations of workers as the instrumentation for measuring weak electric fields continue to improve. As all molecular and intermolecular forces are electromagnetic[5], *electrodynamical forces are sufficient by themselves to mobilize and organize the molecules within the embryonic field into macroscopic pattern.* Prepattern and pattern are but earlier and later stages of the same continuous process.

Electrodynamical forces include polarization fields (which may phase order macromolecular arrays, see Ho and Saunders, this volume), electronic or proton currents, dipole interactions, electromechanical forces and deformations (such as are involved in the interaction of proteins and calcium), electrodiffusion of ions and larger molecules by electrophoresis and dielectrophoresis (see Pethig, this volume) as well as polarization waves, dipole oscillations, phonon and photon exchanges of a resonant nature or otherwise (see also Davydov's solitons, this volume). Given such a wide variety of mechanisms, it is perhaps not surprising that there is as yet no electrodynamical *model* of pattern formation. It remains the major challenge in biocommunication to present specific, realistic, and testable models for pattern formation, which is *the* primary problem of biological organization.

In order to obtain further evidence that the embryonic field is electrodynamical in nature, and to elucidate the mechanisms involved, we decided to investigate the effect of weak external magnetic fields on pattern formation in *Drosophila.* The development of this organism is extremely well characterized in terms of morphological changes and changes in gene expression. Pattern determination processes occur within the first three hours of development, during which time, the body pattern (in the ensuing larva) is sensitive to perturbation from a wide variety of environmental agents[6]. It is also relatively easy to obtain large numbers of embryos that are synchronous to within one minute of development. The homogeneity of the system is important for obtaining reproducible results, which can be further characterized according to developmental time. The reason for using *weak* fields is to avoid thermal and other effects unrelated to the electromagnetic nature of the disturbance.

7.2 Experimental Methods

7.2.1 Methods for Generating Magnetic Fields

Three series of experiments were performed. Two of them involved magnetic fields generated in a Helmholtz coil (300x2 turns in 11 cm; diam. 11cm) wound around an aluminium core, and fitted with an adjustable teflon holder in the centre, which could be turned through 90°, enabling us to expose the embryos in two perpendicular orientations of the field. The first series involved steady magnetic fields generated with a DC power supply (DC-series), the second series involved 60hz AC magnetic fields generated from the ordinary mains supply to the laboratory through an appropriate transformer (AC-series). No attempt was made to cancel out the earth's field as the weakest field applied was at least an order of magnitude above the geomagnetic field.

A third series of experiments (Tor-series) was done using a toroidal coil which generated little or no magnetic field in its vicinity when supplied with a current of 1.5A from a 15V DC power supply. The purpose of this coil is to create 'field-free' (i.e., free of magnetic flux lines generated by the coil) regions where the vector potential does not vanish, and hence the effect of the latter on pattern formation could be investigated. The coil is made of a long thin solenoid, 0.5cm in diameter wound around a flexible plastic tubing - approximately 20 turns per cm. - closed off into a torus 9 cm in diameter. To compensate for the residual magnetic field, a continuous wire is threaded through the core plastic tubing so that current flows in it in the opposite sense to the torus. The experimental embryos were placed in the centre of the torus either before or after the power supply was turned on. Over ten measurments of the fields made on the day of experimentation (with a Bartington 3-axis magnetic field sensor MAG-03MC), the mean and standard deviation of the field in the centre of the torus was almost entirely due to the earth's field, i.e., $36.1\pm2.1\mu$T when the power was turned on, and 36.1 ± 2.3 μT when power was off. (The range of difference between on and off states is -1.5 to $\pm1.5\mu$T, the mean and standard deviation being $0\pm0.3\mu$T.) For comparison, the corresponding values for the field at the control sites were $38.6\pm2.4\mu$T. Thus, the residual magnetic field due to the torus was much smaller than the local spatio-temporal variations in the ambient field.

7.2.2 Methods for Preparing and Rearing Embryos

In a typical experiment, synchronous batches of about 50 embryos were collected on grape-agar impregnated filter-paper discs from 3-7day old female flies previously randomized by pooling them together and cleared of retained eggs[6]. Matched controls and experimental embryos were allowed to develop in a humid atmosphere for 1 hr and in a few cases, 1.5h before exposure to the magnetic field. Control and experimental batches of embryos have to be matched because the fecundity of the females and viability of the embryos vary with the nutritional status of the female flies as well as their age, not to mention the uncontrollable ambient electromagnetic environment, which can vary from day to day. Exposure time was set as stated because we previously found that exposures to DC magnetic fields had no effect on pattern formation after 3.5h of development[7]. This time-window of sensitivity is the same as that exhibited in experiments with a variety of physical factors such as heat or cold shocks, or exposure to organic solvents, which are also capable of perturbing body pattern[6].

For the series of experiments on DC magnetic fields, the embryos were allowed to develop on a shelf in a 25°C incubator (range 21 to 25°C) where the static field is not unusually above or below that of the earth, and the AC field (measured with ELF Sense, Expan Test, Inc.) was the lowest at 1.3μT. In order to reduce the ambient level of AC field, for the AC and Tor-series of experiments, the embryos were allowed to develop in the open laboratory, where the stocks were also kept, near a source of heat from a thermostated heating block. At that rearing site, the ambient AC field is never greater than 0.2μT (measured with Holaday Electromagnetic Field Survey Meter) and the static magnetic field is not substantially different from that of the earth. The temperature, however, fluctuated overnight between the range of 18 to 24.5°C. This was found to improve the hatching rate in the controls, and to reduce the level of 'nonspecifc early arrests' and other abnormalities, except for 'short-segments' which is significantly increased (see below).

7.2.3 Methods of Exposure to Magnetic Fields

For the DC and AC-series, magnetic fields (0.5 to 7mT) were pre-set with a Gaussmeter (Model 9200, F.W. Bell), and the experimental embryos, contained in a moist, covered petri-dish (diam. 2.5cm), were placed in the holder at the centre of the coil before the preset field was activated. The field was uniform throughout the area of the petri dish, and, with a DC power supply, the AC field was never more than 1 μT. For the Tor-series, two sets of experiments were done: one with the embryos in place *before* the pre-set field was switched on, and the second with the embryos placed *after* the field was switched on. Exposure period was 30min for all field strengths in the DC- and AC-series, and 60min for the Tor-series, during which time, the controls were placed in an identical petri-dish on the bench sufficiently far from the coil so that its field had attenuated essentially to zero. The temperature in the petri-dish was measured at the beginning and the end of the exposure period (thermocouple probe, Digitron Electronics, Ltd.). There was no heat generated by the coil at field strengths below 3.5mT in the DC and AC-series, nor in the null field of the Tor-series. In no case was the temperature increase in the petri-dish containing the embryos more than one degree Centigrade above the ambient room temperature and no gradient was detected across the radius of the dish. The temperature changes were well within the range of fluctuation at the sites where the embryos were normally kept, and are not known to induce abnormalities. After exposure, both experimental and control embryos were returned to the rearing site and allowed to develop within a humid perspex chamber for 24h before they were dechorionated, cleared, mounted on a microscope slide and examined under dark-field[6]. The scoring was performed 'blind' after initial training sessions to prepare embryos appropriately and to recognize the abnormalities. Further details are reported elsewhere for the DC experiments[7]. The same protocol was used in the AC-series and the Tor-series.

7.3 Results

7.3.1 Effect on Hatching Rates and Total Patterning Abnormalities

The effect on hatching rates and total patterning abnormalities over all treatments are compiled in Table 7.1. Significant decrease in hatching rate occur only in treatments with both DC and AC fields but not with the null field of the Tor series. This suggests that hatching rates are, at least to some extent, independent of pattern abnormalities. All field exposures, possibly with the exception of the 0.82mT AC field, result in significant increases in overall patterning abnormalities. For both DC and AC fields, exposures along the anterior/posterior axis tend to give greater percentages of abnormalities than exposures along the dorsal/ventral axis for the same field intensities.

7.3.2 The Categories of Pattern Abnormalities

The abnormalities were classified into 8 categories (see Figs. 7.1 and 7.2). 'Nonspecific early arrests' (nea) appear as empty vitelline membranes. They are almost certainly a heterogeneous assemblage including unfertilized eggs as well as embryos arrested before any cuticle has been laid down. The other categories are as follows. 'Headless' (hl) typically fail to form head and other anterior segments (Fig. 7.1 a-c); included in this category are a subclass which can be named 'featureless' as they lack any discernable cuticular structure, appearing as faint ghostly masses within the vitelline membrane (Fig. 7.1 a). They represent embryos in which patterning has failed in the very initial stages. 'Crooked' larvae (cr) are those with one or more ventral denticle bands which are crooked. 'Short-segments' (ss) are those in which one, or a few, or all the segments are truncated (see the remaining segments in Fig. 7.1b and e). 'Head and or tail reduced' (h/t red.), have either head, or tail, or both reduced (most of the embryos in Fig. 7.2), or with tail entirely absent (Fig. 7.1 d,e). 'Twisted' larvae (tw) are by far the most interesting. They exhibit various rotations or twisting of the segments around the anteroposterior axis. Most of the rotations involve either the anterior or posterior ends to varying extent (Fig. 7.2), in one

Table 7.1: Summary of hatching rates and abnormalities on exposure to weak magnetic fields

	Field (mT)	No.	% hatched	% Normal
A. DC d/v	7.00	458	67.3[a]	54.4[a]
	control	428	80.5	80.1
	3.00	355	63.7	56.1[a]
	control	302	61.3	67.9
	2.00	508	65.4	59.7[a]
	control	400	70.5	69.3
	0.50	383	63.2[c]	59.3[b]
	control	370	71.1	69.2
a/p	7.00	282	61.4	44.3[a]
	control	235	67.7	66.4
	3.50	424	72.6	51.7[a]
	control	271	70.5	70.1
	2.00	318	61.3	43.7[a]
	control	255	66.7	64.7
	All DC	2729	65.4[a]	53.5[a]
	control	2261	69.3	70.4
b. AC d/v	2.36	220	77.3	47.3[a]
	control	275	83.6	72.7
	1.76	357	85.4	49.0[b]
	control	395	82.8	60.3
	0.82	214	79.4	68.7
	control	244	68.7	73.4
a/p	2.36	280	68.9[a]	27.9[a]
	control	253	85.8	61.3
	1.50	366	75.4	50.0[a]
	control	332	77.7	71.4
	All AC	1437	77.5[c]	47.8[a]
	control	1499	86.1	67.3
C. Toroid	On after	672	70.2	43.8[a]
	control	624	73.9	67.8
	On before	303	81.0	49.1[a]
	control	273	74.0	64.1
	All Tor	975	73.5	45.4[a]
	control	897	73.9	66.7

[a]Significant at 0.1%, [b]1.0% , and [c]5.0% level respectively, 2x2 chi-squared contingency test.

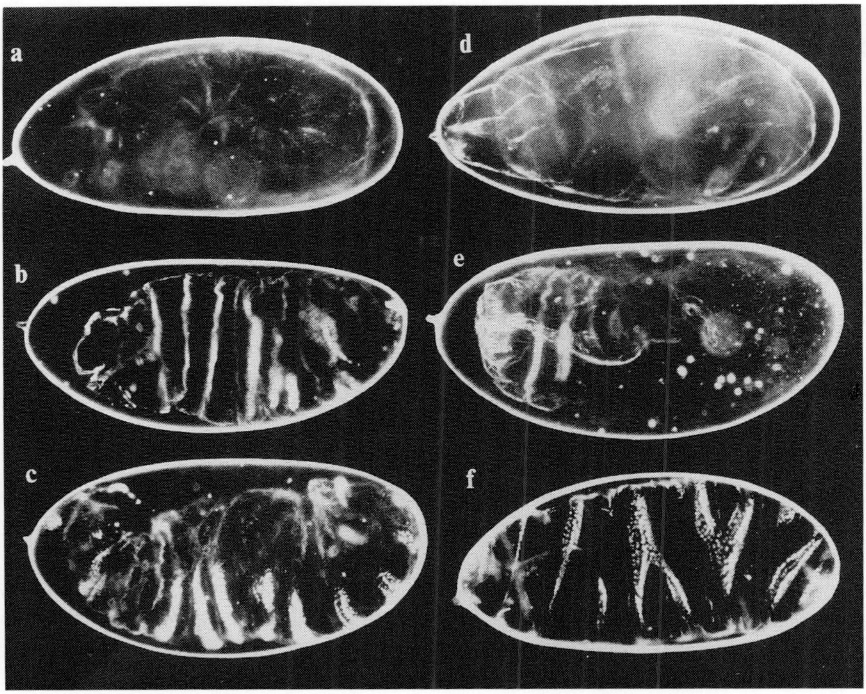

Figure 7.1: *Abnormalities among unhatched larvae exposed to static magnetic fields. a. 'Headless', subcategory 'featureless'; b, c, 'headless' with 'short segments' and reduced tail; d, tailless larva with head and very little trace of segments posterior to the head; e, tailless larva with reduced and abnormal head plus a couple of truncated or short segments; f, 'Twisted' larva with two superposed helices if opposite handedness.*

case, a complete conversion of the normal consecutive segmentation pattern into a helical configuration is observed (Fig. 7.2h). The unhatched embryo in Figure 7.1f shows two superposed helixes of opposite handedness. 'Strong dorsal denticles' (sdd) are larvae showing thicker than normal dorsal denticles in the thorax and anterior abdominal region (see Fig. 7.2c-h, especially 7.2e), as though they have taken on the characteristics of the ventral denticles. 'Tl reduced' (Tl red.) are the the least abnormal, and are those in which the first thoracic denticle band is reduced to a small ventral patch, or is absent altogether (this may also be seen in Fig. 7.2g,h).

The categories of abnormalities almost always overlap, as most embryos have multiple defects. We used two criteria to place embryos in a particular category: (i) according to the category of defect most specific to the magnetic field exposure; (ii) according to the most pronounced, or extensive abnormality. Thus, the order of precedence is 'twisted', 'head/tail reduced', 'crooked' or 'strong dorsal denticles', 'T1 reduced', 'headless' and 'short segments'. The same order applies in all three series.

7.3.3 Different Categories of Abnormalities in DC and AC Fields

Although significant increases in overall patterning abnormalities occur in all three series of experiments, the categories of abnormalities obtained are quite different. A comparison of the DC and AC-series shows that the latter contains far fewer categories of abnormalities. Thus, the categories, 'twisted' , 'strong dorsal denticles', 'T1 reduced', and to a large extent, 'head/tail reduced', turn up fairly often in the DC series, but seldom in the controls or in exposures to AC fields (see Table 7.2). These four categories together comprise an increase of 9.9 percent over the matched controls for the entire DC series, whereas the corresponding figure for the AC-series is 0.4 percent (see Tables 7.3).

In the DC exposures, all categories, except 'headless' and 'short segments', show significant increases over controls (Table 7.3). The zero-flux controls of the three series, obtained by placing the experimental embryos in their usual place within the coil without turning on the power supply, are given in Table 7.4. As mentioned above, the controls in the AC- (and Tor-) series also show a different pattern of abnormalities from the controls in the DC series, as they were reared under different conditions. There is a decrease in 'non-specific early arrests', as well as 'headless', but 'short segments' is substantially increased. Despite this, 'short segments' is significantly *further* increased as the result of exposure to AC fields, and this single category is perhaps the most consistent and characteristic effect of AC fields.

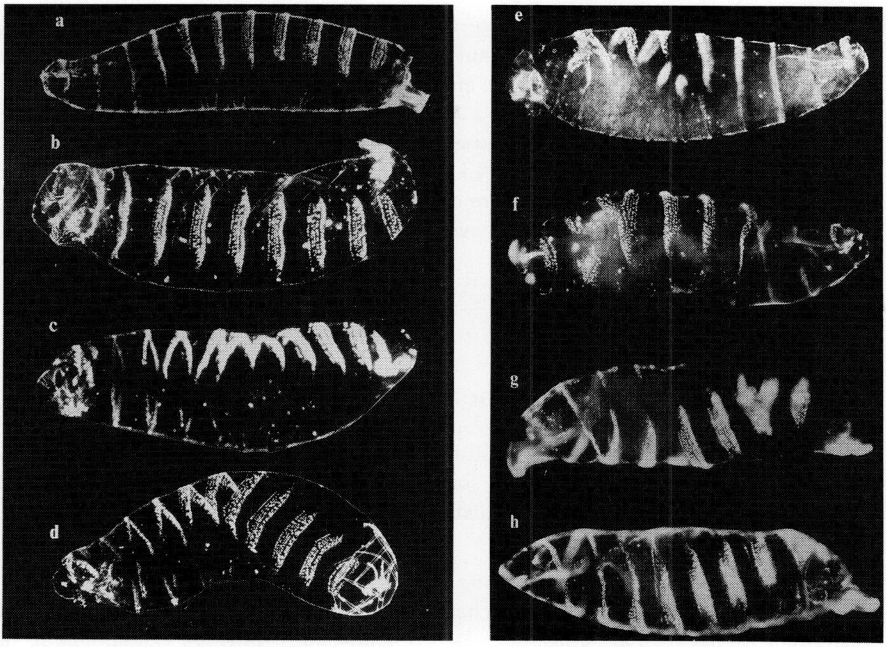

Figure 7.2: *Abnormalities among hatched larvae exposed to static magnetic fields.*

a. Normal first instar larva; b. 'twisted' larva with head and the first two thoracic segments all twisted up; c. 'head and tail reduced' larva also showing tendency of twists and crooked segments as well as strong dorsal denticles; d. 'twisted' larva with head, thorax and anterior abdominal segments twisted into a helical configuration, also strong dorsal denticles; e - h, various 'twisted' phenotypes: e, posterior abdominal segments and tail twisted and fused, and strong dorsal denticles; f. localized helical conliguration involving T2, T3 and Al, posterior abdomen also rotated, tail reduced; g, posterior abdomen rotated, tail reduced, head reduced, head and thorax twisted, helical configuration involves T2,T2 and Al, Tl is absent; h, a continuous helical configuration involving all segments, head and tail reduced, Tl absent.

7.3.4 Tor-series Abnormalities Resemble the DC-series

The results with the toroidal coil are quite tantalizing. Despite the fact that the magnetic field is negligible, significant increases in abnormalities are found over matched controls, and both when the embryos are in place before or after the power supply is switched on. Apart from a significant increase in 'short-segments', the pattern of abnormalities resemble those of the DC magnetic field exposures in that 'twisted', 'strong dorsal denticles', 'T1 reduced' and 'head/tail reduced' all occur whether the embryos are placed before or after the field is turned on. Together, they comprise an increase of 12.1 percent over the matched controls (see Table 7.3).

7.4 Discussion

7.4.1 The Effect of Magnetic Fields on Body Pattern

Our results confirm the findings of Ramirez *et al*[8], demonstrating in *Drosophila* a decrease in hatching rate (and hence viability) as the result of exposure to weak DC and AC magnetic fields. Our identification of specific morphological abnormalities, however, is new.

In general, the pattern defects resulting from both DC and AC magnetic field exposures are quite limited in kind, and characteristically different from those obtained with other agents such as exposure to heat or cold shocks, or ether vapour[6]. Gaps in segmentation, which occur in abundance and in diverse forms in the latter treatment, are almost entirely absent with magnetic field exposure, as are other fairly common defects, such as *homeotic* transformations which convert specific body structures to their homologues normally present elsewhere on the body (e.g., the conversion of the halteres into wings).

There is good evidence that successive bifurcation reactions are involved in segmentation which can be arrested by ether exposure at any time and anywhere along the body, giving rise to a wide variety of gaps in segmentation[9]. The effects of the magnetic fields, however, seem not to involve inhibition of the bifurcation reactions once they have been initiated, hence the conspicuous lack of gap phenotypes. For DC fields, the effects include the following. (a) Inhibition of the *initiation* of the bifurcation series, resulting in the absence of any patterning (represented by the 'featureless' defects, and probably, the excess in the 'non-specific early arrests' over controls). (b) Disturbance to either or both anterior and posterior extremities, resulting in 'headless', and 'head and/or tail reduced'. (c) Twisting and/or rotation of the segmental pattern, which in the extreme case, converts a stack of consecutive segments into a continuous helix. Most of the twists are confined to the extremities and are thus probably related to (b). All in all, the DC magnetic fields appear to cause specific global and local disturbances to the developmental field itself which

Table 7.2: Pattern abnormalities in exposure to different magnetic fields

	n	nea	hl	ss	cr	h/tred	tw	sdd	Tlred	other
DC 7.0 d/v	54.4	25.6	2.6	0.9	2.6	2.0	5.7	1.5	2.6	2.1
control	80.1	14.3	3.5	0.0	0.0	0.5	0.5	0.2	0.2	0.7
exp. - cont.	-25.7[a]	11.3[a]	-0.9	0.9	2.6[a]	1.5[c]	5.2[a]	1.3	2.4[b]	1.4
DC 3.0 d/v	56.1	26.2	3.7	3.9	0.9	2.3	2.8	1.1	2.3	0.7
control	67.9	22.2	4.0	1.3	0.0	0.0	0.0	0.0	0.0	4.6
exp. - cont.	-11.8[a]	4.0	-0.3	2.6[c]	0.9	2.3[b]	2.8[b]	1.1	2.3[b]	-3.9[a]
DC 2.0 d/v	59.7	27.2	3.2	0.4	1.2	2.8	1.6	0.4	0.4	3.1
control	69.3	25.5	2.8	1.0	0.0	0.3	0.5	0.0	0.3	0.3
exp. -cont.	-9.6[a]	1.7	0.4	-0.6	1.2	2.5[b]	0.9	0.4	0.1	2.8[b]
DC 0.5 d/v	59.3	31.1	2.4	0.5	0.8	3.1	1.8	0.0	1.3	0.0
control	69.2	26.0	2.4	0.0	0.0	1.1	0.0	0.3	0.5	0.5
exp. - cont.	-9.9[b]	5.1	0.0	0.5	0.8	2.0	1.8[b]	-0.3	0.8	-0.5
DC 7.0 a/p	44.3	33.0	2.1	0.7	3.9	4.3	5.3	0.7	3.9	1.8
control	66.4	28.1	3.4	0.4	0.4	0.9	0.4	0.0	0.0	0.0
exp. - cont.	-22.1[a]	4.9	-1.3	0.3	3.5[b]	3.4[c]	4.9[b]	0.7	3.9[b]	1.8
DC 3.5 a/p	51.7	23.3	1.7	2.8	1.9	1.2	5.0	2.1	9.7	0.6
control	70.1	24.7	1.9	2.2	0.4	0.0.	0.0	0.0	0.7	0.0
exp. - cont.	-18.4[a]	-1.4	-0.2	0.6	1.5	1.2	5.0[a]	2.1[c]	9.0[a]	0.6
DC 2.0 a/p	43.7	33.3	3.1	0.6	6.3	1.6	6.6	1.6	1.9	1.3
control	64.7	27.8	3.2	1.6	0.8	0.0	0.4	0.4	0.3	0.8
exp. - cont.	-21.0[a]	5.5	-0.1	-1.0	5.5[a]	1.6	5.8[a]	1.2	1.6	0.5
AC 2.36 d/v	47.3	18.6	2.3	25.9	2.3	0.5	0.0	0.0	0.0	3.2
control	72.7	14.6	0.4	11.6	0.0	0.0	0.0	0.0	0.0	0.8
exp. - cont.	-25.4[a]	4.0	1.9	14.3[a]	2.3	0.5	0.0	0.0	0.0	2.4
AC 17.6 d/v	49.0	13.7	0.3	34.5	1.7	0.0	0.0	0.0	0.0	0.8
control	60.3	15.7	0.0	23.0	1.0	0.0	0.0	0.0	0.0	0.0
exp. -cont.	-11.3[b]	-2.0	0.3	11.5[a]	0.7	0.0	0.0	0.0	0.0	0.8
AC 0.82 d/v	68.7	16.8	0.9	13.1	0.5	0.0	0.0	0.0	0.0	0.0
control	73.4	11.5	2.9	11.1	0.4	0.4	0.4	0.0	0.0	0.0
exp. - cont.	-5.7	5.3	-2.0	2.0	0.1	-0.4	-0.4	0.0	0.0	0.0
AC 2.36 a/p	27.9	25.4	2.9	40.0	0.7	0.0	0.0	0.0	0.0	0.7
control	61.3	11.9	1.2	22.9	0.8	0.0	0.0	0.0	0.0	2.0
exp. - cont.	-33.4[a]	13.5[a]	1.7	17.1[a]	-0.1	0.0	0.0	0.0	0.0	-1.3
AC 1.50 a/p	50.0	19.1	1.1	11.5	10.1	1.1	0.6	0.0	0.0	6.6
control	71.4	16.9	0.6	7.2	1.8	0.3	0.0	0.0	0.0	2.1
exp. - cont.	-21.4[a]	2.2[a]	0.5	4.3[a]	8.3[a]	0.8	0.6	0.0	0.0	4.5
Tor On after	43.8	25.7	0.3	11.0	4.2	0.9	3.6	8.9	0.6	1.0
control	67.8	22.4	1.1	7.2	0.2	0.2	0.2	0.2	0.0	0.8
exp. - cont.	-24.0[a]	3.5	-0.8	3.8[c]	4.0[a]	0.7	3.4[a]	8.7[a]	0.6	0.2
Tor On before	49.1	15.2	1.2	12.6	9.6	2.9	4.1	2.3	0.6	2.4
control	64.1	21.6	1.8	7.0	3.3	0.4	0.8	0.0	0.0	1.2
exp. - cont.	-15.0[a]	-6.4[c]	-0.6	5.6[c]	6.6[b]	2.5[c]	3.3[c]	2.3	0.6	3.1

[a]Significant at 0.1%, [b]1.0% , and [c]5.0% level respectively, 2x2 chi-squared contingency test.

Table 7.3: Comparison of pattern abnormalities resulting from the three series

	n	nea	hl	ss	cr	h/tred	tw	sdd	T1red	other
DC - d/v	57.4	27.4	2.9	1.3	1.4	2.5	3.4	0.8	1.6	1.6
control	72.1	21.7	3.1	0.7	0.0	0.5	0.3	0.1	0.3	1.3
exp. - cont.	-14.7[a]	5.7[a]	-0.2	0.6	1.4[a]	1.5[a]	3.1[a]	1.3[b]	2.4[a]	1.4
DC - a/p	47.2	29.1	2.3	1.6	3.8	2.2	5.6	1.6	3.7	1.2
control	66.4	28.1	3.4	0.4	0.4	0.9	0.4	0.0	0.0	0.0
exp. - cont.	-19.2[a]	1.0	-1.1	1.2	3.4[a]	1.3[a]	5.2[a]	1.6[b]	3.7[a]	1.2
DC - total	53.5	28.0	2.7	1.4	2.3	2.4	4.0	1.1	3.1	1.5
control	70.4	24.7	1.9	2.2	0.4	0.0	0.0	0.0	0.7	0.0
exp. - cont.	-16.9[a]	3.3[a]	0.8	-0.8	1.9[a]	2.4[a]	4.0[a]	1.1[a]	2.4[a]	0.5
AC - d/v	53.9	15.9	1.0	26.3	1.5	0.1	0.0	0.0	0.0	1.3
control	67.5	14.2	0.9	16.4	0.6	0.1	0.1	0.0	0.0	0.2
exp. -cont.	-13.6[a]	1.7	0.1	9.9[a]	0.9[c]	0.0	-0.1	0.0	0.0	0.9
AC - a/p	40.4	21.8	1.9	23.8	6.0	0.6	0.3	0.0	0.0	4.0
control	67.0	14.7	0.9	14.0	1.4	0.2	0.0	0.0	0.0	2.1
exp. - cont.	-33.4[a]	13.5[a]	1.7	9.8[b]	4.6[a]	0.4	0.0	0.0	0.0	1.9[c]
AC - total	47.8	18.6	1.4	25.2	3.6	0.4	0.1	0.0	0.0	2.5
control	67.3	14.4	0.9	6.9	0.9	0.1	0.0	0.0	0.0	0.9
exp. - cont.	-19.5	4.2	0.5	18.3[a]	2.7[a]	0.3	0.1	0.0	0.0	1.6[b]
Tor - On after	43.8	25.7	0.3	11.0	4.2	0.9	3.6	8.9	0.6	1.0
control	67.8	22.4	1.1	7.2	0.2	0.2	0.2	0.2	0.0	0.8
exp. - cont.	-24.0[a]	3.5	-0.8	3.8[c]	4.0[a]	0.7	3.4[a]	8.7[a]	0.6	0.2
Tor - On before	49.1	15.2	1.2	12.6	9.6	2.9	4.1	2.3	0.6	2.4
control	64.1	21.6	1.8	7.0	3.3	0.4	0.8	0.0	0.0	1.2
exp. - cont.	-15.0[a]	-6.4[c]	-0.6	5.6[c]	6.6[b]	2.5[c]	3.3[c]	2.3	0.6	3.1
Tor - total	45.4	22.5	0.6	11.5	5.9	1.5	3.7	6.9	0.6	1.4
control	66.7	22.2	1.3	7.1	1.1	0.2	0.3	0.1	0.0	0.9
exp. - cont.	-21.3[a]	0.3	0.7	4.6[b]	4.8[a]	1.3[b]	3.4[a]	6.8[a]	0.6	0.5

[a]Significant at 0.1%, [b]1.0% , and [c]5.0% level respectively, 2x2 chi-squared contingency test.

Table 7.4: Zero-field controls for the three series

		n	nea	hl	ss	cr	h/tred	tw	sdd	Tlred	other
DC (0)	(412)	73.5	21.8	1.0	1.5	0.5	0.5	0.0	0.0	0.0	1.2
control	(363)	75.0	20.4	1.7	2.5	0.0	0.0	0.0	0.0	0.3	0.0
AC (0)	(403)	73.2	16.4	1.2	8.9	0.2	0.0	0.0	0.0	0.0	0.0
control	(339)	73.7	15.0	0.8	8.6	0.3	0.6	0.3	0.0	0.0	0.6
Tor(0)	(323)	72.4	19.2	1.6	5.7	0.9	0.2	0.0	0.0	0.0	0.0
control	(420)	66.7	22.4	3.3	4.8	2.4	0.3	0.2	0.0	0.0	0.0

are *distinct from those resulting from other environmental agents.* What could be the mechanisms involved? There are two kinds of possibilities. Static magnetic fields can act via moving charges within the embryo, or by orientation effects on phase-ordered membrane lipids and proteins, similar to those occurring in liquid crystalline mesophases.

One candidate for moving charges are the transembryonic ionic currents found in *Drosophila*[10] and all other species examined[11]. As reported by Overall and Jaffe[10], the transembryonic ionic current at the preblastoderm stages in *Drosophila* typically flows into the anterior pole and out around the posterior pole, and is believed to be carried mainly by Na^+. The extraembryonic current is confined within the highly insulating vitelline membrane, and the average thickness of the pervitelline space is one micron The total current is estimated to be 1nA as determined with the *vibrating probe* technique, although this may be an underestimate due to the incomplete permeabilization of the vitelline membrane. Nevertheless, we could take a maximum current density of $1\mu Acm^{-2}$, or $10^{-2}Am^{-2}$ as a reasonable figure, as this appears to be typical of many other embryos[10,11]. To estimate the Hall effect field in the perivitelline space, we assume that the current is carried by a single species of univalent ion of number density $6\times10^{25}m^{-3}$ (corresponding to a 0.1molar solution). The drift velocity of the carriers is then of the order $10^{-9}ms^{-1}$, and for a perpendicular magnetic field of 10mT, the Hall effect field is, $10^{-11}Vm^{-1}$. This is miniscule compared to the electric field across the embryonic membrane, which is of the order of 10^7 Vm^{-l} . Another candidate for moving charges are electrons or proton currents flowing in or along the membrane. In order to affect patterning to the global extent exhibited in the abnormalities (see Figs. 7.1 and 7.2), such currents would have to be coherent over the whole embryonic field. For example, currents flowing in loops as shown in Figure 7.3, could easily be deflected by the external field into a helical configuration similar to that exhibited in the category of abnormality, 'twisted'. There is no evidence at present that such currents exist in the embryonic field, although they have been theoretically predicted and demonstrated in artificial systems[5].

Similarly, if one were to envisage a direct orientation influence of the magnetic field on individual magnetic dipoles in the cell membrane, the resulting magnetic energy

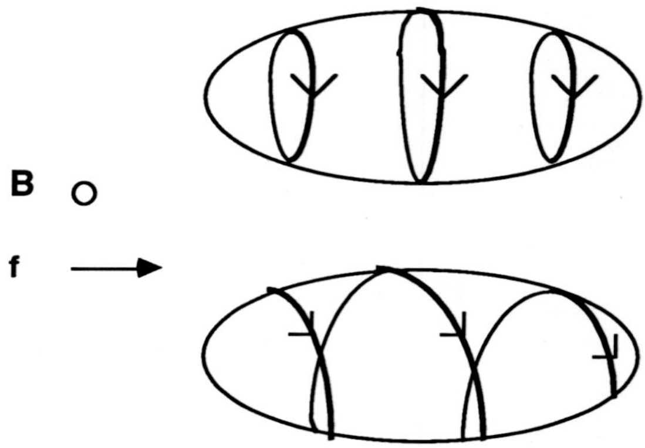

Figure 7.3: *Deflection of loop currents by external magnetic field can result in a helical pattern similar to that in the 'twisted' category of abnormalities (see text).*

would be of the order of μ_BB, where μ_B is the Bohr magneton. At 10mT, $\mu_B B \sim 10^{-25}$J, which is several orders of magnitude below the thermal background at room temperature, $kT \sim 4 \times 10^{-21}$J. This is fairly typical of weak electromagnetic field effects in biological systems reported so far (see Edmonds, and Liburdy, this volume). It seems therefore, that there can be no significant effect unless the magnetic field is acting via a high degree of cooperativity among the molecules involved in the processes of pattern determination. One form of cooperativity is Fröhlich's 'coherent excitation', involving a non-equilibrium phase transition to dynamic order under energy pumping from metabolism (see Wu, this volume).

The cell membrane and cortical layer of all embryos are intimately involved in determining body pattern, which may in turn depend on the phase ordering of liquid crystalline membrane lipids and proteins (see Ho and Saunders, this volume). Liquid crystals can be phase ordered by electric or magnetic fields. One way in which the external magnetic field could affect pattern is by interacting with an endogenous polarizing electric field. Membranes are extremely sensitive to external fields during phase-transition (see Liburdy, this volume). The sensitivity of the anterior and posterior extremities of the *Drosophila* embryo to perturbation by external magnetic fields is also consistent with this interpretation. Tenforde and Liburdy[12] have developed a model of magnetically induced membrane effects in which domains of correlated lipid molecules form in the bilayer membrane at prephase transition temperatures. These produce an instability in the elastic properties of the membrane when the local radius of curvature approaches the critical value R/6, where R is the radius of the spherical liposomes. The anterior and posterior extremities of the *Drosophila* embryo are domains of increased curvature and are hence expected to be particularly sensitive to perturbation by external magnetic fields.

The failure of 60hz AC fields to produce the same kinds of abnormalities as DC fields is consistent with the hypothesis that the DC fields act via changes in orientation of coherent, phase-ordered domains in the embryonic membrane. It would be of interest to investigate the effect of lower frequency fields as well as pulsed fields in order to estimate the relaxation time of the orientation processes involved.

One major caveat in all these hypothesis is that there is no *definitive* evidence for electrodynamical coherence in the embryo, despite several lines of indirect evidence which suggest that it does indeed exist[7,13,14]. The most definitive evidence would be if the *Aharonov-Bohm effect*[15] could be demonstrated in our embryos.

7.4.2 Embryos as Possible Detectors for the Aharonov-Bohm Effect

The Aharanov-Bohm effect describes a paradoxical situation in which electrons are influenced, in the form of a phase shift, by electromagnetic fields which they do not experience. This arises because of the electromagnetic potentials in quantum theory. In classical electrodynamics, potentials are merely a mathematical tool for calculating electromagnetic fields. The fundamental equations of motion can always be expressed in terms of fields alone. In quantum mechanics, however, potentials enter explicitly into Schrödinger's equation and cannot be eliminated. Nevertheless, these equations are all gauge (or scale) invariant; so that it may seem that even in quantum mechanics, the potentials do not have any physical significance. However, Aharonov and Bohm[15] conjectured that this may not be the case, and furthermore, proposed experiments aimed at clarifying how potentials would affect electrons passing through field-free regions.

They showed that it is possible to arrange for an electron to be confined in a Faraday cage, so that it experiences the potential but not the electromagnetic field. When the generator is turned on, the electron undergoes a phase shift, which cannot be detected under ordinary circumstances. These phase shifts could be detected, however, if a coherent electron beam were to be split into two, and each of which were to shift its phase differently, so that on recombining, an interference pattern would result. One way to achieve that is to pass the split coherent electron beams to either side of a long thin solenoid carrying a magnetic flux inside, but with the region outside the solenoid essentially free from magnetic fields lines. The vector potential, however, does not vanish, and differ on the two sides of the solenoid. When the beam is recombined, an interference pattern results. The scale of this effect depends on the characteristic wavelength of the electron, which is 0.3 A, and the coherence length that can be achieved experimentally, which is of the order of tens of microns. The experiment has since been done by a number of workers under increasingly stringent conditions, and the Aharonov-Bohm effect has been conclusively demonstrated[16].

Thus, if dynamically coherent electrons, protons, or oscillating charges are present in the embryonic field that somehow determine pattern, then pattern formation should be sensitive to the vector potential, as it would perturb the phase of the charged particles. (More specifically, it should be sensitive to $S = \int A.dl$, the integral of the vector potential, A, over the dimension of the embryo, l, in which pattern formation processes are occurring.) The embryos, in this case, serving as detectors for the Aharonov-Bohm effect. Our experimental set-up in the Tor-series is designed for this purpose. The magnetic flux is almost completely trapped within the torus, while the region outside has a nonvanishing vector potential that is dependent on the magnetic flux inside the torus. Our results suggest that indeed, the embryos may be sensitive to the vector potential in an essentially field-free region. The same kinds of abnormalities specific for exposure to static magnetic fields are obtained in the Tor-series. That the effect occurs whether the embryos are placed before or after the current is switched on further indicates that it is not primarily due to the transient electric field generated when the coil is switched on. We consider our results in the Tor-series extremely preliminary, although the rationale for attempting the experiment essentially sound. The Aharonov-Bohm effect is so fundamental that it would require a great deal more experimentation under the most stringent conditions to be able to establish it for biological systems. If it is eventually confirmed, it would be by far the most definitive proof that the embryonic field is a highly coherent macroscopic quantum domain. Only such a domain would be sensitive to the phase shifts induced by the vector potential.

Acknowledgment

We are grateful to our colleagues, Stephen Swithenby and Ray Macintosh, for advice, support, and stimulating discussions. Experiments on the Tor-series were possible thanks to Bartington U.K. who provided us, at considerable discount, the necessary magnetic field sensor MAG-03MC for measuring very weak magnetic fields. Frazer Robertson of the Electronics Workshop gave us invaluable help with the instrumentation. Part of this research is supported by a small grant from the Fetzer Foundation.

Bibliography

[1] Turing, A. M. (1952). The chemical basis of morphogenesis. *Phil. Trans. R. Soc. Lond. B.* **237**, 37-72.

[2] Oster G.F. and Odell, G.M. (1983). The mechanochemistry of cytogels. In *Ronts, Interfaces and Patterns* (A. Bishop, ed.), North Holland, Amsterdam.

[3] Goodwin, B.C. and Trainor, L.E.H. (1985). Tip and whorl morphogenesis in Acetabularia by calcium-regulated strain fields. *J. theor. Biol.* **117**, 79-106.

[4] Thompson, D'A. W. (1917). *On Growth and Form*, Cambridge University Press, Cambridge.

[5] Ho, M.W. (1993). *The Rainbow and the Worm, The Physics of Organisms*, World Scientific, Singapore.

[6] Ho, M.W., Matheson, A., Saunders, P.T., Goodwin, B.C., and Smallcomb, A. 1987 Ether-induced disturbances to segmentation in *Drosophila melanogaster. Roux Arch. Devl . Biol.* **196** 511-21

[7] Ho, M.W., Stone, T.A., Jerman, I., Bolton, J., Bolton, H., Goodwin, B.C., Saunders, P.T. and Robertson, F. (1992). Brief exposures to weak static magnetic field during early embryogenesis cause cuticular pattern abnormalities in *Drosophila* larvae. *Phys. Med. Biol.* **37**, 1171-1179.

[8] Ramirez, E., Monteagudo, J.L., Garcia-Gracia M., Delgado, J.M.R. (1983). Oviposition and development of Drosophila modified by Magnetic fields. *Bioelectromagnetics* **4**, 315-326.

[9] Ho, M.W. 1990 An exercise in rational taxonomy *J. Theor. Biol.* **147** 43-57

[10] Overall, R. and Jaffe, L.F. 1985 Patterns of ionic currents through Drosophila follicles and eggs *Developmental Biology* **108**, 102-19.

[11] Nuccitelli, R. (1988). Ionic currents in morphogenesis. *Experientia* **44**, 657-666.

[12] Tenforde, T.S. and Liburdy, R.P. (1988). Magnetic deformation of phospholipid bilayers: effects on liposome shape and solute permeability at prephase transition temperatures. *J. theor. Biol.* **133**, 385-396.

[13] Ho, M.W., Xu, X., Ross, S. and Saunders, P.T. (1992). Light emission and rescattering in synchronously developing populations of early Drosophila embryos: evidence for coherence of the embryonic field and long range cooperativity. In *Recent Advances in Biophoton Research and its Applications*, (F.A. Popp, K.H. Li and Q. Gu, eds.), pp. 287-306, World Scientific, Singapore.

[14] Ho, M.W., Ross, S., Bolton H., Popp, F.A. and Li, X.X. (1992). Electrodynamic activities and their role in the organization of body pattern. *J. Scient. Expl.* **6**, 59-77.

[15] Aharonov, Y. and Bohm, D. (1959). Significance of Electromagnetic Potentials in the Quantum Theory. *Physical Review* **115**, 485-491.

[16] Peshkin, M and Tonomura, A. (1989). *The Aharonov-Bohm Effect*, Lecture Notes in Physics, Springer-Verlag.

Chapter 8

Liquid Crystalline Mesophases in Living Organisms

Mae-Wan Ho and Peter T. Saunders

8.1 Liquid Crystals and Living Matter

It has been widely noted that biological molecules such as membrane lipids and proteins possess properties of liquid crystals at physiological temperatures[1,2]. Indeed, as George Gray[3] points out recently, liquid crystals are ideal for living matter as they combine order with mobility, giving rise to "tunable responsive, self-assembly systems". Living matter is maintained far from thermodynamic equilibrium by energy flow, and one of its most distinguishing characteristics is the rapid and appropriate way in which it can respond to contingencies[4]. That involves maximizing the *potential* degrees of freedom (keeping all possibilities open) so that the *actual* single degree of freedom required for coherent action can be instantly realized[5]. Studies of red cell membranes have shown that all the double-bonds in the long-chain fatty acids of membrane lipids are in the *cis*-configuration, i.e., not in their lowest energy state, and that the smectic A phase is adopted instead of the more stable smectic C phase. This has the effect of maximizing the potential degrees of freedom, which in turn allows for the expression of the highly correlated motions[6] that are required for their proper functioning. As John Lydon says, when referring to the special characteristics of liquid crystals which make them good at mixing: it works so long as the thermal motions of the different molecules match, it is "the way they dance together" that is crucial[6].

Despite these excellent insights into the properties of liquid crystals which make them ideally suited for living matter, there have been no serious investigations on liquid crystal mesophases in living organisms. Huxley and Taylor[7] initiated the study on changes in birefringence of isolated, contracting muscle fibres (see Irving *et al*[8] for recent developments), but no attempt has been made to observe *dynamically* phase ordered regimes *within* living organisms.

8.2 Coherence and Phase Ordering of Liquid Crystals

We were motivated to try to observe dynamically phase ordered regimes in living organisms by the large number of experimental observations - including those from our own laboratory[9–11] (see also Ho *et al*, this volume) - suggesting the existence of electrodynamical coherence at all levels within living systems. Forty to fifty years ago, a number of scientists, including Schrödinger[12] and Szent-Györgi[13] have already remarked how the 'anti-statistical' behaviour of organisms is reminiscent of the collective modes of activity, such as superconductivity and superfluidity, that aggregates of molecules can exhibit at low temperatures. The next important step was taken by Fröhlich[14,15], who presented a theoretical model showing how living systems made up predominantly of dielectric macromolecules packed rather densely together, will exhibit a non-equilibrium phase transition to dynamically coherent regimes under the most general conditions of energy pumping from metabolism. This has been confirmed by Wu (see this volume) and more recently by Duffield[16] who show that the 'Fröhlich state' is an asymptotically stable, global attractor.

Electrodynamical coherence, if it exists, ought to be reflected in the ordering of biological molecules as liquid crystalline mesophases. In the course of investigating this possibility, a novel technique was discovered which optimizes the detection of small birefringences in phase ordered biological molecules, enabling us to obtain for the first time, high contrast and high resolution colour images of entire, living, moving organisms[17]. Since then, we have been able to analyze the optics involved in colour generation, and to gain further insights into the biological implications of the technique for the study of coherence in living organisms.

8.2.1 The Technique

The technique, as reported in brief elsewhere[13], involves viewing the organisms between crossed polars with an asymmetrically placed compensatory wave-plate - a birefringent quartz crystal whose thickness is sufficient to cause a relative retardation of 560nm between the fast and slow components of the light ray, corresponding to a phase difference of 2π, or a full wavelength of yellow light. (Another innovation in our technique is that we can differentially illuminate various internal organs and tissues by adjusting the level of the condenser.) In conventional crystallographic work, the wave-plate is placed at 45° between the crossed polars to line up with the polarizing directions of the crystal under observation, so that the birefringence of the crystal can be estimated from the colour changes in the specimen by addition or subtraction from that due to the wave-plate. In our technique, the wave-plate is at a small angle from *either* of the crossed polars (between 2 - 15°, with the optimum about 4 - 7.5° depending on the species of organism). These include first instar larva of *Drosophila*, nauplius larva of the brine shrimp, hatchling of the zebra fish, larva of the crested newt, *Daphnia*, rotifers, protozoa, human buccal epithelial cells, as well as fresh frozen sections of pig skin, bovine cornea, sclera and lens, chick heart and brain. In the asymmetric placement of the wave-plate, full colours are generated in all *live* organisms and fresh sections of tissues examined thusfar. For all specimens, chromatic response under the conventional conditions of a symmetically placed wave-plate is either nonexistent or very poor. Conversely, for strongly birefringent materials such as rock and mineral crystals, our technique offers little or no improvement in either colour contrast or chromaticity.

Figure 8.1a gives the vertical arrangement of the essential parts of the polarizing microscope and the position of the living organism/tissue imaged. The polarizing directions and amplitudes of the plane polarized light emerging from the polarizing entities other than the organism are shown in Figure 8.1b. The detailed angle settings for the polarizer and analyzer are not important so long as they are crossed, ie, at 90°, and the wave-plate is at a small angle with respect to either of them. Furthermore, the wave-plate could be positioned above or *below* the organism. The reference axis of the organism is the anteroposterior, which is usually aligned E-W on the rotating stage. Again, this is arbitrary, as it is impossible to orientate very rapidly mobile organisms.

As can be seen from Figure 8.1b, the optical system consisting of the polarizers and the wave-plate is such that there is a highly unequal resolution of the plane polarized light POP' from the polarizer into the slow wave, SOS', and the fast wave, FOF', orthogonal to it. The slow component is retarded relative to the fast by 560nm on emergence from the wave-plate. This is equivalent to one wavelength of yellow light, but for the other constituents of white (halogen) light, the retardation will be more or less than one wavelength depending on whether the wavelength is shorter or longer than 560nm. These phase differences are added to or subtracted from the phase differences caused by birefringences in the organism, which are expected to be many and variously oriented with respect to the polarizers. The slow and fast components then recombine through the analyzer to give equal and opposite contributions, AOA', so that slow and fast wave components at each frequency can interfere with each other, either destructively or constructively according to the phase differences. White light is thereby resolved into light of many colours (ie, white light missing different spectral components and/or with certain components dominating) in accordance with the birefringent characteristics of the molecular constituents of the specific tissues. A typical image obtained - that of a first instar *Drosophila* larva - is shown in Figure 8.2.

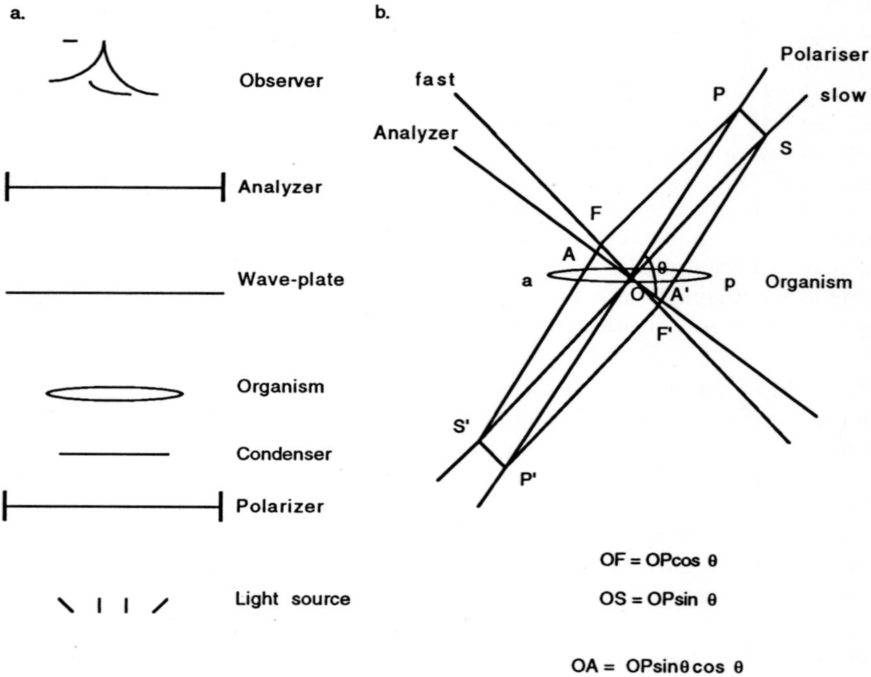

Figure 8.1: *Optical system for Interference Colour Vital Imaging. a. Schematic diagram of the essential parts of the Zeiss universal photomicroscope used in our investigations. b. Polarizing directions and amplitudes of the plane polarized light emerging from each polarizing entity other than the organism. The light is coming perpendicularly out of the page. a-p, the organism oriented with anteroposterior axis in a E-W direction; POP' polarizing direction and amplitude of the plane polarized light from the polarizer; SOS', polarizing direction and amplitude of the slow wave emerging from the wave-plate; FOF', polarizing direction and amplitude of the fast wave emerging from the wave-plate; AOA', polarizing direction and amplitude of the recombined waves emerging from the analyzer.*

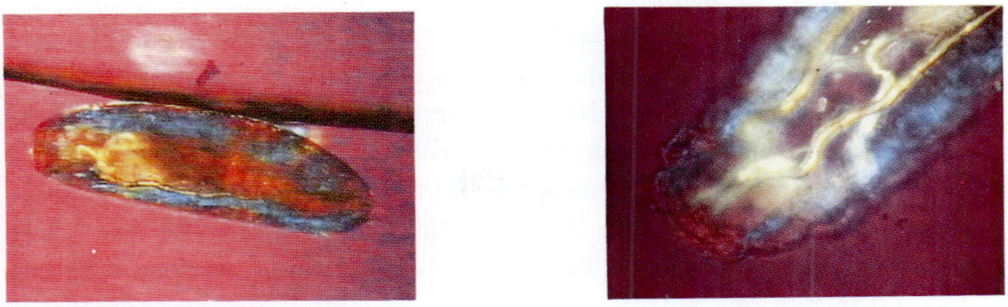

a. b.

Figure 8.2: *a. Live, first instar Drosophila larva shortly before hatching magnification 125x. b. Posterior end of first instar larva shortly after hatching, magnification 250x. Both with anteroposterior axis oriented E-W on the microscope. Recording was done on U-matic video with a Hitachi ccd KD7000 colour camera. The stills are obtained by video transfer and photosync from the video recording.*

8.3 The Nature and Biological Significance of the Colours

8.3.1 The Colours Give Anatomical Details Noninvasively

The colours are tissue-specific, following the anatomy quite precisely and independently of the geometry of the organism (which is fortunate, for organisms, especially mobile ones, are notoriously ungeometric). This is because to a first approximation, we can consider the organism to be non-refractive everywhere - being predominantly aqueous - except when the polarized light passes through layers of phase ordered birefringent molecules. Given that the ordered regimes are in membranes or tissue layers, it is perhaps not surprising that the colours should be independent of body geometry. Thus, we have a technique that give anatomical details noninvasively without chemical staining.

8.3.2 The Colours Are Informative of Molecular Structure and Dynamic Molecular Phase Order

The generation of colours works specifically for *living* biological tissues, although freshly fixed cryostat sections are as chromatic as the fresh tissue when the phase ordering has been preserved. Ordinary fixed histological sections are not chromatic, or only weakly so, presumably because the phase ordering of the molecules does not survive dehydration and embedding procedures. This suggests that the colour generated is dependent on the *degree* of phase ordering.

The colours also carry information on molecular structure (birefringence) as well as phase ordering, not just of the static kind, but of dynamic phase changes and changes in phase orientation accompanying activity. This is very important, as the frequency of visible light is about 10^{14}hz, coherent modes in the molecules of the tissues - as a rule much smaller that 10^{10}hz - will still appear highly ordered. Thus, contracted muscle bands in the *Drosophila* larva change from bright blue to red in certain orientations as the wave of contraction passes along the body (see Fig. 8.2a). That means our technique is capable of distinguishing between coherent and random molecular motions within a wide range of time scales as characteristic of living tissues.

8.3.3 The Colours Give Information Concerning Global Coherence of the Organism

It is of great significance that in *all* organisms examined, the anteroposterior axis is also the major polarizing axis for *all* the tissues within the body. In other words, like ordinary interference colours, the colours of all polarizing tissues within the organism change according to the orientation of its anteroposterior axis with respect to the crossed polars. DC fields have been identified in all organisms (including developing embryos of many species) examined by different workers using a variety of measurement techniques (see Chapter 1, this volume), and liquid crystals are, of course, subject to orientation by electric and magnetic fields. Our observation is the most suggestive evidence, by far, in support of the age-old hypothesis that the organizing embryonic field is global in character, right down to individual macromolecules, and that its major axis is electrodynamical in nature (see ref. 10). The mesophases revealed by our imaging technique in living organisms may indeed be ordered *dynamically* by endogenous global electrodynamical fields.

8.3.4 The Colours are Correlated with the Energetic and Physiological Status of the Organism

These endogenous fields may be further associated with particular energetic regimes. For example, chromaticity (the intensity of colours) waxes and wanes in the course of development. In *Drosophila* for example, there are chromatic stages early in development (between 2 and 4h), when from other studies, we know that pattern determining processes are active[14], but very little colour is subsequently observed until some hours before the first instar larva hatches (about 23-24h at 25°C). At around 17-18h of development, the colour of the segmental muscle bands changes rapidly within 10 mins. from a faint, dull blue to a bright blue *as the embryo starts to move.* Thus, two adjacent embryos may appear structurally indistinguishable, and as indicative of the same developmental stage, and yet only the one which has begun to move will have the intense colour. Thereafter, more and more colours develop in other tissues until at hatching, the full spectrum is present. After hatching, the colours continue to intensify as the larva becomes more active.

As consistent with the intensification of colours with activity, dead organisms can be distinguished by their lack-lustre colours, which may gradually fade away altogether. *Drosophila* larvae exposed to dehydration or to low temperatures also lose their colours concomittantly as they become immobile, but on being revived (by rehydration and by warming up respectively), regain their full colour and and mobility within 15 min.

All these observations suggest that the colour and colour changes carry important structural information concerning the degree of coherent dynamic order, as well as the degree of birefringence, which can change according to the energetic, developmental and physiological status of the organism. In order to be more precise, we need to examine the optical basis of the technique.

8.4 The Optical Basis of the Technique

8.4.1 The Technique is Optimized for Visualizing Biological Crystals

One major factor responsible for generating colours under our conditions is the maximization of *effective amplitude modulation* in the asymmetric placement of the wave-plate. This arises because the organism is a mixture of tissues each with its own polarizing axes. Within each tissue, the biological molecules typically have complicated shapes, and one would expect the presence of many different minor polarizing axes oriented differently from the major axes (all oriented antero-posteriorly) (see Fig. 8.3).

Ignoring dispersions, optical activities and other effects, two main mechanisms are involved in generating colour as far as the organism is concerned. First, the intrinsic birefringences of the biological molecules, which introduce phase differences that add to or subtract from that due to the wave-plate. Second, for the same range of birefringences, different polarizing directions in the molecules will make different vectorial contributions to the resultant composition of light passing through the wave-plate and then on to the analyzer. It is the second factor that is crucial in an organism - as opposed to a uniform mineral crystal with only one or at most a few polarizing axes - and which is affected by the asymmetric placement of the wave-plate. In contrast to the symmetrical placement of the wave-plate, the asymmetric placement maximizes subtractions and additions of phase differences over all orientations of polarizing directions in the molecular arrays of the organism giving *increased colour contrasts.* In addition, for a given intrinsic birefringence, it evens

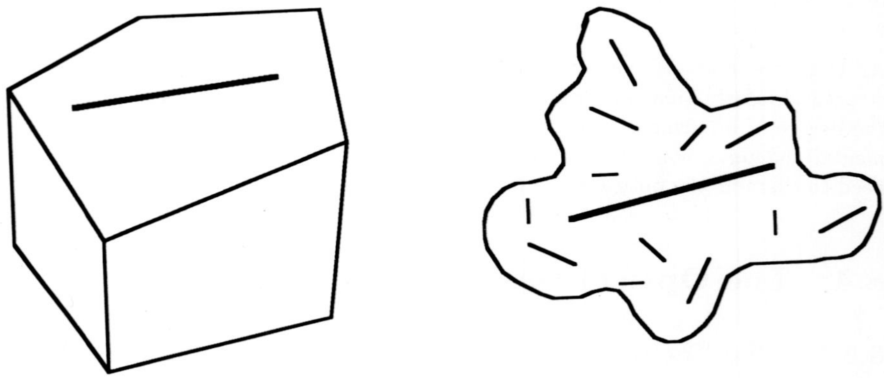

Figure 8.3: *Diagrammatic representation of a solid crystal (left) with a liquid crystal (right). The thick line represents the major polarizing axis of the crystal, the thin lines, the minor axes oriented differently from the major one.*

out the *absolute* intensities of light over all polarizing directions and across all phase differences introduced by the wave-plate within the visible spectrum, so that we effectively 'see' a wider range of birefringences simultaneously; in other words, our technique also *optimizes overall chromaticity* of the organism as a whole.

8.4.2 Mathematical Analysis in Terms of Two Superposed Crystal Plates

The effects are more precisely illustrated by considering the intensity of light emerging from two superposed crystal plates at different orientations between crossed polars[18],

$$
\begin{aligned}
I \;=\; & -\sin 2(\psi_2-\psi_1)\sin 2\psi_1 \cos 2\psi_2 \sin^2 \delta_1/2 \\
& +\sin 2(\psi_2-\psi_1)\cos 2\psi_1 \sin 2\psi_2 \sin^2 \delta_2/2 \\
& +cos^2(\psi_2-\psi_1)\sin 2\psi_1 \sin 2\psi_2 \sin^2(\delta_1+\delta_2)/2 \\
& -sin^2(\psi_2-\psi_1)\sin 2\psi_1 \sin 2\psi_2 \sin^2(\delta_1-\delta_2)/2
\end{aligned} \tag{8.1}
$$

where δ_1 and δ_2 are the phase differences respectively of the wave- plate and the biological 'crystal', and ψ_1, ψ_2, the angles that their slow (or fast) wave makes with the polarizer. Using these equations, we can work out the variation in intensity with orientation angle in the biological crystal, ψ_2, keeping δ_2 constant at representative angles, viz., 36°, 72° and 180°, for δ_1 from 250° to 500° (spanning the visible range of the optical spectrum) and ψ_1 at 45° and 7°. The graphs are shown in Figure 8.4.

A number of interesting features emerge. First, most of the full range of intensities (0 to 1) is available for the symmetrically placed waveplate at all values of δ_2, the intensity varying considerably across the visible spectrum, introducing systematic biases into the colours which differ for different values of ψ_2. Thus, for δ_2 at 36°, effects due to the green and yellow parts of the spectrum will be relatively diminished compared with the blue and red ends. As δ_2 increases, the biases shift across the spectrum without evening out. For the asymmetric case, on the other hand, the range of intensity is more level across the visible spectrum for any given ψ_2. As δ_2 increases, the absolute intensity also increases, but there is even less variation across the spectrum so that at 180°, the intensity is almost contant for any given ψ_2. This explains why overall chromaticity is optimized with our technique.

The optimization in chromaticity proves to be very sensitive to ψ_1; thus at angles $> 10°$, the intensity contours begin to resemble those at 45°. This agrees very well with our empirical observations, lending support to the explanation offered. The increase in intensity with δ_2 under our conditions means that in practice, *both the specific colour (spectral composition) and brightness (amplitudes) could be used to*

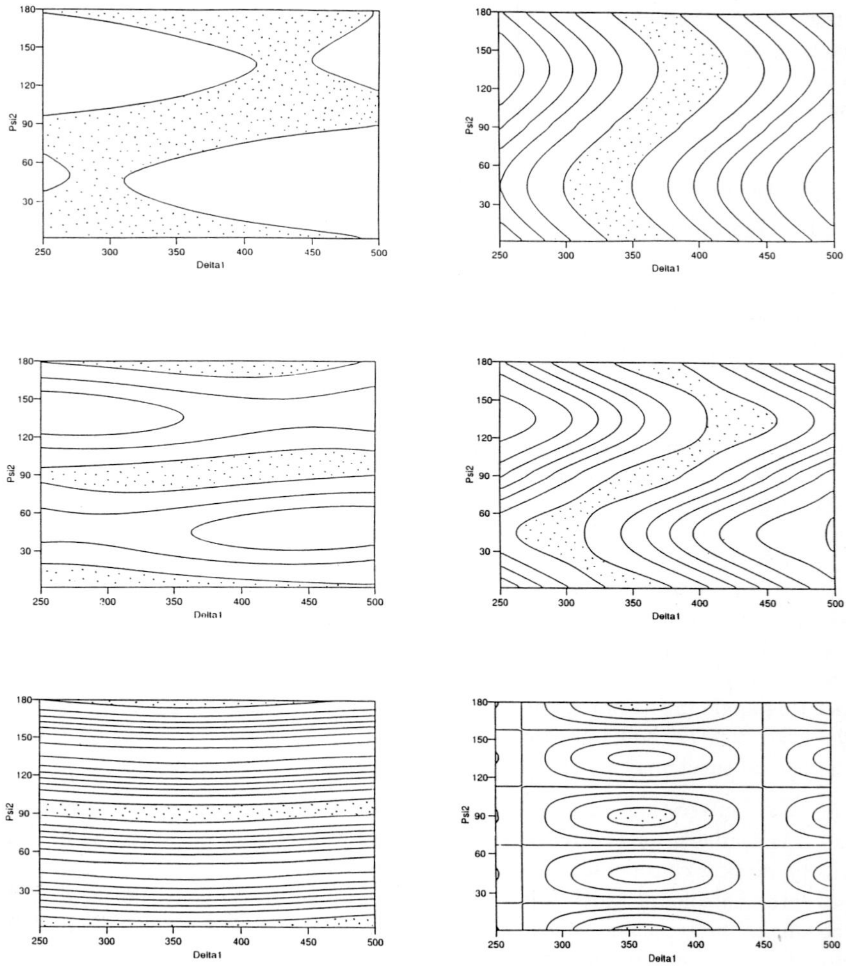

Figure 8.4: *Intensity contours on the* $\frac{\psi_2}{\delta_1}$ *plane for* δ_2 *at* 36° *(top diagrams)* 72° *(middle) and* 180° *(bottom) and* ψ_1 *at* 45° *(right) and* 7° *(left). Stippled areas are less than 5%, thereafter, the contours go up at intervals of 15%.*

estimate birefringence. Such changes in brightness are often observed in rapid muscular activities.

In contrast to the relative lack of variation in intensity across δ_2 for $\psi_1 = 7^o$ compared to $\psi_1 = 45^o$, the range of variation in intensity over ψ_2 is greater at the small angle. We can express this parameter as *the degree of modulation,*

$$M = \frac{I_{max} - I_{min}}{I_{max} + I_{min}}$$

The contours of M on the $\frac{\delta_1}{\delta_2}$ plane for the two angle settings are shown in Figure 8.5.

As can be seen, degrees of modulation above 0.95 occupy a narrow diagonally crossed ridge on the plane with ψ_1 at 45°. In other words, modulation is maximum when δ_1 and δ_2 are precisely matched. At 7° on the other hand, a broad plateau above 0.95 covers most of the plane. This is certainly consistent with the greatly increased overall colour contrast obtained at the small angle. The degree of modulation turns out not to be as sensitive as the intensity variation across the visible spectrum, and significant deterioration only occurs for $\psi_1 > 20^o$. Nevertheless, it offers yet another explanation which is in line with the empirical observation that the best results are obtained at small values of ψ_1.

8.5 Conclusions

The imaging technique described in this Chapter has many potential applications. It gives detailed anatomical information as in other forms of imaging, with the added advantages that it is relatively nonivasive, and the colours offer physical information concerning the shape and dynamic phase ordering of the molecules making up the tissues. As the colours are tissue specific, it enables us to distinguish one tissue from another in the absence of staining and other more invasive procedures. Organisms or cultures could be monitored continuously in testing the short and long term effects of drugs, for example. We are currently developing computer software to quantify colour and intensity changes of tissues in different developmental stages or physiological states for diagnostic purposes. With the aid of video-recording techniques, accurate analysis of muscular and other movements can be made in order to identify malfunction. As mentioned above, the most important aspect of the image is that the specific colours obtained for each tissue reflect the physical organization of the constituent macromolecules, and hence it may enable us to recognize coherent changes in protein conformations accompanying functional activities. This technique - suitably adapted for use with fibre optics - will have important applications in clinical diagnosis of organ/tissue malfunction.

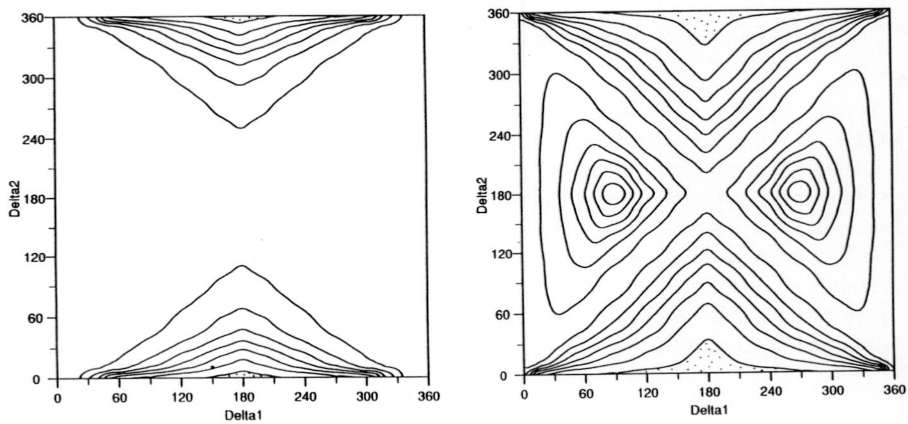

Figure 8.5: *Contours of the degree of modulation, M, on the* $\frac{\delta_1}{\delta_2}$ *plane for the two angle settings of* ψ_1, *45° (right), and 7° (left). Stippled areas are less than 5%, thereafter, the contours go up at intervals of 15%.*

Acknowledgment

We are grateful to Steve Swithenby for very helpful comments on an earlier draft, and to to E.R. Pike, Alan Durrant, J.G. Walker, Tom Smith, Ray Mackintosh and Graham Dunn for stimulating discussions. This work was carried out in the laboratory of Andy Tindle in Earth Sciences Department who kindly allowed us access to the polarizing microscopes, and gave us advice on using them. Michael Lawrence, co-discoverer of the technique with M.W.Ho, gave invaluable help and advice in microscopy and video-recording. Excellent technical assistance was provided by Katerina Zdrahalova, Julian Haffegee and Adrian French in preparing the embryos, by Brian Sloam, Trevor White, Brian Rowley, David Wilson and John Petriwlo (BBC.OUPC) for video transfer and photosync, by Denise Taylor for video tape editing and by Jackie Brown for cryosectioning. Rachel Bourne and Vicky Stirling provided us fixed and fresh-frozen sectioned materials of the chick. Keith Meek and YiFei Huang brought us bovine cornea, sclera and lens. Verina Waights gave us the newt larva and Brian Gasking and Nigel Holder, the zebra fish fry.

Bibliography

[1] Special issue on liquid crystals, *Physics Today,* May, 1982.

[2] Hartshorne, N.H. and Stuart, A. (1970). *Crystals and the Polarizing Microscope,* Edward Arnold, London, especially chapters 14 and 15.

[3] Gray, G. (1993). Liquid crystals -molecular self-assembly. *British Association for the Advancement of Science, Chemistry Session: Molecular Self-Assembry in Science and Life, Presidential Address,* Sept. 1, Keele.

[4] Ho, M.W. (1993). *The Rainbow and the Worm: The Physics of Organisms,* World Scientific, Singapore.

[5] Ho, M.W. (1993). What is negentropy? . (submitted for publication).

[6] Lydon, J. (1993). The dynamic nature/dimension of the cell membrane. *British Association for the Advancement of Science, Chemistry Session: Molecular Self-Assembry in Science and Life,* Sept. 1, Keele.

[7] Huxley, A.F. and Taylor, R.E. (1955). Activation of a single sarcomere. *J. Physiol* **130**, 49-50.

[8] Irving, M., Maylie, J., Sizto, N.L. and Chandler, W.K. (1987). Intrinsic optical and passive electrical properties of cut frog twitch fibers. *J. Gen. Physiol* **89**, 1-40.

[9] Ho, M.W. and Popp, F.A. (1992). Biological organization, coherence and light emission from living organisms. In *Thinking About Biology* (W.D. Stein and F. Varela, eds.), Addison Wesley, New York.

[10] Ho, M.W., Xu, X. Ross, S. and Saunders, P.T. (1992a). Light emission and rescattering in synchronously developing populations of early Drosophila embryos. In *Recent Advances in Biophoton Research and its Applications* (F.A. Popp, K.H. Li and Q. Gu, eds.), World Scientific, Singapore.

[11] Ho, M.W., Stone, T.,A., Jerman, I., Bolton, J., Bolton, H., Goodwin, B.C., Saunders, P.T. and Robertson, F. (1992b). Brief exposure to weak static magnetic fields during early embryogenesis cause cuticular pattern abnormalities in *Drosophila* larvae. *Physics in Medicine and Biology* 37, 1171-1179.

[12] Schrödinger, E. (1944). *What is Life?* Cambridge University Press, Cambridge.

[13] Szent-Györgi, A. (1960). *Introduction to a Submolecular Biology,* Academic Press, New York.

[14] Fröhlich, H. (1968). Long range coherence and energy storage in biological systems. *Int. J. Quantum Chem.* **2**, 641-649.

[15] Fröhlich, H. (1980). The biological effects of microwaves and related questions. *Adv. Electronic and Electron. Phys.* **53**, 85-152.

[16] Duffield, N.G. (1988). Global stability of condensation in the continuum limit for Fröhlich's pumped phonon system. *J. Phys. A: Math. Gen.* **21**, 625-41. I thank Geoffrey Sewall for drawing my attention to this work.

[17] U.K. patent, Nov. 1992; see also Ho, M.W. and Lawrence, M. (1993). Interference colour vital imaging: A novel noninvasive microscopic technique. *Microscopy and Analysis* September, 26. 18. From Hartshorne and Stuart (1970) p. 302.

Chapter 9

Dielectric and AC Electrodynamic Properties of Cells

Ronald Pethig

9.1 Introduction

The uses of direct-current (DC) electrical fields, such as those employed in the electrophoretic characterisation of cells, in the Coulter method for cell counting and sizing, and in the permeabilisation and electrofusion of cells, are well appreciated by biologists. Not so well known, perhaps, are the effects arising from the interactions of cells with alternating-current (AC) fields. Current studies of such phenomena indicate that they may find application in new assays and techniques for the characterisation, selective manipulation and separation of cells.

If a non-uniform AC field is applied to a cell, it experiences a force which can cause it to move. This effect is known as dielectrophoresis and its application to the study of cells was largely pioneered by Pohl[1]. As shown by Masuda *et al*[2], electrical traveling-waves can also be used to impart linear motion on cells. In certain circumstances, non-uniform fields can induce a torque on a cell, causing it to spin. An early report of such an effect was given by Teixeira-Pinto *et al*[3] and later described in more detail by Pohl[1], but the underlying influences were not fully understood. The controlled way to induce cellular spin is to subject the cell to a rotating electrical field, and the first reports of this were given by Arnold and Zimmerman[4] and Mischel *et al*[5].

In this chapter an outline is presented of the unifying theory that has been developed to link together the electrokinetic phenomena of dielectrophoresis, electrorotation and traveling wave effects as applied to cells. Examples will also be given of practical results of relevance to applications in the biotechnological and biomedical sciences.

9.2 Theoretical Background

9.2.1 Dielectric Properties of a Cell:

When an A.C. field is applied to any particle, including a cell, electrical energy is both consumed and stored by the particle to an extent determined by its dielectric properties. These properties can be characterised by measurement of the cell's effective electrical conductance G(units Ohm^{-1} or Siemens) and electrical capacitance C (units Farads). The conductance is a measure of the ease with which free electrical charges can move within the cell's structure under the influence of an applied electric field, whilst its capacitance is a measure of the amount of electrical charge that must be given to it to raise its electrical potential by one unit (a capacitance of 1 Farad requires 1 Coulomb of charge to raise its potential by 1 Volt). The capacitance of a cell is primarily associated with the accumulation of induced electrical charges at membrane surfaces.

If measurements are made on the same cell using the same experimental arrangement, then its conductance and capacitance can be defined by two equations:

$$G = k\sigma, \text{ and } C = k\epsilon_o\epsilon \tag{9.1}$$

where k is a geometry factor (units meter) and the conductivity σ(units Siemens/meter) is the proportionality factor between the induced electric current density and the applied electric field. For cells, the conductivity is mostly associated with mobile hydrated ions and protons, and with displacement currents arising from the induced motion of molecular dipoles and electrical double layers occuring at membrane surfaces. The factor ϵ_o is the dielectric permittivity of free space (of value 8.854×10^{-12} Farad/meter) while ϵ is the permittivity of the cell relative to that of free space. This relative permittivity is a measure of the extent to which molecular dipoles and localized charge distributions within and at the surface of the cell can be polarized by the applied field.

A cell, to an electrical engineer at least, is a very complicated particle. In order to begin to appreciate how it responds to an imposed electric field we shall first have to strip away these complications. In fact, we will first consider a cell to have the form of an uncharged homogeneous sphere! This approach allows us to make use of a well known problem set to students of electrostatics, namely that of deducing the effect of applying an electric field to a system comprising a spherical particle of radius **a** and relative permittivity $\in_p$ suspended in a medium of relative permittivity $\in_s$. The result, as described in many textbooks on electrostatics, is that electrical charges are induced to appear on the surface of the particle so that is acts like an electric dipole (see figures 9.1a and 9.1b). The value of the associated induced dipole moment **m** is given by

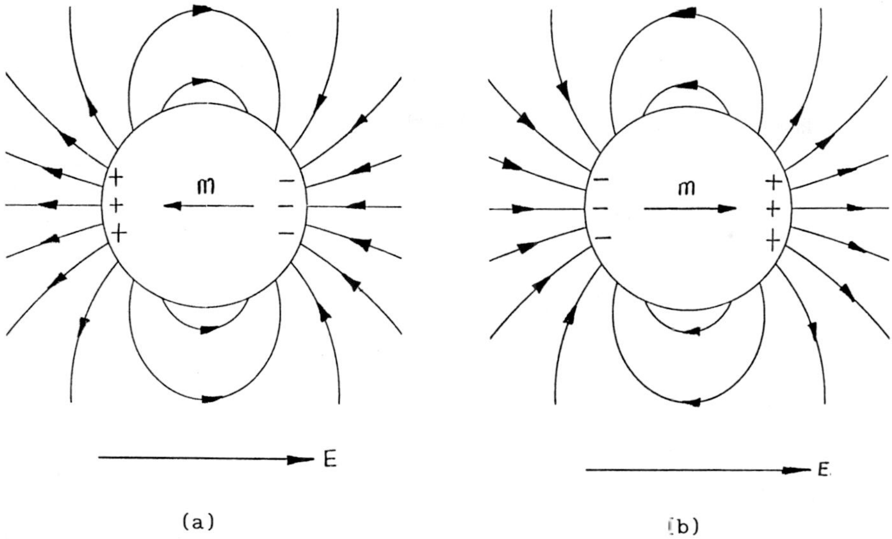

Figure 9.1: *The induced electric charge distribution, resulting dipole moment and associated electric field for a spherical particle whose polarisability is either (a) less than, or (b) greater than that of the surrounding medium. E is the externally applied electric field.*

$$\mathbf{m} = 4\pi\epsilon_o\epsilon_s\left(\frac{\epsilon_p^* - \epsilon_s^*}{\epsilon_p^* + 2\epsilon_s^*}\right)a^3\mathbf{E} \tag{9.2}$$

where **E** is the applied field.

In Eq. (9.2), the usual situation considered in textbooks has been modified to include the general case where the field varies sinusoidally in time and where both the particle and suspending medium are not perfect loss-free dielectrics (defined by a permittivity value only) but also exhibit energy losses associated with a finite conductivity (σ_p and σ_s, respectively). To allow for this, the permittivity has to be expressed mathematically as a complex quantity (i.e. having real and imaginary components). For the particle, we then have

$$\epsilon_p^* = \epsilon_p - j\sigma_p/\omega$$

where ω is the angular frequency of the applied electric field ($\omega = 2\pi f, f$ being the frequency). The symbol $j(=\sqrt{-1})$ identifies the imaginary component (σ_p/ω) of ϵ_p^*

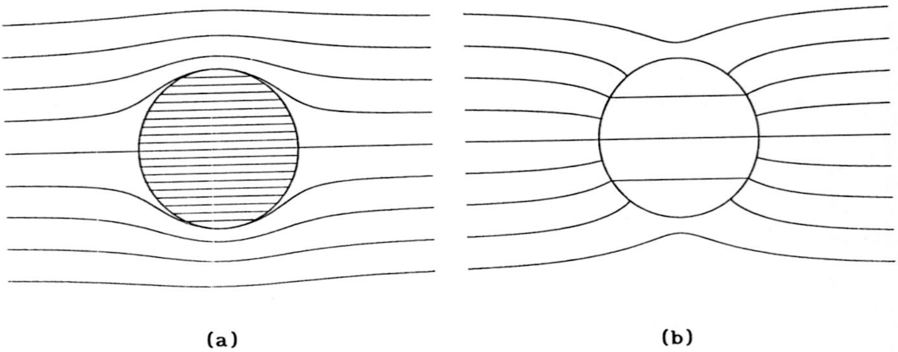

Figure 9.2: *Resultant field patterns for the particles of Figure 9.1*

and also signifies that the current associated with the conductivity σ_p lags by a phase angle of 90° the (displacement) current associated with the (real) component of the permittivity ϵ_p. Evaluation of Eq. (9.2) reveals that the magnitude and polarity of the induced dipole moment depends in quite a complicated manner on the frequency of the applied electric field and on the relative magnitudes of $\epsilon_p$and ϵ_s, σ_p and σ_s^6. For example, if the magnitude of ϵ_p^* is less than that of ϵ_s^* the effective polarisability of the particle is less than that of its suspending solution and the induced charges on the particle are as shown in Figure 9.1a, with the induced moment **m** directed against the applied field. Conversely, as shown in Figure 9.1b, if the polarisability of the particle exceeds that of the suspending medium (i.e. $| \epsilon_p^* | > | \epsilon_s^* |$) then the arrangement of the induced charges produces a dipole moment directed along the same direction as the applied field. The resultant electric field patterns within and around a particle for the cases where its effective polarisability is either less or greater than the suspending medium, are shown in Figures 9.2a and b.

Another important fact to note is that the surface charge that appears at the interface between the particle and the surrounding medium is not created immediately. The charges, and hence the induced dipole moment, appear at a rate which is exponential with time after application of the field. The characteristic time constant τ describing this is given by Equation (9.3):

$$\tau = \left(\frac{\epsilon_p + 2\epsilon_s}{\sigma_p + 2\sigma_s} \right) \tag{9.3}$$

Equations (9.2) and (9.3), which taken together describe the magnitude, polarity and time response of the dipole moment induced in a particle subjected to an external AC electric field, represent the key to understanding the phenomena of dielectrophoresis, electrorotation and traveling wave motion exhibited by particles. To understand these effects for cells, we need to consider a more realistic model for them.

9.3 Cells in Electric Fields

Cells are not homogeneous in their electrical and physical properties, but it turns out that the basic concepts we have just described for homogeneous particles in electric fields are of relevance.

A simple cell to model is the red blood cell (erythrocyte), which in most species (including Man but not in chickens, for example) does not contain a nucleus. As first shown in 1913 by the dielectric measurements of Höber[7], an erythrocyte appears electrically as a conducting entity (the cytoplasm) surrounded by a poorly conducting membrane ("eine dielektrische Hülle", in the words of Höber). If, for the present, we neglect the negatively charged neuraminic acid residues projecting from their membranes, then erythrocytes in their spherical form can be modelled as a conducting sphere surrounded by a thin and poorly conducting shell. Such a single-shell model is shown in Figure 9.3a, which also depicts how the field-induced interfacial charges are distributed across the resistive membrane of the cell.

Maxwell[8] demonstrated that such a concentric system can be replaced by a homogeneous sphere of the same outer radius having an effective specific resistance r_p given by:

$$r_p = r_2 \frac{(2r_1 + r_2)a^3 + (r_1 - r_2)a_1^3}{(2r_1 + r_2)a^3 - 2(r_1 - r_2)a_1^3}. \tag{9.4}$$

where, as shown in Figure 9.3a, r_1 and r_2 are the specific resistances of the inner sphere (radius a_1) and outer shell (radius a), respectively. When placed in an electrical field the "smeared-out" uniform sphere of Figure 9.3b of equivalent resistance r_p can be substituted for the heterogeneous sphere of Figure 9.3a, without altering the form of the resulting external electric field pattern.

The membrane charging effect shown in Figure 9.3a manifests itself as a dielectric dispersion, whereby the capacitance of a suspension of cells falls as the frequency of measurement is increased. For suspensions of erythrocytes this fall in capacitance takes place within the frequency range of 50 kHz to 50 MHz, and is termed the β-dispersion[9,10]. The theories of relevance to dielectric measurements on cells and other biological materials were developed principally by Fricke[11], Cole[12] and Dänzer[13], and led to Fricke's[14] major achievement in deducing the ultrathin nature of erythrocyte membranes.

The process of calculating the effective complex permittivity of two concentric spheres in terms of a single homogeneous sphere can be repeated endlessly. So, by placing such a smeared-out sphere inside another sphere we can, for example, model

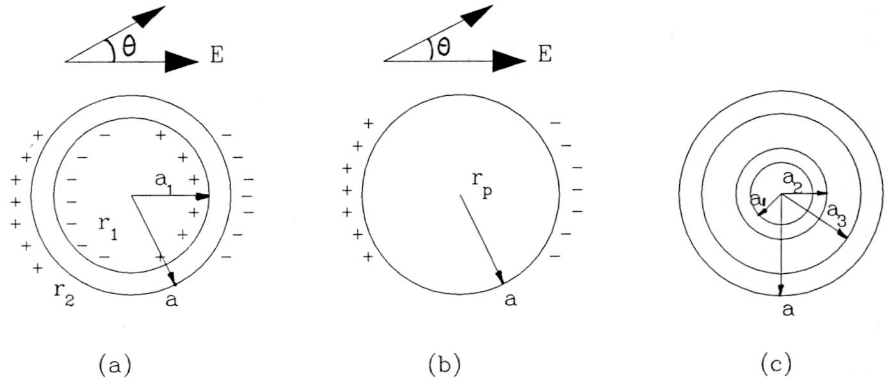

Figure 9.3: *(a) Single-shell model of a cell and the distribution of induced charges at its interfaces. (b) The homogeneous sphere equivalent model. (c) The double-shell model of a cell having either a cell wall or nucleus.*

a cell with a nucleus (as in Fig. 9.3c) or model a bacteria having a cell wall and a cytoplasmic membrane. Such a procedure has been extensively developed by Irimajiri et al[15,16], Asami and Irimajiri[17], and Hanai[18] in particular contributed greatly to extending the theory of heterogeneous dielectrics beyond that originally developed by Maxwell for inanimate particles.

For a simple cell structure consisting of a thin, resistive, membrane surrounding a cytoplasmic electrolyte (e.g. Fig. 9.3a) subjected to an electric field $\mathbf{E}$, the radial distribution of the fully established field acting across the membranes thickness d is given to a good approximation by the formula

$$\mathbf{E}_m(\theta) = 1.5(a/d)\mathbf{E}\cos\theta \tag{9.5}$$

This equation shows that the greatest field stress is created across the membrane region that lies in a radial direction parallel with the imposed field. There is also an "amplification" of the field by a factor of the order 10^3 related to the ratio (a/d) of the cell radius to the membrane thickness. The full transmembrane stress will not occur instantantly, but will follow a time course given by

$$\mathbf{E}_m(\theta) = 1.5(a/d)\mathbf{E}\cos\theta[1 - \exp(-t/\tau)] \tag{9.6}$$

where t is the elapsed time after application of the field $\mathbf{E}$ and τ is the characteristic time constant of Eq. (9.3). For an erythrocyte suspended in an aqueous medium, τ is of the order 10^{-7} seconds. At low frequencies (i.e. $f << (2\pi\tau)^{-1}Hz$) the full transmembrane stress is developed during each half-cycle of the applied field $\mathbf{E}$ and this voltage stress is depicted in Figure 9.4a for a simple cell of the form of Figure 9.3a. However, as the frequency is increased above the value $f = (2\pi\tau)^{-1}$ the total electrical charging of the membrane is reduced. This is why the capacitance of a suspension of erythrocytes (and for many other cells) steadily decreases as the frequency is increased above 1 MHz. Another result of this effect is that the cell structure of Figure 9.3a appears as a particle of low conductivity at low frequencies and the field pattern around it takes the form of Figure 9.2a. With increasing frequency the membrane does not develop a voltage stress across itself and the external field is drawn inside the cell as depicted in Figure 9.2b. If the cell contains a nucleus then at high frequencies, depending on the physico-chemical of its membrane, the greatest voltage stress may occur across the nuclear membrane as depicted in Figure 9.4b.

So far we have neglected the fact that cells and micro-organisms usually carry a net negative surface charge of around $1\mu C/cm^2$. As a result of such surface charge, counter-ions will be attracted to the cell so as to electrically neutralise it. This creates an electrical double-layer around the cell, which can be polarized by an external field. Field-induced relaxations of electrical double-layers around charged particles, together with ion diffusion at the particle surface and within the double-layer, produce low-frequency dielectric dispersions which have been designated as α-dispersions[9,10]. We can model the α-dispersion in terms of an effective permittivity and conductivity, and introduce this as an extra outer shell (or shells if we wish to model the finer details of the α-dispersion)[6]. It should also be noted that the magnitude of the permanent membrane charge is around one-thousand times larger than that of the field-induced charges depicted in Figures 9.1 and 9.2. It is because they are asymmetrically distributed around the cell (and thus form an electric dipole) that the effect of the induced charges is not masked by the (symmetrically distributed) natural charge.

To summarise, the basic dielectric and field-induced phenomena that are understood for homogeneous spherical and ellipsoidal particles can be extended to describe the dielectric properties of biological particles. In principle "all" that is required is to calculate the effective value for ϵ_p^* in Eq. (9.2), for example. As already described, the complex relative permittivity of the cell can be written in the form:

$$\epsilon_p^* = \epsilon_p - j\sigma_p/\omega$$

and this can be generalised as

$$\epsilon_p^* = \epsilon_\infty + \sum_{k=1}^{n+1} \frac{\Delta\epsilon_k}{1 + j\omega\tau_k} - j\sigma_o/\omega \qquad (9.7)$$

where ϵ_∞ is the limiting high-frequency permittivity, σ_o is the limiting low-frequency conductivity and $\Delta\epsilon_k$ is the dielectric dispersion associated with its own characteristic relaxation time τ_k. Thus, a suspension of "multi-shelled" particles in general

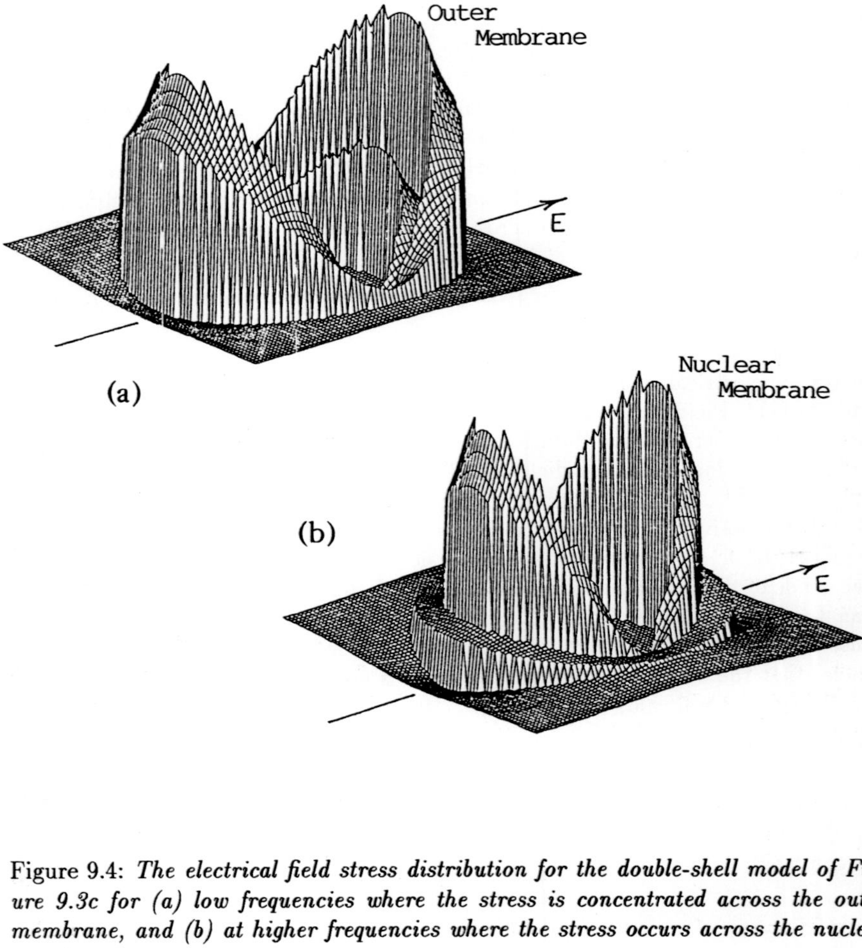

Figure 9.4: *The electrical field stress distribution for the double-shell model of Figure 9.3c for (a) low frequencies where the stress is concentrated across the outer membrane, and (b) at higher frequencies where the stress occurs across the nuclear membrane.*

gives rise to a composite dielectric spectrum, with the maximum number of individual dispersions $\Delta\epsilon_k$ corresponding to the number of interfaces, each of which demarcates the dielectrically distinguishable substructures of the particles. At low frequencies (typically up to a few kHz) ionic conduction and electrical double-layer effects at the outer membrane determine the nature of the dielectric dispersions. As the frequency is increased, and the field penetrates further into the cell structure, each newly "explored" membrane interface contributes to a new dispersion $\Delta\epsilon_k$. Since each of these dispersions is controlled by the physico-chemical and geometric properties of each substructure, we can expect that different cell types, or similar cells of different viability or stages of development, will exhibit different dielectric spectra. This in turn leads to characteristic dielectrophoretic, electrorotational and traveling-field responses for different cells.

9.4 Dielectrophoresis and Electrorotation

As described in the introduction, dielectrophoresis is the motion imparted on electrically polarized particles subjected to non-uniform electric fields and electrorotation occurs as a result of rotational torque exerted on polarized particles subjected to rotating electric fields. These two effects are closely related through the field-induced dipole moment $\mathbf{m}$ of the particle, which from Eq. (9.2) can be seen to be a complex quantity composed of a real (*Re*) and an imaginary (*Im*) component. This in turn reflects the fact that the dipole moment is not created immediately on application of the applied electric field, and thus can be resolved into a component (*Re*) in-phase, and another (*Im*) 90° out-of-phase, with the applied field. As discussed more fully elsewhere[19], the time-averaged dielectrophoretic force $F(\omega)$ is related to the in-phase component of $\mathbf{m}$, whilst the rotational torque $\Gamma(\omega)$ is related to the out-of-phase component, through the following equations:

$$F(\omega) = Re\{\mathbf{m}(\omega)\}\nabla\mathbf{E}^2/2\mathbf{E} \tag{9.8}$$

$$\Gamma(\omega) = -Im\{\mathbf{m}(\omega)\}\mathbf{E} \tag{9.9}$$

where ∇ is the del vector operator and $\nabla\mathbf{E}$ is a measure of the field non-uniformity.

Equations (9.8) and (9.9) indicate that the polarity of the dielectrophoretic force and the sense of electrorotation depend on the polarity of the induced dipole moment. If the polarisability of the cell exceeds that of the surrounding medium (i.e. $\mathbf{m}(\omega)$ is $+ve$) the dielectrophoretic force $F(\omega)$ is positive and thus directed towards regions of high field strength, whilst the torque exerted will be negative and cause the cell to rotate in a sense that opposes that of the rotating field. If the polarisability of the cell is less than that of the surrounding medium, the cell will be directed to low field regions and spin in the same sense as the rotating field.

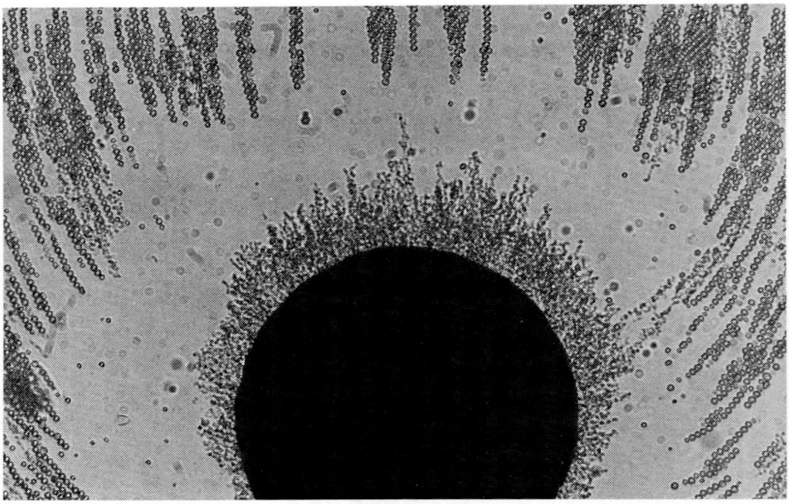

Figure 9.5: *This photograph shows the result of applying a 10 kHz voltage to a pin electrode that is surrounded by a mixture of sheep erythrocytes and bacteria (Micrococcus luteus) suspended in an aqueous electrolyte of conductivity* $10mS/m$. *Within a few seconds the bacteria have collected at the electrode under the action of a positive dielectrophoretic force, whilst the blood cells are pushed away under the influence of negative dielectrophoresis.*

The two types of dielectrophoretic behaviour are demonstrated in Figure 9.5 for a mixture of sheep erythrocytes and bacteria. The different behaviour exhibited by the cells in Figure 9.5 arises from the fact that the erythrocytes are surrounded by a resistive lipid membrane, whereas the cell walls of the bacteria contain polysaccharides and are relatively conducting. At a frequency of 10 kHz the erythrocytes are less, and the bacteria more, polarisable than the surrounding medium of conductivity $10mS/$m.

Non-uniform fields can be generated using a variety of electrode geometries apart from the pin-type shown in Figure 9.5. One example, of the form of interdigitated, castellated microelectrodes constructed using conventional photolithography[20,21] is shown in Figures 9.6 and 9.7. By appropriate scaling of the characteristic dimensions of such electrodes, cells can be collected under the influence of either positive or negative dielectrophoretic forces[22]. Cells collected into "pearl chains" under the influence of positive dielectrophoresis (see Figure 9.6) are more firmly immobilised to the electrode system than those collected under negative dielectrophoresis into the triangular and diamond-shaped aggregations shown in Figure 9.7. Thus, with an ap-

Figure 9.6: *The interdigitated, castellated, design of microelectrodes. Yeast cells are shown collecting in "pearl chain" formations at the electrode edges under the influence of positive dielectrohoresis.*

propriate suspending medium conductivity and applied voltage frequency, different cell types can first be separated locally into pearl chain and triangular aggregations and then by using either fluid flow or gravitational forces (with the AC signal still applied) they can be completely isolated as separate suspensions[23,24].

The cells forming the triangular-shaped patterns in Figure 9.7 have in fact been directed into regions of isolated electric field minima. Another design of electrode, of the so-called polynomial geometry[25], can also be used to collect cells away from electrode edges into energy wells associated with field minima, and an example of this is shown in Figure 9.8.

The electrodes shown in Figure 9.8 have been energised such that adjacent ones are of opposite polarity[25]. If the electrical connections are altered such that opposite pairs of electrodes have opposing polarity (with four electrodes this is equivalent to their being phased 90° apart) a rotating electric field is generated. The polynomial electrodes enable near simultaneous measurements to be made of both dielectrophoresis and electrorotation effects and data obtained for yeast cells in this way are shown in Figure 9.9.

The anti-field rotation peak centred around 50 kHz (for viable yeast cells suspended in $3mS/m$ medium) is a characteristic feature exhibited by many viable cells, and

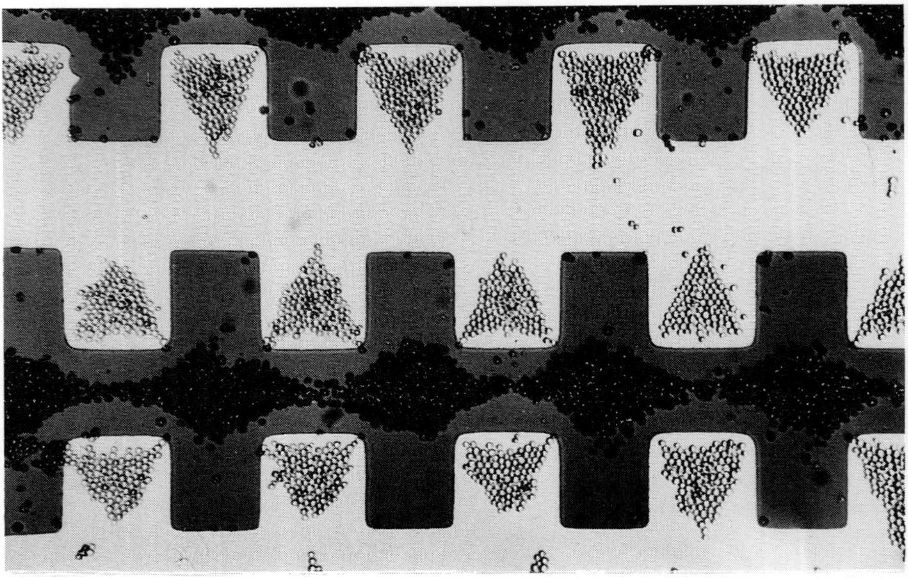

Figure 9.7: *Yeast cells collecting under the influence of negative dielectrophoresis into triangular and diamond-shaped aggregations using interdigitated microelectrodes. See reference*[22] *for details.*

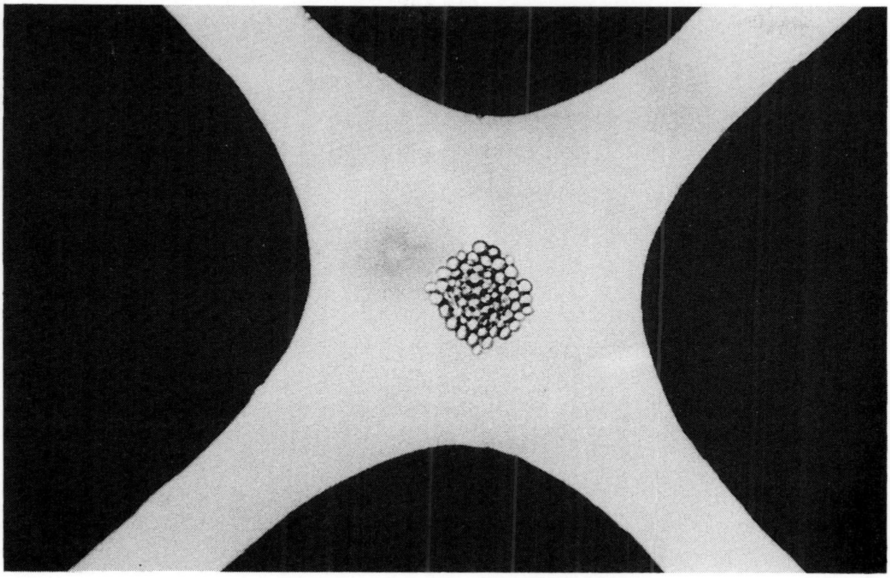

Figure 9.8: *Yeast cells directed by negative dielectrophoresis into an energy well associated with an electric field minimum*[23,25]. *By changing the electrical connections this "polynomial" electrode geometry can also be used to electrorotate cells.*

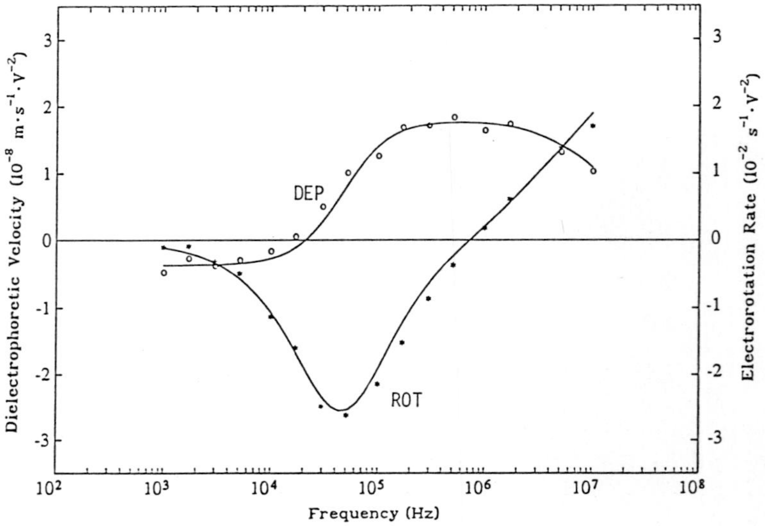

Figure 9.9: *Dielectrophoretic (DEP) and electrorotation (ROT) spectra for viable yeast cells (Based on reference[19]).*

also occurs where the corresponding dielectrophoretic behaviour changes polarity from negative to positive as the frequency is increased[6,26]. When the yeast cells are heated, so as to make them non-viable (and degrade the resistive properties of the plasma membrane) this anti-field rotation peak disappears[19]. Indeed, a large anti-field rotation peak can be taken as a reliable indicator of the degree of viability for many cell types, and appears to be related to the integrity of the plasma membrane. The fact that a particle should rotate in the opposite sense to that of the imposed rotating field seems counter-intuitive, but a simple explanation is given in Figure 9.10 and based on the fact that an intact plasma membrane lends to the cell the properties of a poorly conducting particle at low frequencies.

As demonstrated by the pioneering work of Arnold and Zimmermann[4,28,29] electrorotation measurements provide a very sensitive method for monitoring the physiological state of cells and their sensitivity to exposure to chemicals and other influences. If instead of cells, small beads are used that are coated with active chemical agents such as antibodies or nucleic acids, the electrorotation technique can also be used for such purposes as a sensitive assay of the viability and concentration of toxic organisms in drinking water, or for DNA sequencing, for example[30,31].

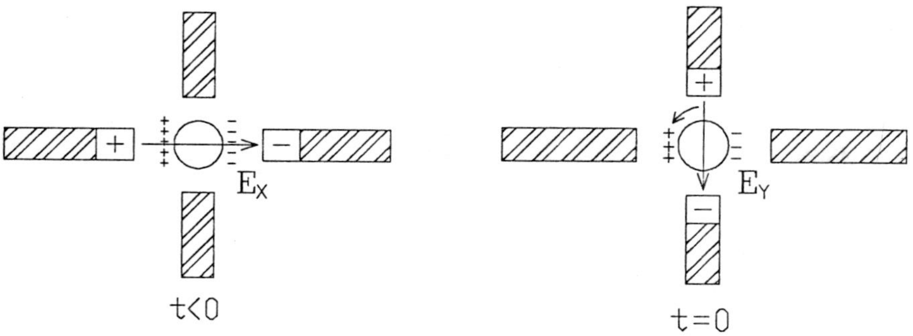

Figure 9.10: *Schematic representation of the anti-field electrorotation effect. (a) In the instant* ($t < 0$) *before voltage polarity switching occurs the induced dipole moment in the cell opposes the applied field. (b) At the instant* ($t = 0$) *of clock-wise polarity switching the resulting interaction between the new field and the remanent dipole moment induces anti-clockwise rotation (from reference[27]).*

9.5 Travelling-Wave Effects

The first reports on the manipulation of biological particles using travelling electric fields appear to be those of Masuda *et al*[2,32] who found that three-phase voltages of frequency between $0.1Hz$ and $100Hz$ could be used to induce controlled translational motion of erythrocytes and lycopodium particles. At these low frequencies electrophoretic forces were largely responsible for the effects observed, and so it was proposed that different particles could be separated according to their size or net electrical charge. In more recent work, Fuhr *et al*[33,34] have demonstrated that travelling fields of frequency between 10 kHz and 30 MHz can be used to induce linear motion of pollen and cellulose particles. At these frequencies, dielectrophoretic forces, rather than electrophoretic ones, are dominant and Fuhr *et al* derived a theoretical model that took into account interactions between the multipoles induced in the particles and the travelling field to explain the observed "travelling-wave dielectrophoresis" effects.

Travelling electric fields can be generated using a comb-like electrode arrangement such as that shown in Figure 9.11, in which the electrodes are sequentially addressed with sinusoidal voltages of phase separation 90° apart, keeping directly opposing electrodes on either side of the channel phase-shifted from each other by 180°.

For the electrode arrangement shown in Figure 9.11, it has been shown by Huang *et al*[27] that the time-averaged force $F_{TF}(\omega)$ acting on a particle in the centre of the channel is given by:

$$F_{TW}(\omega) = -\frac{\pi}{\lambda} Im\{\mathbf{m}(\omega)\}\mathbf{E} \tag{9.10}$$

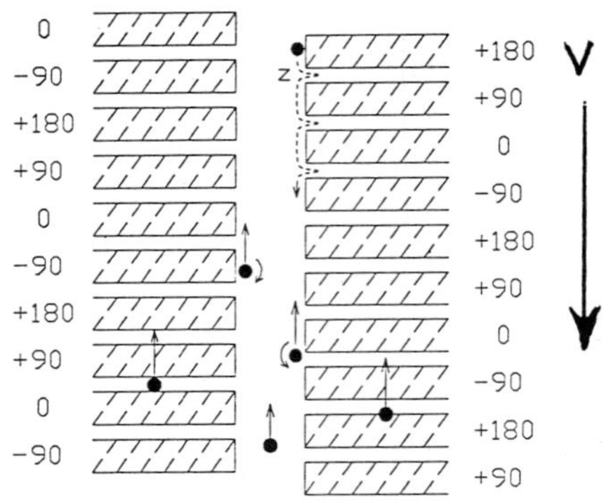

Figure 9.11: *Travelling electric fields, of propagation direction indicated by the arrow V, are created by addressing the electrodes with the voltage phase sequences (in degrees) shown. Cells move over the electrodes and along the channel in the direction opposing that of the travelling field, and when moving near the sides of the channel they also spin. In a fundamentally unstable (FUN) mode of behaviour, the cells can also travel in a zig-zag manner down the edge of the channel in the same direction as the travelling field. (From reference[27])*

where **E** is the field strength across the channel and λ is the wavelength of the travelling field of value equal to the repetitive distance between electrodes of the same phase. Thus, although a translational motion is imparted on the particle and is termed "travelling-wave dielectrophoresis" by Fuhr *et al*[33,34], unlike conventional dielectrophoresis the travelling field effect is related to the *imaginary*, rather than to the *real*, component of the induced dipole moment $\mathbf{m}(\omega)$. We can understand this from the fact that the travelling field effect and electrorotation share the common experimental feature of the applied field undergoing translational and angular motion, respectively, so that at any instant the interaction of importance is that between the field and the *remanent* dipole moment.

The minus sign in Eq. (9.10) indicates that under the conditions when $Im\{\mathbf{m}(\omega)\}$ is positive the particle will move in a direction opposing that of the travelling field. As shown in Figure 9.12 this interesting effect can be understood in rather the same way as the anti-field electrotation effect described in Figure 9.10.

Referring to the dielectrophoresis and electrorotation results for viable yeast cells shown in Figure 9.9, we can also understand some other important aspects of the travelling field effect. For the particles to be capable of translational motion over the electrodes or along the channel, they must not be attracted to the electrodes by a positive dielectrophoretic force. For the conditions applicable to the results shown in Figure 9.9 this means that the travelling field effect will not occur above a frequency of around 20 kHz. Further more, below about 1 kHz the magnitude

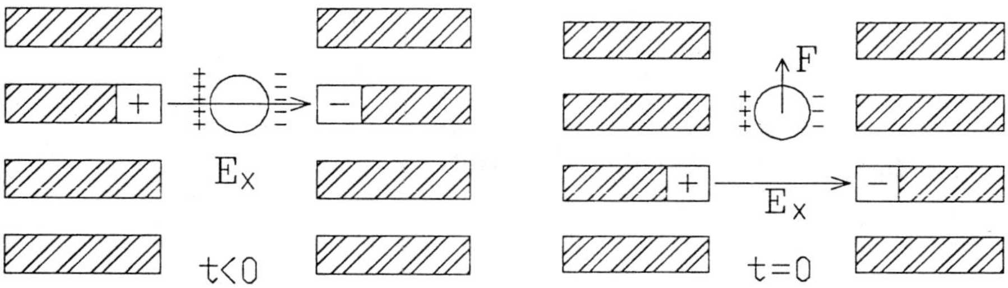

Figure 9.12: *Schematic representation of the travelling field effect for a poorly conducting particle in a conducting solution. In the instant* ($t < 0$) *before electrode switching occurs the induced charge distribution around the particle is as shown. On switching the electrode voltages at time* $t = 0$, *the resulting interaction between the field across the channel and the remanent induced charges pushes the particle in the opposite direction to that of the travelling field (from reference[27]).*

of $Im\{\mathbf{m}(\omega)\}$(as mirrored in the magnitude of the electrotation rate) reduces to a very low value and so according to Equation (9.10) the travelling field effect will cease below this point. We can also deduce from Figure 9.9 that in the frequency range between 1 kHz and 20 kHz the parameter $Im\{\mathbf{m}(\omega)\}$ is positive (this follows from Equation 9.9 and the fact that the rotation is anti-field), which in turn means (from Equation 9.10) that the particle will move in a direction opposing that of the travelling field, as depicted in Figures 9.11 and 9.12. The fundamentally unstable (FUN) regime of behaviour depicted in Figure 9.11 occurs in the narrow frequency range where the dielectrophoretic behaviour changes polarity from negative to positive and $Im\{\mathbf{m}(\omega)\}$ continues to increase in magnitude with increasing frequency. In the FUN regime a particle experiences two counter-acting forces, a positive dielectrophoretic force directing it towards high field regions at the electrode tips and a travelling-wave force attempting to push it along the channel.

Because the frequency dependencies of $Re\{\mathbf{m}(\omega)\}$ *and* $Im\{\mathbf{m}(\omega)\}$ will differ markedly for different cell types, we can expect to be able to use the travelling-wave effect to good advantage in the separation of cells, and this has been demonstrated by Huang *et al*[27] for the case of mixtures of bacteria and yeast. From Equation (9.10) it is possible to deduce how changes in electrode dimensions and spacing can lead to enhanced selectivity of particle manipulation and separation.

9.6 Concluding Remarks

There is now a good understanding of the basic factors that influence the A.C. electrokinetic behaviour of cells and bioparticles. From Equations (9.8), (9.9) and

(9.10) we can see that the common link between the dielectrophoretic, electrorotation and travelling-wave effects is the magnitude and time response of the induced dipole moment. This dipole moment represents the net effect of a variety of ionic, dipolar and interfacial polarisations and is therefore a very sensitive function of the frequency of the applied electric field. It is also a sensitive function of the dielectric properties of the suspending medium. These factors, taken together with the range of non-uniform, rotating and travelling-wave electric fields that can be applied to particles in suspension, provide us with a wide range of exploitable AC electro-kinetic effects.

The principal characteristics of cell electrorotation, namely the sense and maximum rate of rotation, are expressed over a relatively narrow frequency range of the applied electric field and are very sensitive to changes in the physico-chemical properties of the cell and its environment. This makes the technique particularly exploitable for the real-time detection and monitoring of pharmacological agents and toxins, or as an assay when using beads coated with a specific binding agent such as an antibody or antigen.

The differences in dielectrophoretic or travelling-wave behaviour of different cell types, or of chemically treated beads, can be enhanced through judicious choice of such parameters as the conductivity and/or permittivity of the suspending medium, the frequencies of applied fields (more than one signal can be applied at the same time) and the geometric design of the electrodes. By controlling these various parameters a variety of effects can be achieved, such as the selective manipulation and separation of bioparticles, the controlled interaction between chemically active particles, and the monitoring of the heterogeneity or viability of cell cultures, for example.

Acknowledgments

I wish to acknowledge the valuable collaborations and work of my colleagues, without which this article could not have been written. These include Julian Burt, Ka Lok Chan, Ralph Hölzel, Ying Huang, Mike Hughes, Gerard Markx, Mark Talary, John Tame, Xiao-Bo Wang and Xiao-Feng Zhou at Bangor, who like me enjoy a valuable collaborative programme of work with Frederick Becker and Peter Gascoyne at the University of Texas M. D. Anderson Cancer Center, Houston. We also enjoy stimulating interactions with Mike Arnold in Würzburg, Ted Grant in London, Karan Kaler in Calgary, Thomas Jones in Rochester and Masao Washizu in Tokyo. The work at Bangor has been supported by the National Foundation for Cancer Research, the AFRC and SERC Research Councils, British Technology Group Ltd, ICI Ltd, Leverhulme Trust, National Grid Co., Scientific Generics Ltd. and by the award of British Council, CVCP, EC and SERC scholarships.

Bibliography

[1] H. A. Pohl, *Dielectrophoresis*, Cambridge University Press, Cambridge, 1978.

[2] S. Masuda, M. Washizu and I. Kawabata, Movement of blood cells in liquid by nonuniform traveling field, *IEEE Trans. Ind. Appl.* **24** (1988) 217 - 222.

[3] A. A. Teixeira-Pinto, L. L. Nejelski, J. L. Cutler and J. H. Heller, The behavior of unicellular organisms in an electromagnetic field, *Exp. Cell Res.* **20** (1960) 548 - 564.

[4] W. M. Arnold and U. Zimmermann, Rotating-field-induced rotation and measurement of the membrane capacitance of single mesophyll cells of *Avena sativa*, *Z. Naturforsch*, **37c**, (1982) 908 - 915.

[5] M. Mischel, A. Voss and H. A. Pohl, Cellular spin resonance in rotating electric fields, *J. Biol. Phys.* **10** (1982) 223 - 226.

[6] X-B. Wang, Y. Huang, R. Hölzel, J. P. H. Burt and R. Pethig, Theoretical and experimental investigations of the interdependence of the dielectric, dielectrophoretic and electrorotational behaviour of colloidal particles, *J. Phys. D: Appl. Phys.* **26** (1993) 312 - 322.

[7] R. Höber, Messungen der inneren Leitfähigkeit von Zellen, *Pfl. Physiol. Archiv. Mensch. Tiere,* **150** (1913) 15 - 45.

[8] J. C. Maxwell, *A Treatise on Electricity and Magnetism*, 3*rd* ed., Vol.1, Ch.ix, Clarendon Press, Oxford, 1891.

[9] H. P. Schwan, Electrical properties of tissue and cell suspensions, *Adv. Biol. Med. Phys.* **5** (1957) 147 - 209.

[10] R. Pethig, *Dielectric and Electronic Properties of Biological Materials*, J. Wiley, Chichester, 1979.

[11] H. Fricke, A mathematical treatment of the electric conductivity and capacity of disperse systems, *Phys. Rev.* **24** (1924) 575 - 587.

[12] K. S. Cole, Electric impedance of suspensions of spheres, *J. Gen Physiol.* **12** (1928) 29 - 36.

[13] H. Dänzer, Über das Verhalten biologischer Körper im Hochfrequenzfeld, *Ann. Physik* **20** (1934) 463 - 480.

[14] H. Fricke, The electric capacity of suspensions of red corpuscles of a dog, *Phys. Rev.* **26** (1925) 682 - 687.

[15] A. Irimajiri, T. Hanai and A. Inouye, Evaluation of a conducto- metric method to determine the volume fraction of the suspensions of biomembrane-bounded particles, *Experientia* **31** (1975) 1373 - 1375.

[16] A. Irimajiri, T. Hanai and A. Inouye, A dielectric theory of "multi- stratified shell" model with its application to a lymphoma cell, *J. Theor. Biol.* **78** (1979) 251 - 269.

[17] K. Asami and A. Irimajiri, Dielectric analysis of mitochondria isolated from rat liver II. Intact mitochondria as simulated by a double-shell model, *Biochim. Biophys. Acta* **778** (1984) 570 - 578.

[18] T. Hanai, Theory of the dielectric dispersion due to the interfacial polarization and its application to emulsion, *Kolloid Z.* **171** (1960) 23 - 31.

[19] Y. Huang, R. Hölzel, R. Pethig and *X*-B. Wang, Differences in the A.C. electrodynamics of viable and non-viable yeast cells determined through combined dieletrophoresis and electrorotation studies, *Phys. Med. Biol.* **37** (1992) 1499 - 1517.

[20] J. A. R. Price, J. P. H. Burt and R. Pethig, Applications of a new optical technique for measuring the dielectrophoretc behaviour of micro-organisms, *Biochim. Biophys. Acta* **964** (1988) 221 - 230

[21] J. P. H. Burt, T. A. K. Al-Ameen and R. Pethig, An optical dielectrophoresis spectrometer for low-frequency measurements on colloidal suspensions, *J. Phys. E: Sci. Instrum.* **22** (1989) 952 - 957.

[22] R. Pethig, Y. Huang, *X*-B. Wang and J. P. H. Burt, Positive and negative dielectrophoretic collection of colloidal particles using interdigitated castellated electrodes, *J. Phys. D: Appl. Phys.* **25** (1992) 881 - 888.

[23] X-B. Wang, Y. Huang, J. P. H. Burt, G. H. Markx and R. Pethig, Selective dielectrophoretic confinement of bioparticles in potential energy wells, *J. Phys. D: Appl. Phys.* **26** (1993) 1278-1285

[24] P. R. C. Gascoyne, Y. Huang, *R* Pethig, J. Vykoukal and F. F. Becker, Dielectrophoretic separation of mammalian cells studied by computerized image analysis, *Meas. Sci. Technol.* **3** (1992) 439 - 445.

[25] Y. Huang and R. Pethig, Electrode design for negative dielectrophoresis, *Meas. Sci. Technol.* **2** (1991) 1142 - 1146.

[26] X-B. Wang, R. Pethig and T. B. Jones, Relationship of dielectrophoretic and electrorotational behaviour exhibited by polarized particles, *J. Phys. D: Appl. Phys.* **25** (1992) 905 - 912.

[27] Y. Huang, X-B. Wang, J. A. Tame and R. Pethig, Electrokinetic behaviour of colloidal particles in travelling electric fields: studies using yeast cells, *J. Phys. D: Appl. Phys.* **26** (1993) 1528-1535.

[28] W. M. Arnold and U. Zimmermann, Electro-rotation: Developments of a technique for dielectric measurements on individual cells and particles, *J. Electrostatics,* **21** (1988) 151 - 191.

[29] X. *Hu*, W. M. Arnold and U. Zimmermann, Alterations in the electrical properties of T and B lymphocyte membranes induced by mitogenic stimulation. Activation monitored by electro-rotation of single cells, *Biochim. Biophys. Acta* **1021** (1990) 191-200.

[30] A. Coghlan, Sticky beads put bacteria in a spin, *New Scientist*, No:1873 (15 May 1993) p.21.

[31] T. Beardsley, Putting a spin on parasites, *Scientific American*, **269** (July 1993) p.87

[32] S. Masuda, M. Washizu and M. Iwadare, Separation of small particles suspended in liquid by nonuniform traveling field, *IEEE Trans. Ind. Appl.* **23** (1987) 474 - 481.

[33] G. Fuhr, R. Hagedorn, T. Müller, B. Wagner and J. Gimsa, Asynchronous traveling-wave induced linear motion of living cells, *Studia Biophys.* **140** (1991) 79 - 102.

[34] R. Hagedorn, G. Fuhr, T. Müller and J. Gimsa, Traveling-wave dielectrophoresis of microparticles, *Electrophoresis* **13** (1992) 49 - 54.

Chapter 10

Dynamic Cell-Membrane Events Following the Application of Single-Pulse Electric Fields

J.J. Chang, Y. Liu, and K.J. Kong

10.1 Introduction

Exogenous molecules which are normally excluded from living cells, can be made to pass through cell membranes under the influence of a high voltage pulsed electrical field (PEF) with a short duration. This phenomenon, involving the electrical modification of cell membrane permeabilities, has been known as electropermeabilization or more commonly, electroporation. Since Zimmermann[1] described the principle and applications of electrofusion and electroporation in 1982, both which are based on the interactions between the cell membrane and PEF, more and more biologists are becoming interested in the effects of PEF on the cell membrane, as it is important, not only as a new approach in biotechnology, but also as a new topic in theoretical studies on cell membranes.

Until now, however, there have only been a few publications[2–6] on the dynamics of the process. In order to study the dynamic events of the effects of PEF on the cell membrane, some experiments[7,8] have been performed in our laboratory using fluorescence measurements. In this chapter we summarize our recent findings and propose a mechanism for electroporation of cell membrane. We also discuss electric forces and electrical interactions in living systems.

*

*Supported by The National Nature Science Fundation of China

10.2 Materials and Methods

Human erythrocyte ghosts and bone marrow cells of rats were used. $TbCl_3$ (Terbium chloride hexahydrate) was purchased from Aldrich, and DPA (Pyridine-2, 6-dicarboxylic acid) from Fluka Company, EB (ethidium bromide) from BBH chemical company. All other chemicals are analytical pure and commercially available in China.

10.2.1 Preparation of Human Erythrocyte Ghosts

Human erythrocyte ghosts were prepared according to modifications of previous methods[9–10]. Fresh human red blood cells were obtained from the local blood bank, washed three times with PBS buffer (pH 8.0) and lysed in 5mM phosphate buffer pH 8.0 for 15min. After washing the lysed erythrocytes membranes three times by centrifugation, the membranes were sealed in 7.5 mM T_bCl_3. Sealed ghost membranes were kept in the ice bath until the application of PEF. Before pulsation treatment, the ghosts were washed twice with Hepes and suspended in 280 mM sucrose containing 0.15 mM DPA in deionized water. The final concentration of ghosts were kept at 10^7 cells/ml[7].

In order to study the mechanism of electroporation, some ghosts were pretreated with ethanol in concentration of 20mM to 40mM and with 10-20mM glutaraldehyde[8].

10.2.2 Preparation of Bone Marrow Cells

The bone marrow of Wister rats cells were washed out gently from leg bones with 0.9% NaCl. After washing the cells were suspended in the is-osmotic solution of 280mM sucrose with added 25 μg/ml ethidium bromide (EB). The final concentration of cells was kept at 25^7 cells/ml[8].

10.2.3 Electropulsation and Fluorescence Measurements

Electropulsations were applied by the Gene Transformation Electroporation System ZA 2025 PDS Ins, which gives nearly square pulses.

For each pulsation, the same volume of cell suspension, of about 120 μl, was placed into the stainless steel discharge chamber (Fig. 10.1), and a single pulse of 1.0 kV/cm to 2.4 kV/cm and 10 μs to 50 μs duration was applied. The same volume of ultrasonicated ghost suspension served as the control, the fluorescence intensity of which is always referred to as the maximum (100 %). The same volume of suspension of bone marrow cell lysed by hypotonic solution was used as the control for bone marrow cell measurements.

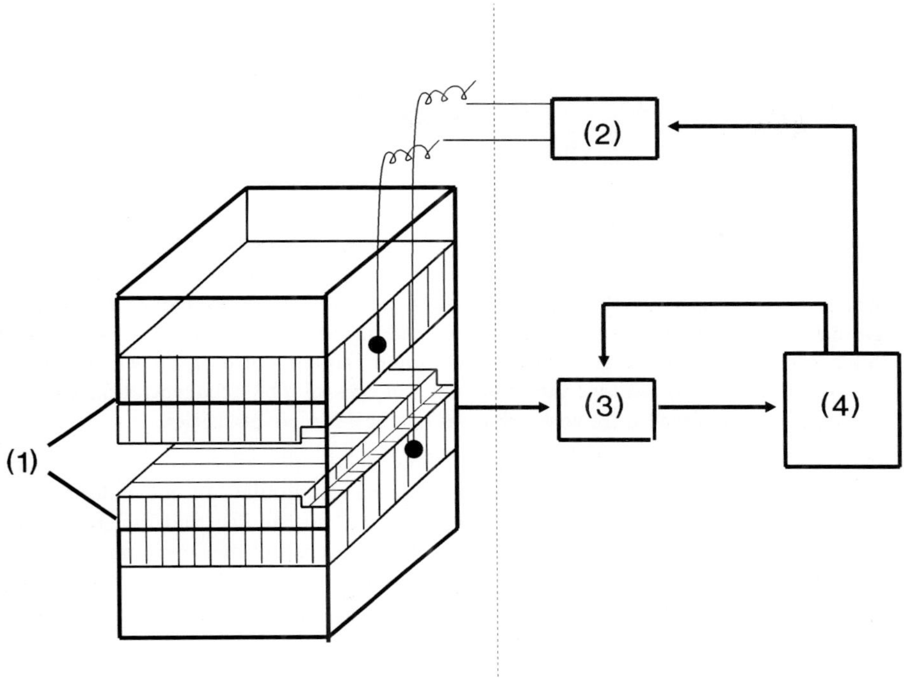

Figure 10.1: *Electropulsation chamber and the measurement systems: 1-electrodes, with adjustable distance, 2-pulse generator, 3-fluorescent spectrometer, 4-computer system.*

All the fluorescence measurements were performed in a PERKIN ELMER LS-5 fluorescence spectrometer. A special pulsation chamber consisting of two parallel electrodes with adjustable distance was fixed in the fluorescence spectrometer. It was connected with a computer system and with the PEF generator, as shown in Fig. 10.1. The pulse generator was connected synchronously with a fluorescence spectrometer and the computer, which registered the data at 0.1 sec. intervals. In some experiments, continuous recordings of up to 7 min. were made, otherwise the recordings were 5 seconds long.

10.3 Results

10.3.1 Dynamic Process of Electropermeabilization of Human Erythrocyte Ghosts

Under our experimental conditions and without external electrical pulses, the ions of Tb^{3+} (159 Da.) loaded in the ghosts were impermeable to the cell membrane. Therefore, the fluorescence intensities of the suspensions remained very low for a long period of several minutes. However, as soon as a pulse was applied to the suspension, the fluorescence intensities immediately increased rapidly with time, which indicated that Tb^{3+} ions leaked out of the ghosts and complexed with DPA, giving thus rise to the increase in fluorescence intensities. Under our experimental conditions, both the fluorescence intensities and the amount of Tb^{3+} which leaked out were linearly correlated, therefore, the changes of Tb^{3+} concentration can be quantitatively estimated from the changes of fluorescence intensities.

Our results, presented in Fig. 10.2-10.4, show that under all conditions, the time course of the increase in the fluorescence intensities is similar, except under rather weak electrical fields. From Fig. 10.2 to Fig. 10.4 we can see that when a pulse is applied to a ghost suspension, the fluorescence intensities increases immediately, reaching a maximum dependent on the parameters of the electrical field.

Fig. 10.5 shows the relationship between the intensities changes and electric field. A threshold of about 2.2 kV/cm can be identified in Fig. 10.5, below which and above which the electropores have different relationship with electrical field intensities. This is because at below this threshold, the ghost membranes are not porated, whereas above it, poration takes place.

10.3.2 Effects of Ethanol and Glutaraldehyde on the Electropermeability of Tb^{3+} Loaded in Human Ghosts

Rols[11] has shown that treatment of CHO cells with 20mM to 400mM ethanol can inhibit the electropermeabilization of cells. Ethanol is considered to penetrate into bilayer of cell membrane lipids and change the orientation of lipid molecules. At low concentrations, ethanol reduces membrane order, thus increasing the energy barrier for membrane transitions. As a consequence, it can inhibit electropermeabilization of cells. Glutaraldehyde is known to denature proteins. It affects membrane proteins and membrane cytoskeleton proteins. In addition, it can inhibit electropermeabilization of cell membranes. In order to study the mechanism of electropermeabilization, some ghosts were pretreated in ethanol and in glutaraldehyde before pulsation.

Our experimental results, displayed in Figs. 10.6, shows that ethanol at 20mM to 40mM and glutaraldehyde at 10mM to 20mM both inhibit electropermeabilities of

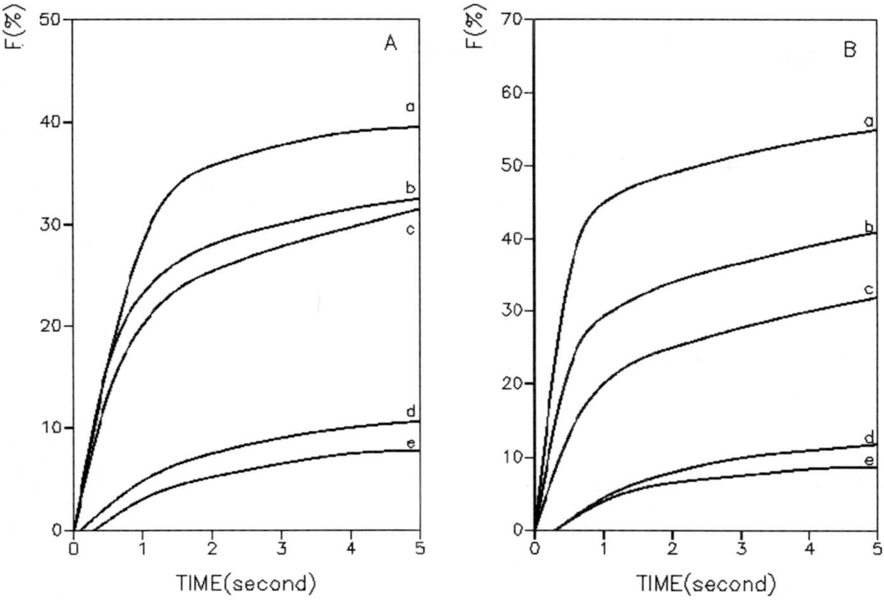

Figure 10.2: *Fluorescence intensities as a function of pulse duration and pulse strength. Curves in A were obtained with a constant field strength of 2.4 kV/cm (a, 50 μs; b, 40 μs; c, 30 μs; d, 20 μs; e, 10 μs). Curves in B were obtained at a constant duration of 50 μs (a, 2.4 kV/cm; b, 2.2 kV/cm; c, 2.0 kV/cm; d, 1.8 kV/cm; e, 1.6 kV/cm).*

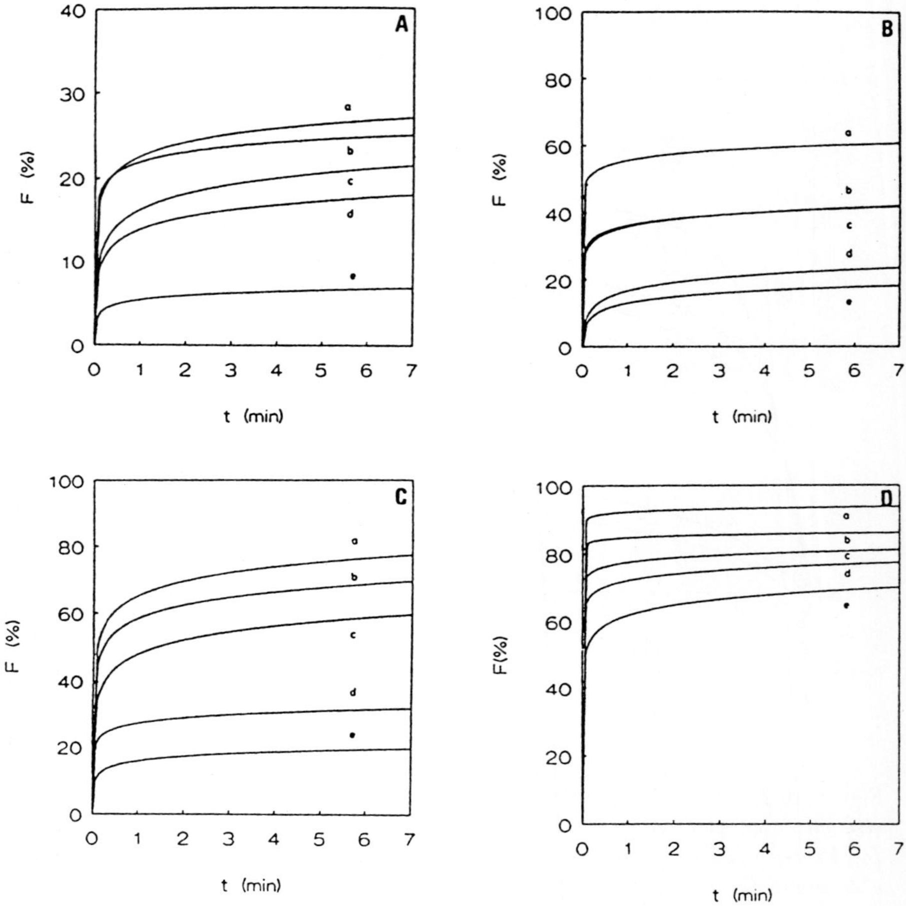

Figure 10.3: *Fluorescence changes measured continuously for 7 minutes. Fig. A at pulse strengths 1.8 kV/cm; Fig. B at 2.0 kV/cm; Fig. C at 2.2 kV/cm; Fig. D at 2.4 kV/cm; curve a: 50 μs, b: 40 μs, c: 30 μs, d: 20 μs, e: 10 μs.*

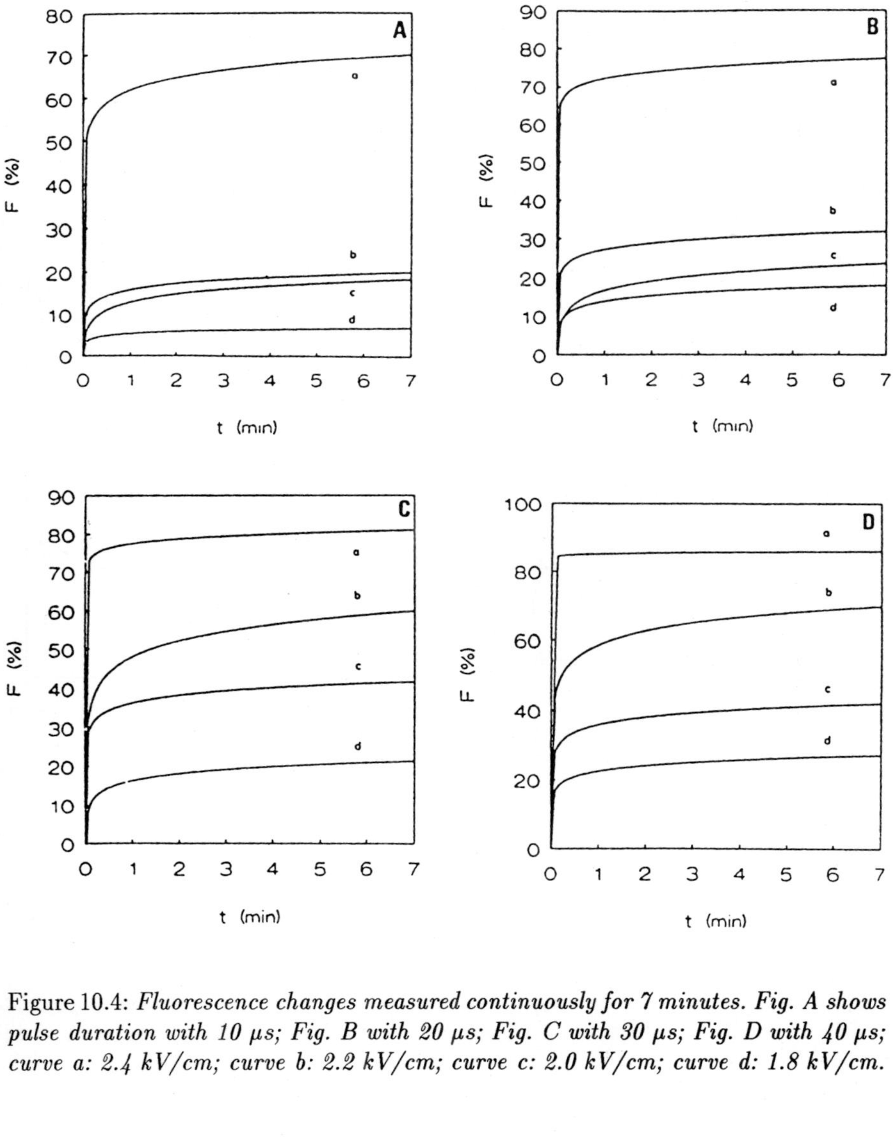

Figure 10.4: *Fluorescence changes measured continuously for 7 minutes. Fig. A shows pulse duration with 10 μs; Fig. B with 20 μs; Fig. C with 30 μs; Fig. D with 40 μs; curve a: 2.4 kV/cm; curve b: 2.2 kV/cm; curve c: 2.0 kV/cm; curve d: 1.8 kV/cm.*

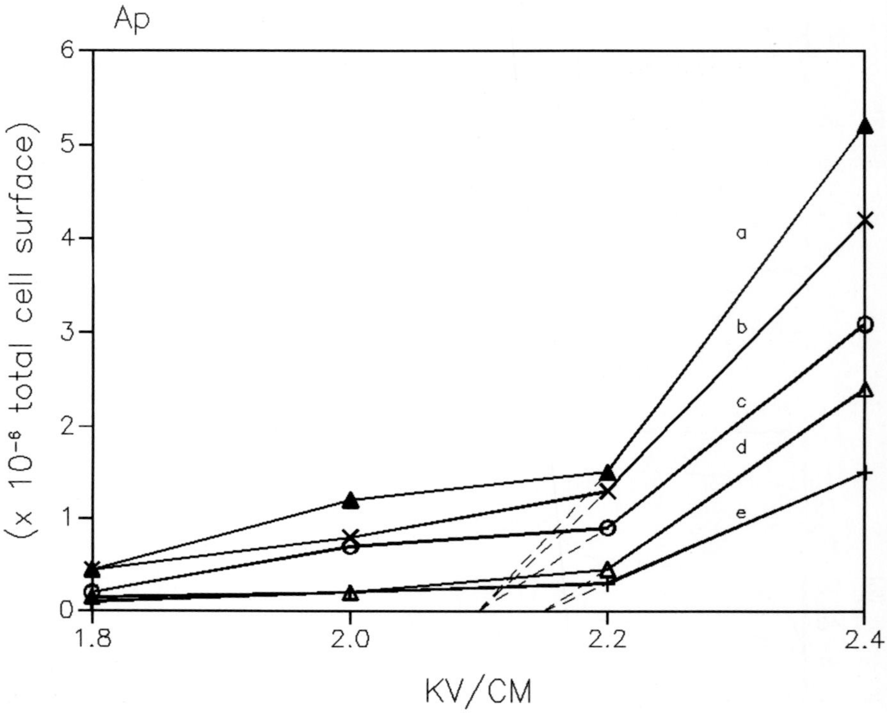

Figure 10.5: *Relationship of electropore area to the parameters of pulsed electrical fields in erythrocyte ghosts. Curve a, 50 μs; b, 40 μs; c, 30 μs; d, 20 μs; e, 10 μs.*

human ghosts, with the effect of glutaraldehyde being stronger than that of ethanol. This suggests that membrane proteins and membrane skeletal proteins may play the main role in electropermeabilization of cell membranes.

10.3.3 Dynamic Changes of Electropermeable Areas of Human Ghosts Membranes

According to theoretical analysis, the leakage of Tb^{3+} may be mainly due to diffusion mechanism, although electrophoresis of Tb^{3+} also takes place[7] during electropulsation. We have,

$$C_i(t) = C_0 e^{-Kt}\,, \tag{10.1}$$

where C_0 and $C_i(t)$ are the concentrations of Tb^{3+} within the ghost at time t_0 and at time t of pulsation, respectively.

(1) can be written as

$$F_i(t) = F_0 e^{-Kt} \tag{10.2}$$

where $F_i(t)$ is the fluorescence intensity at time t and F_0 at time t_0.

$$K = \frac{DA_p}{Vd}\,, \tag{10.3}$$

where D is diffusion coefficient of Tb^{3+} in aqueous solution, A_p is the total electroporated area of the ghost membranes, V is the total volume of ghost membranes and d is the thickness of the membrane.

From Equations (10.2) and (10.3) we have

$$A_p = -\frac{V \cdot d}{D \cdot \Delta t}\{\ln[1 - F(t_1)] - \ln[1 - F(t_2)]\} \tag{10.4}$$

After applying Equation (10.4) to the curves in Fig. 10.6A and Fig. 10.6B we get the curves in Fig. 10.6C and 10.6D which show that the total electroporated area of the ghost membranes changes with time. When a pulse is applied to a ghost suspension, the porated area increases immediately, which means that the electroporated pore forms suddenly and expends rapidly. It reaches a maximum within 0.2 seconds to 0.3 seconds, and then decreases rapidly from 0.3 to 0.8 seconds, as the pore reseals. However, weak electrical fields below the threshold do not give these dynamic changes, implying that there is no electropore formation.

From Fig. 10.6C and Fig. 10.6D, we can also see that the maximum areas of electropores are related to the parameters of E.F. as well as to the pretreatments with

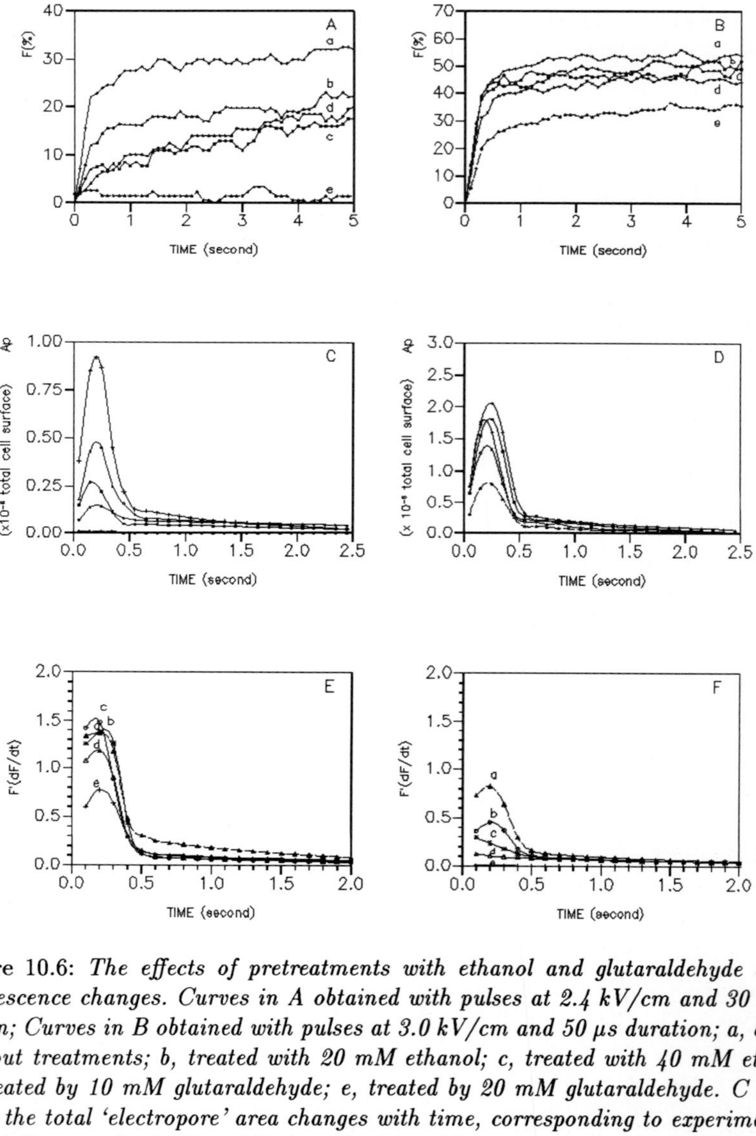

Figure 10.6: *The effects of pretreatments with ethanol and glutaraldehyde on the fluorescence changes. Curves in A obtained with pulses at 2.4 kV/cm and 30 μs duration; Curves in B obtained with pulses at 3.0 kV/cm and 50 μs duration; a, control without treatments; b, treated with 20 mM ethanol; c, treated with 40 mM ethanol; d, treated by 10 mM glutaraldehyde; e, treated by 20 mM glutaraldehyde. C and D show the total 'electropore' area changes with time, corresponding to experiments in A and B, respectively. E and F show the rate of fluorescence changes as the function of time, corresponding to experiments in A and B respectively.*

ethanol and glutaraldehyde. However, the period during which the electropores expand to their maximum is independent of the E.F. or of pretreatments.

10.3.4 Dynamic Process of Electropermeabilization of Bone Marrow Cell Membrane

The dynamic process of electropermeabilization of bone marrow cells was studied by using ethidium bromide (EB) as fluorescence probe of DNA. Because EB molecules are impermeable to living cell membranes, the fluorescence intensities of the bone marrow cell suspension remained unchanged for a considerable period of time when electropulsation was not applied. As soon as pulses were applied, EB molecules permeated into the cell and complexed with DNA, resulting in an increase of fluorescence.

The experimental results are displayed in Figs. 10.7 and 10.8, which show that the dynamics of electropermeabilization for bone marrow cells are similar to that of erythrocyte ghosts, although the time course differs considerably.

The total electropore areas expends to a maximum within 0.4 seconds to 0.9 seconds. After that, the decrease takes place more slowly than the increase. The minimum is reached after 3 seconds to 5 seconds. As shown in Figs. 10.7 and 10.8, a significant difference exists between bone marrow cells and human erythrocyte ghosts in the period of time taken for the electropore to reach a maximum, which is related to both the strength and the duration of the PEF. A broadly similar set of biphasic curves are obtained for the electropore area as a function of field strength for different pulse durations (see Fig. 10.9). However, the threshold for electropore formation is lower than for erythrocyte ghosts (about 1.2 kV/cm compared to about 2.1 kV/cm). The differences between erythrocyte ghosts and bone marrow cells may be attributed to specific membrane properties in the two systems; chemical properties of the permeating species and the direction of electropermeation. The last factor is especially relevant in accounting for the different kinetic behaviour of 'electropore expansion'. In the case of erythrocyte ghosts, the faster response may be due partly to a release of the inorganic ions from membrane surface sites on electropulsation. This process will not be involved in the case of ethidium bromide permeation *into* the cell. Further research with faster time resolution will have to be done in order to clarify this point. However the main reason may be due to the different sizes of cells. Because of the distribution in sizes the unseparated bone marrow cells are subject of different processes under the same electric field. Actually, the experimental results of bone marrow cells display some average behaviour of the ensemble.

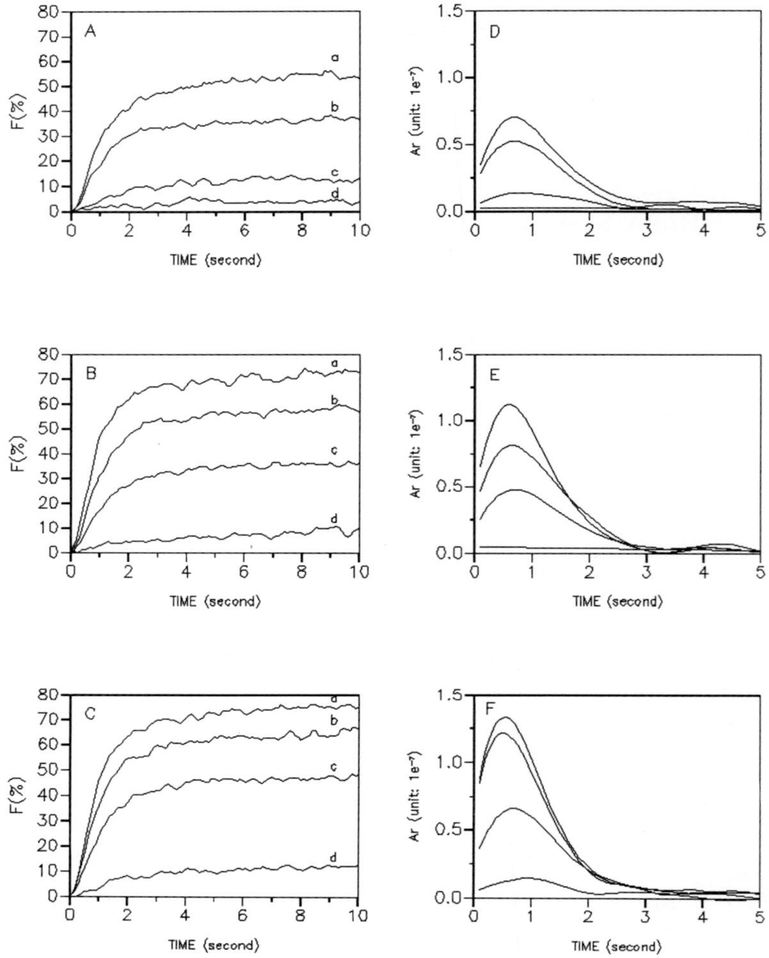

Figure 10.7: *Fluorescence changes of bone marrow cells as a function of pulse strengths at fixed durations. A, at a fixed duration of 10 μs; B, 30 μs; C, 50 μs. Curve a, at 1.6 kV/cm; curve b, 1.4 kV/cm; curve c, 1.2 kV/cm; curve d, 1.0 kV/cm; D, E, F, show the changes of total electropore area corresponding to experiments in A, B and C respectively.*

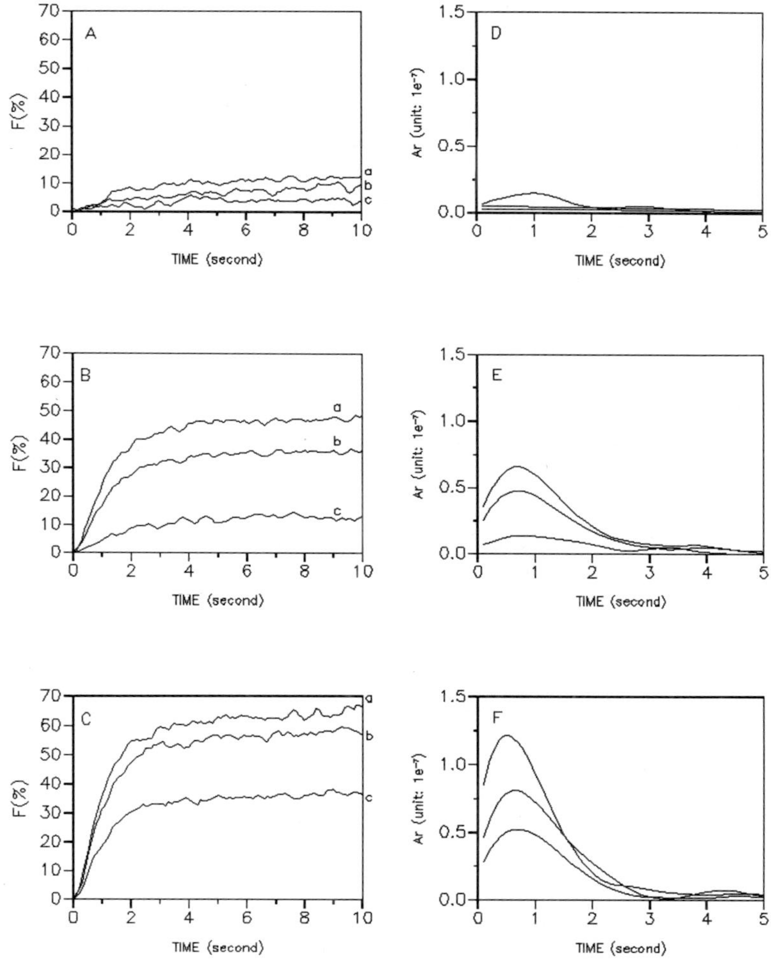

Figure 10.8: *Fluorescence changes of rat bone marrow cells as a function of pulse durations and pulses strengths. A, at 1.0 kV/cm; B, at 1.2 kV/cm; C, at 1.4 kV/cm. Curve a, 50 μs; curve b, 30 μs; curve c, 10 μs. D, E, and F show the changes of the total electropore area of bone marrow cells corresponding to experiments in A, B and C respectively.*

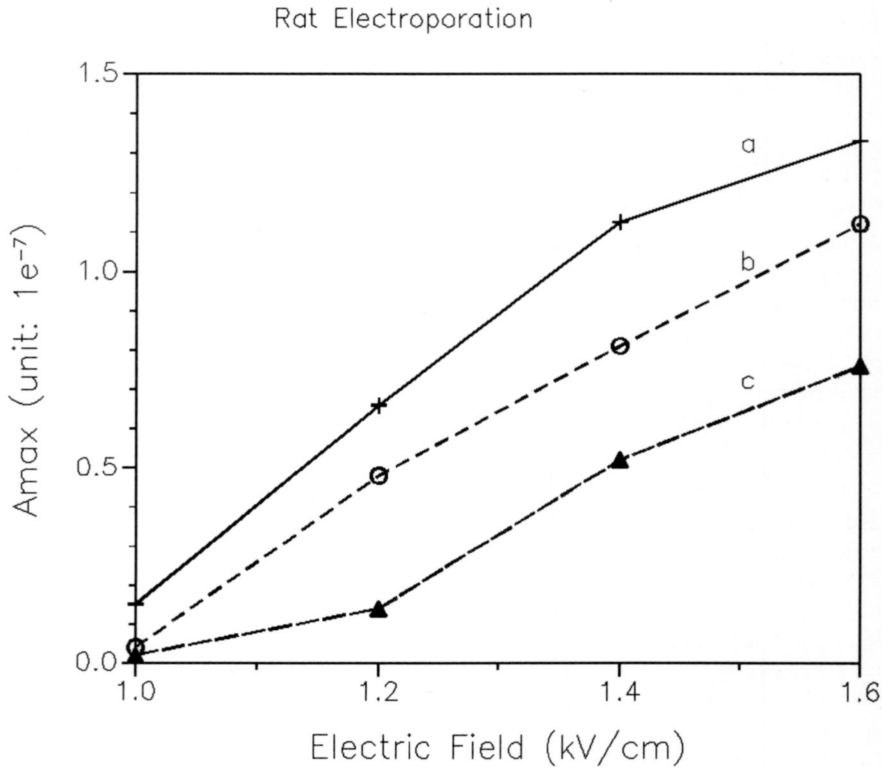

Figure 10.9: *The relationship of electropore area to the parameters of electrical fields in bone marrow cells. Curve a, 50 μs; b, 30 μs; c, 10 μs.*

10.4 Discussion

In summary, by using an electroporation system constructed in our lab, we have studied the dynamic processes of human erythrocyte ghosts membrane and bone marrow cells exposed to external single-pulse electrical fields. Our results show that subsequent to a single high-voltage pulse of short duration, electropores form and expand to their maximum rapidly and then reseal. The total areas and speed of expanding and resealing of electropores are related to parameters of the external electrical fields as well as the specific cell membrane properties.

Although we can at present achieve a time resolution of only 100ms, the results are significant. For applications in biotechnology, they give an indication of the best conditions for electrogenetransformation and electrocellfusion. Besides, dynamic processes which take place within and between cells and their environments are always significant for our understanding of the nature of life, as life is the whole result of combining coherently all the processes and movements which take place within or between cells and their surroundings at a submolecular, molecular, subcellular and cellular level[12].

Electric forces and electrical interactions or electrodynamic processes within cells and between cells and their surroundings may be the most significant for the understanding of biological systems[13]. Although it is technically difficult at present to study dynamical electrical interactions within cells it provides an alternative method to study the interactions of biological systems and external electrical fields by investigation of the intracellular electrofeatures. Furthermore, it is particulary important for understanding biocommunication mechanisms to study the effects of external electrical field on cell membranes. Up to now, it is not yet clear how external physical signals, for instance, electrical signals, cross the cell membrane. We have shown elsewhere that Na^+, K^+-ATPase on cell membrane is activated by pulsed external electrical field[14]. Tsong[15,16] has reported similar results (see also Tsong and Gross, this volume). It is interesting to study whether some membrane proteins are involved in a physical signal correlating to chemical receptor signals.

There are several proposals for the mechanism of electroporation at present. Our results suggest that membrane proteins (and membrane cytoskeletal proteins) may play the major role. This was confirmed by Chang et.al.[17] in freeze-fracturing studies of human erythrocyte membranes exposed to external pulsed electrical field, in which we found that membrane proteins and membrane skeleton proteins can escape from pulsed cells, and that changes in protein-protein interactions and protein-lipid interactions caused by external electrical field may be the dominating mechanism of cell membrane electroporation.

What is the mechanism of exchange of mobile molecules under electropermeabilization? According to our theoretical analysis, diffusion is the main process contributing to electropermeabilization. However, from our experimental curves of porated areas, we found that they are not symmetric. The increasing part takes place more rapidly than the resealing part, suggesting that other processes, for instance, electrophoresis of Tb^{3+} may be involved.

On account of methodological limitation, we have lost the information for the process within the pulse duration of 50 μs, as shown in Figs. 10.6E and 10.6F. We can see that, at the beginning of the pulse the electropore area reaches a maximum in a very short time. If we could follow the dynamic process within the pulse period, it would provide new insights into the interactions of cell membranes with external electrical fields.

Although we describe the dynamic changes as electropore formation, whether there are real pores on cell membranes, is still unknown[18]. There may also be dynamic changes of areas in the membrane. Therefore the so-called 'pore area' is only a descriptive term which means that molecules can pass through this area. One need not think that there are permanent pores during the whole period.

Acknowledgment

The authors would like to express their thanks to Prof. Bei Shi Zhang for his long term guidance in our research. We are very grateful to F.A. Popp and M.W. Ho for valuable discussions, comments and suggestions for improving the paper. Thanks are also due to Miss J. Deny for typing the manuscript.

Bibliography

[1] U. Zimmermann: Electric field-mediated fusion and related electrical phenomena. *Biochim. Biophys. Acta,* **694**, 1982; 227-277.

[2] K. Schwister and B. Deuticke: Formation and properties of aqueous leaks induced in human erythrocyte by electrical breakdown. *Biochim. Biophys. Acta,* **818**, 1985; 332-348.

[3] W. Mehrle, U. Zimmermann and R. Hamps: Evidence for asymmetrical uptake of fluorescent dyes through electro-permeabilised membranes of Avena mesophyll protoplasts. *FEBS Lett.,* **185**, 1985; 89-94.

[4] D.A. Stenger and S.W. Hui: Human erythrocyte electrofusion kinetics monitored by aqueous cotents mixing. *Biophys. J.*, **53**, 1988; 883-888.

[5] D.S. Dimitrov and A.E. Sowers: Membrane electroporation fast molecular extrange by electroosmosis. *Biochim. Biophys. Acta,* **1022**, 1990; 381-392.

[6] M. Hibino, H. Hoh and K. Kinosita: Time courses of cell electroporation as revealed by submicrosecond imaging of transmembrane potential. *Biophys. J.* **64**, 1993, 1789-1800.

[7] K.J. Kong, Y. Liu and J.Z. Zhang (J.J. Chang): Dynamic studies on electropermeabilities of Tb^{3+} loaded in human erythrocyte ghosts. *Biophysica Sinica (Chinese)*, 1994 vol. 10 No.1.

[8] Y. Liu, K.J. Kong and J.J. Chang: Dynamic studies on electroporation of cell membrane by fluorometric method. *Acta Biophysica Sinica (Chinese)*, in press 1994.

[9] T.L. Steck and J.A. Kant: Preparation of impermeable ghost and inside-out vesicles from human erythrocyte membranes: *Methords enzymol,* **31**, 1974; 172-180.

[10] D.D. Hoekstra, J. Wilschut and G. Scherphot: Fusion of erythrocyte ghosts induced by calcium phosphate-kinetic characteristics and the role of Ca^{3+}, phosphate and calcium-phosphate complexes. *Eur. J. Biochem.* **146**, 1985; 131-140.

[11] M.P. Rols, F. Dahhou, K.P. Mishra and J. Teissie: Control of electric field induced cell membrane permeabilization by membrane order. *Biochem.*, **29**, 1990; 2960-2966.

[12] J.J. Chang: Progress and perspective on cell biophysics: progress in *Biochem. Biophs, (Chinese)* 21, 1994: 55-60.

[13] F.A. Popp: Electromagnetism and living systems: in this volume.

[14] J.J. Chang, T. Sun, Y.J. Chen and S.Z. Pang: NMR studies on the changes of intracellular Na^+ concentration in human erythrocytes effected by pulsed electrical field. In *Abstracts of papers of the 5th Meeting of Chinese Society for Cell Biology.* Hangzhou 1992, p. 256.

[15] T.Y. Tsong: Electrical modulation of membrane protein. *Anna. Rev. Biophys. Chem., vol. 19*, 1990; 83-106.

[16] T.Y. Tsong: On electroporation of cell membranes and some related phenomena. Bioelectrochem. *Bioenergy*, **24** (3), 1990; 271-296.

[17] J.J. Chang, K.J. Kong, Q.W. Lu and R.J. Dong: Freeze-fracturing studies on human erythrocyte membranes effected by external pulsed electrical field. *Exper. Biol. Acta (Chinese)* 27, 1994: No. 2.

[18] D.C. Chang and T.S. Reese: Changes in membrane structure induced by electroporation as revealed by rapid-freezing electron microscopy. *Biophys. J.,* **58**, 1990: 1-12.

Chapter 11

On the Biological Nature of Biophotons

Wei-Ping Mei

We all know what light is, but it is not easy to tell what it is.
— Samuel Johnson, from Vol.3 of Boswell's Life —

11.1 Introduction

Longman's dictionary[37] defines "photon" as follows:

> A quantum of electromagnetic radiation, the magnitude of which is obtained from the product of Planck's constant h and the frequency ν of the radiation. A photon can be regarded, in some contexts, as an elementary particle; it is then considered to have zero rest mass, a constant velocity of 3×10^8m/s and no charge, and to change its momentum by altering the associated frequency. Photons are generated when an electric charge is in collision with nuclei or electrons, and the electric charge changes its momentum. The loss of momentum appears as a photon. Photons are also generated in the decay of radioactive nuclei and in the decay of unstable fundamental particles. Photons are generally considered, in a narrow sense, to be quanta of light; they have the constant velocity of light in vacuum.

For many people there is no need to introduce the special word "biophoton", because photons participate in *all* atomic and molecular interactions in biological systems, a photon is a photon, and we can not get any biological information about and from it.

But that is not the case. So far, two classes of light emission from biological systems have been recognized differing mostly in their intensities. In contrast to the well established bioluminescence research which is based on luciferin-luciferase reactions and mostly also correlated with biological function e.g. circadian rhythm, synchronization etc., low intensity photon emission between 200nm and 800nm detected with a photon counting device, has been found in almost every species of biological tissues examined.

Biophotons, i.e. emission of weak radiation from the UV to the visible range of the electromagnetic spectrum by biological systems, have been described by several authors[49,56–58,86]. Experimental results increasingly give evidence that biophoton emission cannot be decoupled from biological processes such as cell division and death, oxidative metabolism, carcinogenesis, etc., and is highly sensitive to environmental changes. It is really both a product of physiological processes and their actual regulator. So it is necessary to use the key word biophoton to emphasize that it carries bioinformation in connection with physiological processes in organism.

11.2 Principle of Measurement

Very weak photon emission in the visible region (400-700nm) was first clearly detected[15,16,76] in the 1950s using the newly developed photomultiplier (PM), which has proved to be a very sensitive and reliable method for the detection of very weak light. Since then, the photomultiplier is mostly used for measuring biophotons in the mode of "photon counting". Several reviews on the detector systems and the optimization of the measurement have appeared[30,33,38,42,83,89].

It is difficult to find a satisfactory classification scheme for methods in areas of biophoton research. This is essentially because they have evolved into a wide variety of combinations, where the correlations among parameters are sought internally between distinct methods. But technical sophistication is also constantly progressing, notably in extending the limits of sensitivity and time resolution.

The method of processing output signals from PM can be divided into the analog mode and digital mode, depending upon the intensity of incident light. When incident light enters the photocathode of a PM, photoelectrons then travel through the dynodes where they are intensified by high voltages some 10^6 to 10^7 times. These multiplied photoelectrons are then collected by the anode, through which they finally output to an external circuit (e.g preamplifer, discriminator and counter).

The measurement of these discrete pulses is called the photon counting or the digital mode. In the analog mode, the mean value of the AC signal becomes a DC signal, whereas in the photon counting mode each pulse is converted into a binary digit which is counted. The photon counting mode therefore has many advantages, it is considerably more effective, especially in the region of very low-light levels.

The efficiency of collecting photoelectrons is called the collection efficiency. The ratio of the collected value to the number of incident photons is called the *detection efficiency* or the *counting efficiency* (D),

$$D = \frac{N_c}{N_i} = \eta \cdot \beta$$

where N_c is a counted value, N_i is the number of incident photons, η is the quantum efficiency of the photocathode, and β is the collection efficiency.

More detailed descriptions of the measuring components of biophoton-detecting are given by Inaba[30] and Mieg et al.[42].

Some attempts have also been made to develop image acquisition systems to check for possible sources and localization of biophoton emission[28,65,69,70]. Since 1970, a variety of two-dimensional photon counting techniques have been developed, using microchannel plates (MCP) and several types of position-sensitive anodes. The *photon counting image acquisition system* (PIAS) has been developed in 1985 as an integrated system consisting of a photon counting image head, a position analyzer and an image processor[82] for use in two-dimensional analysis of very low-level-light phenomena.

11.3 Nature of Biophotons

Photon emission from biological systems has different aspects for the physicist, the chemist and the biologist. I shall try to present an overview of all the aspects and also describe some new results in our Laboratory.

11.3.1 Spontaneous and Induced Photon Emission

Biophotons are referred to under a variety of expressions by different authors, e.g. "ultraweak photon emission", "low level luminescence", "dark luminescence", "low intensity luminescence", "delayed luminescence" or "ultraweak luminescence".

All of these expressions deal with investigations on weak light emissions from biological systems. When the emission is induced by some external excitation it is referred to as "luminescence". Because the intensity of "luminescence" is very low, words like "weak" or "ultraweak" are added. But generally, the term "biophoton" seems to be more suitable, because there are many sources for biophotons, and they cannot be *only* assigned to specific reactions of oxidation or some radical reactions[54–57,86]. The term "biophoton" has to include all possible sources contributing to photon emission from biological systems.

For investigating biophotons it may be appropriate to distinguish between *spontaneous photon emission* (PE) and *induced photon emission* (IPE), which are mostly used in the literature[12,21,45,48,54,57,62,64,68,81].

Spontaneous photon emission (PE) is photon emission without any external perturbation. For such investigations, super-high sensitivity of the measurement device is required.

By contrast, induced photon emissions (IPE) have been throughly investigated for example in photosynthesis, delayed luminescence from different plants, cellular suspensions and tissues. Most of the investigations use light illumination or some other physiological parameters, e.g. temperature, as the inducer.

11.3.2 Physiological Nature of Biophotons

Recent results of biophoton research have increasingly confirmed that biophoton emission cannot be decoupled from its biological processes, and it is really both a product of physiological processes and also a regulator.

Cell Cycle

The pioneering work on cell cycle and photon emission has been done by Konev et al.[32] in 1960s. They used a culture of *Torula ultilis* which were synchronized by the method of nutrition limitation. In the synchronized culture, all the cells (60%) are in approximately the same metabolic state and hence would be expected to emit a pulse of light which is correlated with the synchronous division step. They detected an UV emission peak (mainly at 250-380nm) which preceded the first wave of cell division by about 1h and a second weaker peak which corresponded in the same way to the second synchronous division step.

The most extensive investigations on biophoton emission in the cell cycle have been done by Chwirot[12,13] in meiosis during pollen grain formation in the anthers of *Larix europaea* and by Mei[39.40] in mitosis of yeast, *Saccharomyces cerevisiae.*

Biophoton emission, as well as biochemical and other parameters, DNA synthesis, GSH-glutathione and cell number of *Saccharomyces cerevisiae*, which were synchronized by the method of mating hormones, were measured as a function of the cell division cycle[39,40]. The investigations show that self emission in yeast follows a characteristic pattern in the course of cell division cycle, increasing in the late S phase to the G2 phase (Fig. 11.1). The spectrum of the emitted photons is continuous with a maximum in the UV- and blue-region.

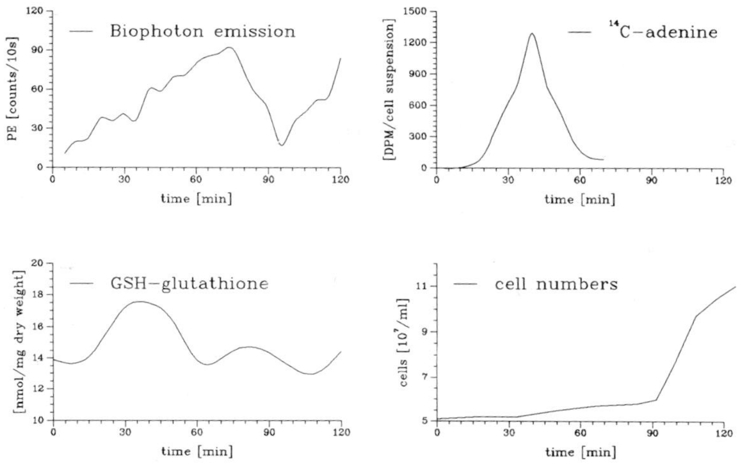

Figure 11.1: *Biophoton emission from synchronized culture of yeast,* **Saccharomyces cerevisiae**, *and its corresponding DNA synthesis, GSH-glutathione level and cell numbers.*

This characteristic cell cycle pattern of photon emission is obliterated only in the presence of the uncouplers of the electrontransport DNP (2,4-dinitrophenol), or FCCP (carbony cyanide *p*-trifluoromethoxyphenylhydrazone), and the inhibitor of protein synthesis cycloheximide. There is no influence on the pattern after addition of the inhibitor of the mitochondrial respiratory chain oligomycine(B), and the inhibitor of mtDNA synthesis of yeast ethidium bromide, indicating that this pattern of photon emission can not be originated from the mitochondria. After excluding the possible contributions of mitochondria to photon emission and considering that the electrontransport can be originated both from mitochondrien and nuclear[66,67], results suggested that cell nucleus may play a very important role in biophoton emission in yeast[39,40,66,67].

Chwirot's works[12,13] show that the meiotic divisions and the development of the microsporocyres are indeed accompanied by considerable changes in the properties of both self emission and light induced photon emission (IPE). The correlation between the changes in the level of IPE and variations in the activity of DNase has been interpreted as an indication of an involvement of DNA in photon emission by the living systems. Also, the existence of hyperbolic decay kinetics and oscillation of the IPE are predicted by Nagl and Popp's model[46] as a consequence of an active storage of photons in DNA dependent on the stage of the development.

Another investigation which has begun since 1974 by Quickenden et al.[60] deal with the exponential- and stationary- phase culture of *Saccharomyces cerevisiae* and

E. coli and its correlated weak photon emission (the so-called "exponential- and stationary- phase photon emission". The results[80] show that the intensity of photon emission for each species examined in the late stationary phase of growth was much higher than in the corresponding exponential phase, demonstrating a close relationship between photon emission and the growth of the cell culture.

Temperature-dependence

In previous work[55], the temperature response of photon emission from plants has been described in terms of dissipative structures of thermodynamically open systems. It has been shown that the proposed model is consistent with the experimental results in cucumber seedlings.

The concept of "dissipative structures of thermodynamically open systems" has been further developed by Slawinski and Popp[73] later on. They described new experimental results which are consistent with the regulatory role of a photon field generated within the living systems. In particular, they used the temperature-dependence of the intensity of photon emission to demonstrate the "temperature hysteresis" and "overshoot reaction" of different plants as an evidence of non-linearity and cooperativity of processes underlying photon emission from plant tissues.

The temperature-dependence of ultraweak photon emission from dark adapted spinach chloroplasts and from leaves has also been reported by Hideg et al.[27]. They found that light emission from both chloroplasts and leaves is highly dependent on the temperature. Whereas the fluorescence of leaves induced by light shows a characteristic maximum in the region of 8-18°C[26,78], ultraweak photon emission from leaf and chloroplasts increase continuously in the region of 5-40°C. They suggested that the reactions resulting in photon emission of chloroplasts are associated with a slow back flow of electrons from NADPH to oxygen via the plastoquinone pool, in other words, chlororespiration[4].

Biophoton emission from cellular systems, for example yeast[39], also shows a temperature-dependence, which is further correlated with the cell cycle. It is well-known that as the temperature is lowered, the cell growth rate decreases, i.e. the cell division cycle is prolonged. The results in yeast show that by lowering the temperature from 30°C to 26°C, the pattern of photon emission from synchronized cell cultures changes also in line with the lengthening of the cell cycle (Fig. 11.2).

The effect of temperature on photon emission from *E. coli* has been examined by Tilbury and Quickenden[79]. It was found that the mean intensities of biophoton emission during the exponential phase decreased by 55(±9%) when the temperature was raised from 33°C to 37°C, with little effect on its spectral distribution. It was suggested that the effects may be attributable to reactions associated with the formation and (or) decomposition of lipid hydroperoxides, and the changes of light emission can be explained in terms of temperature-induced changes in the ratio of

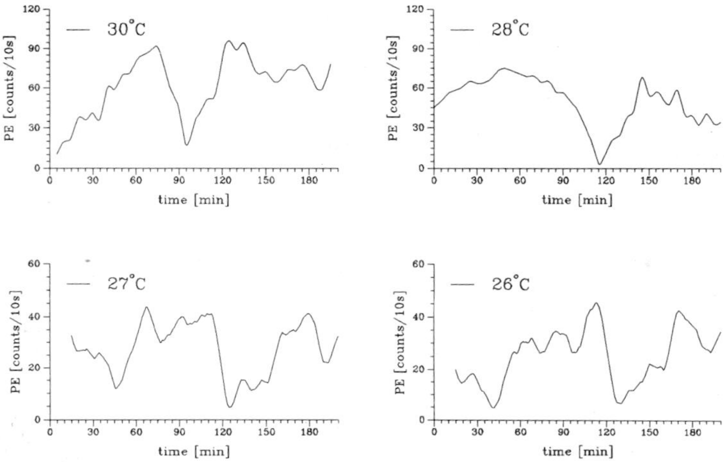

Figure 11.2: *The temperature dependence of the cell cycle and its correlated photon emission.*

unsaturated to saturated lipids.

Fischer et al.[19] have reported the effects of temperature stress on photon emission of *Dictyostelium discoideum.* They also found that starvation of *D. discoideum* induces a photon emission response which was enhanced under heat shock treatment. The emission begins to increase dramatically after about 2h, reaches a peak by about 4h and declines to base level by 8h after the onset of starvation. It was concluded that the increase of temperature enhances the biological process which leads to photon emission by about 10-fold. It is possible that the inducing signal for heat shock is a product, perhaps an oxygen metabolite, of perturbed mitochondrial metabolism in stressed cells. The authors thus suggest that if an oxygen metabolite, such as superoxide or peroxide, is the inducing signal in stressed cells, then photon emission can occur as a response to stress.

Temperature stress not only influences the intensity of photon emission, but also changes its spectral distribution[7,73,75]. The authors claim that these effects are associated with a decrease in the lipid peroxidation rate and a concomitant decrease of generation of electronically excited molecules and subsequent photon emission.

The Localization of Biophotons

Since 1987, Scott et al.[69] first attempted to investigate the biophoton emission with a photon counting image acquisition system (PIAS). Several subsequent reports have also appeared[28,65,70]. These works are mostly concentrated on the investigation of germinating soybean seedlings and cucumber seedlings. All results have shown that although the average count rate of emission is relatively low, remarkable photon emission is observed in the segment of hypocotyl, the junctional region between radicle and plumule, where active cell division is taking place. In contrast, the emission from the plumule itself or from the cotyledon is much less.

In connection with the hypothesis of light piping in plants, the results[65] in cucumber seedlings also indicate that the generation of ultraweak photon emission can be localized at structures different from the localization of apparent light emission, which indicate that the area most probably active as the locus of light generation appears to be the transition zone of the seedling showing the highest mitotic activity[24,25].

These results provide strong evidence that biophoton emission is closely related to the biochemical and biophysical processes of cell multiplication and successive growth which have been described in the previous section.

Oxidative Metabolism

Oxygen is necessary for many cases of photon emission[8–10,43,60,63,74,75,80]. Photon emission from biological systems can originate from the relaxation of excited species or from free radical termination reactions involving oxidative free radicals. Two types are suggested by Cadenas[9] depending on the association of photon emission with membrane lipid peroxidation: a) systems associated with lipid peroxidation and b) reactions not associated with lipid peroxidation.

For photon emission in the visible range, it is generally argued that the sources are excited singlet oxygen and excited triplet state carbonyls[9–11,74]. These species are formed *in vivo* by processes such as lipid peroxidation[47], phagocytosis, enzymatic reactions and interactions of oxygen radicals (especially the superoxide anion) with oxidative metabolites. These processes are often linked with oxidative stress.

Some authors[77] have even stated that biophoton emission results solely from oxidative reactions by the active oxygen species, which consist in singlet excited oxygen ($^1O_2$), superoxide anion radical (O_2^-), hydrogen peroxide (H_2O_2 and its deprotonated species), and hydroxyl radical ($\cdot OH$).

However, work with various scavengers and inhibitors of activated oxygen and spectra in plants suggest[1] that singlet oxygen itself is not a direct source. Similar investigations in yeast[39] showed no direct correlations between photon emission and the radical reactions.

Mammalian Cells and Malignancy

Photon emission by animal tissues and cells has been described for a variety of organs and by many authors[11,85]. Biophotons have been detected from liver[6,72], heart[3,52],

Table 11.1: Light induced photon emission in cell suspensions.

IPE (photons per 10^4 cells)	Cell type
4–8	normal cells
7–36	tumour cells
30-100	fibroblastes

lung[8], nerve[2], muscle[5] and blood[30,31], from mammalian species including cat, Chinese hamster, cow, dog, human, monkey, mouse and rat[21,44,88].

A number of studies focussed on photon emission characteristics of tumor-bearing animals, tumors and isolated tumor cells[48–50,64,68,85–88]. In addition, research activity has increased with respect to the development of biochemical and biophysical models in order to describe the creation of electronically excited states and photons by organisms, the link between them and physiological processes, and the evaluation of their possible informational role.

Most of the results[21,44,48–50,68,85–88] confirm that suspensions of mammalian cells show a weak light-induced photon emission (IPE). The intensity of this emission is dependent on the cell type and its physiological conditions. In general, the lowest IPE values are found for normal, non-mesodermal cells, relatively high IPE values for suspensions of fibroblasts, whereas tumour cells fall in between[88] (see Table 11.1).

Within the last few years, some authors[18,48,71] have reported that most of the induced photon emission originate from cell nuclei. In connection with the results from different investigations[30,39,40,48,53–59,87], it is now clear that the cell nuclei play a fundamental role in photon emission from cellular systems. A typical example is shown in Fig. 11.3.

Further interesting results have been obtained by several authors[64.68,85,86]. In these measurements, cell populations of different densities have been used to study the relation between photon emission and cell number. Tumour cell cultures typically display an exponential increase of photon emission with increasing cell density (Fig. 11.4). A completely different relation between cell density and IPE is exhibited by normal cells where at higher cell densities, the IPE becomes decreased to such an extent that it is not above that due to the empty cuvette.

Using such characteristics, i.e. different behaviours of IPE in different population densities between tumour cells and normal cells Popp suggests that a new detecting system can be developed for the early diagnosis of cancer.

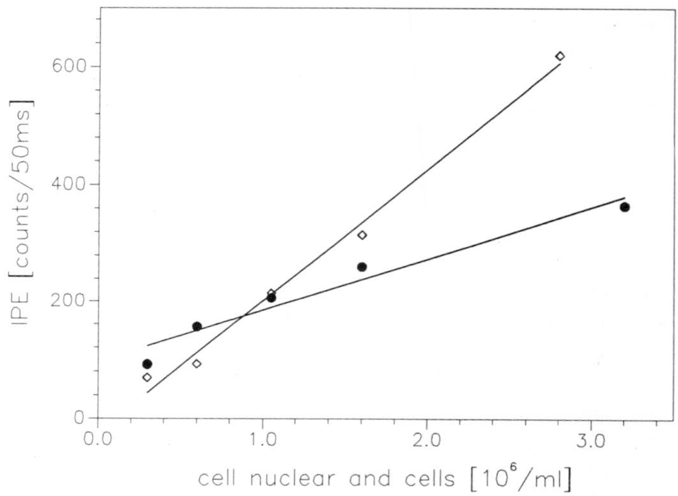

Figure 11.3: *The white-light-induced photon emission of fibroblast cells (◇) and cell nuclear (•) as a function of densities.*

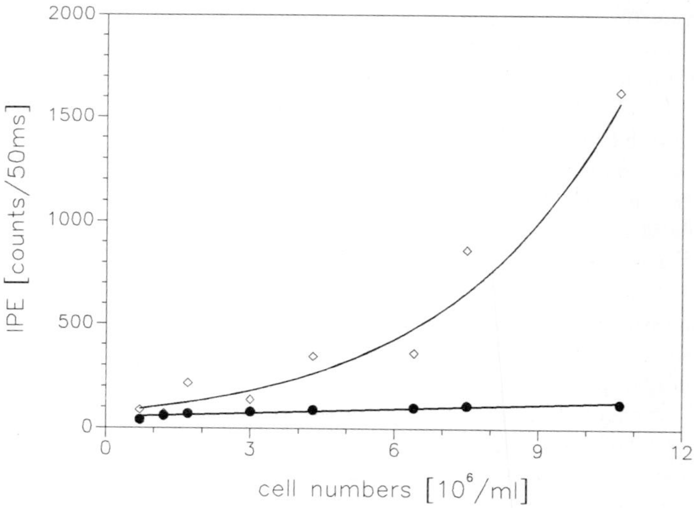

Figure 11.4: *The dependence of biophoton emission on the cell densities from tumour cells (HTC, ◇) and normal cells (•).*

UV-radiation and Biophoton Emission

In the 1920s, Gurwitsch[24,25] claimed that dividing cells emit a very weak UV luminescence which was termed *mitogenetic radiation* due to its ability to stimulate cell division when incident on nearby cells. A review of mitogenetic radiation by Quickenden and Que Hee[61] discusses some of the curious features of its history.

Konev (1966)[32] was the first to employ the UV-sensitive PM tube to detect UV photon emission from living organisms. Also, significant ultraviolet photon emission was observed from cultures of yeast and bacteria[39,40,80].

The ultraviolet wavelengths in the sun's spectrum are known to be a basic cause of human skin cancer. The principal UV damage to DNA is the binding of adjacent bases, particularly when two thymine bases or two cytosines are close to each other[90] and the UV spectrum overlaps the tail of DNA absorption[84]. It has been suggested[29] that ultraviolet photon emission accompanying bodily functions may be just as effective as basic a cause of carcinomas within the body.

Niggli[50] has developed a model system, in which he investigated the short-term response of ultraweak photon emission after irradiation of mitotic factor (MF) and mitonycin C (MMC) induced postmitotic human skin fibroblasts (PMF) with an UV-source. The results have shown that MMC-induced PMF cells tend to lose the capacity to store photons efficiently, which indicated that the UV-induced ultraweak photon emission response may serve as a marker of differentiation.

The effects of different types of background light and UV radiation on growth, photosynthetic pigments, photosynthesis, peroxidation and ultraweak photon emission from plants have been investigated by Panagopoulos et al.[51]. They found that photon emission increases after UV treatment of *Hibiscus* leaves. The highest peroxidase activity and photon emission were observed in leaves from plants grown under white-light(WL)+UVB[1,51]. When UVB radiation was supplemented with UVA, activity decreased markedly, indicating a possible ameliorating effect. It seemed that UVA radiation in general serves to decrease photon emission of WL and UVB treatment. As not all photon emission belongs to the red region of the spectrum, but also to light of wavelengths shorter than 612nm, some processes e.g. lipid peroxidation, which give both kinds of light (red and shorter wavelengths) may exist.

Different cellular systems, e.g. HTC cells, fibroblasts, keratinocytes, yeast and *Acetabularia actubulum*, have been also investigated under UV stress in our laboratory[41]. The results show that white light-induced photon emission after UV-stress and/or UV-induced photon emission are very different for different cellular systems and are also very sensitive to the cellular conditions. UV-radiation has in general "depression effects" on light induced photon emission (IPE) although it works as an activator for self emission (PE)[41].

De Fazi[17] studied in 1924 the effects of UV light on yeast and found that fermentation was greatly stimulated by very weak UV radiation, or by very short exposure to

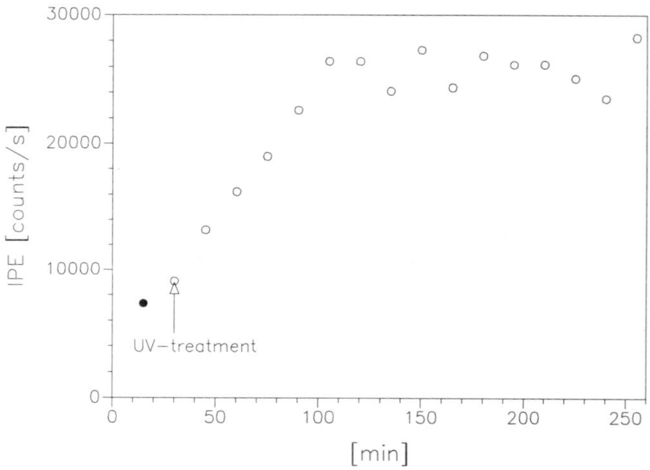

Figure 11.5: *The white-light-induced photon emission from yeast, Saccharomyces cerevisia (5×10^7/ml in 10 ml PBS at 30°C), measured before (•) and after (○) UV-treatment. UV-treatment: 10min (800 J/cm^2) UV-irradiation outside the measuring chamber.*

higher levels of UV radiation. The same phenomenon has been observed in biophoton emission (Fig. 11.5), where white-light-induced photon emission (IPE) increases within a short time (1.5h) to 4-fold after the cells were exposed to 800 J/cm^2 UV-irradiation.

The results in keratinocytes and fibroblast cells also show the very different time course of UV-induced photon emission (Fig. 11.6). With UV applied every 15min at 10 J/cm^2, keratinocytes cells have the ability to survive, but above certain doses they lose their repair ability and increase photon emission until they die, when photon emission once again decreases. In contrast to keratinocytes, fibroblast cells show no changes at all after low doses of UV-irradiation, because they not so sensitive to UV-radiation.

The dynamics of light induced photon emission (IPE) in *Acetabularia acetabulum* (a chlorophyll-contained single cellular system) have been reported by Musumeci et al.[45]. They analysed the decay kinetics of different action spectrum (496nm, 652nm, 695nm, 753nm) before and after the cells were poisoned with different concentrations of atrazine. The UV-induced IPE of *Acetabularia actabulum* was also investigated in our laboratory, which showed that the UV-induced IPE in the first second is strongly dependent on the duration of UV illumination (Fig. 11.7). There is a duration of UV illumination which begins to depress the UV-induced photon emission.

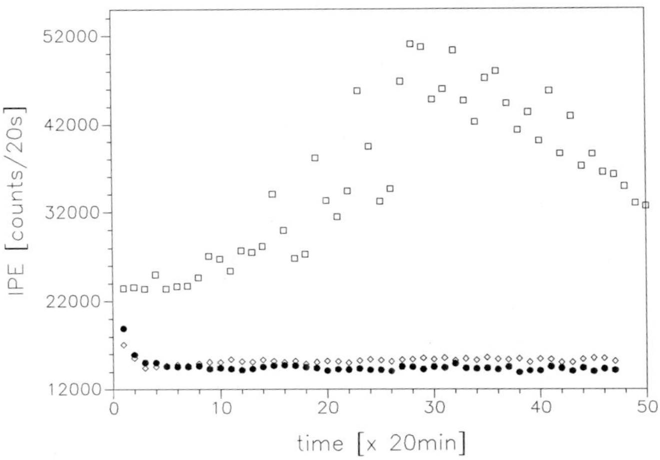

Figure 11.6: *The time dependence of UV-induced photon emission from 10 ml cell suspension of keratinocytes (□), fibroblast (•) and medium (DMEM) alone (◇). Each point represents the IPE-values after 5min UV-irradiation (10 J/cm²) as an inducer.*

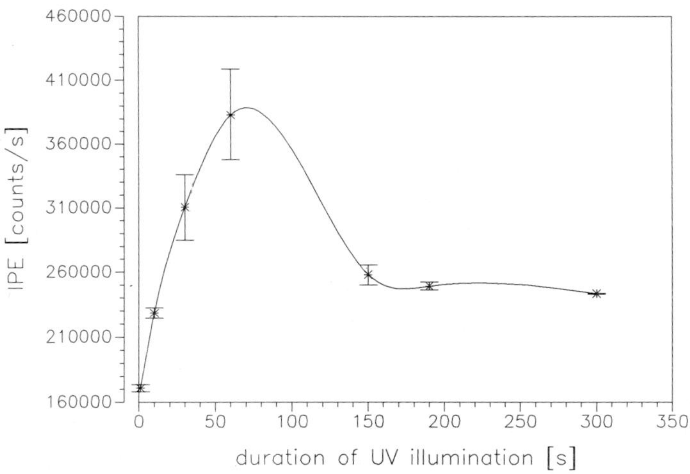

Figure 11.7: *The dependence of UV-induced photon emission of Acetabularia acetabulum on the duration of the UV-illumination. Each point represents the IPE-values after 5min UV-illumination (10 J/cm²) of different duration as an inducer.*

11.3.3 Coherence of Biophotons

The idea of the "coherence" of biophotons was first introduced by Popp[53]. Later, Nagl and Popp[46] postulated that there is a negative feedback loop in living cells which couples states of a coherent field of ultraweak photons (i.e. biophoton) and the conformational state of the cellular DNA (a cellular physical resonance device). They assume a photon transfer or radiationless chemical pumping[14] from the cytoplasmic metabolism which results in changes of the DNA conformation via exciplex/excimer formation.

A lot of experimental results[12,13,49,56,58,62,87] have provided evidence that DNA is a source of biophotons originating from a delocalized coherent electromagnetic field within living matter[20]. As a generator of this field, living matter displays an energy-level distribution characterized by the f_ν = constant-rule, which means that in the ideal case (i.e. enough "pumping" energy is always available) all the relevant excited states of living matter are occupied with about the same probability, independent of the excitation energy[53]. The coherence of biophoton emission can be understood in terms of the emission of the phase-locked and mode-locked biophotons from living systems in their quasi-stationary state operations[23]. A sufficient condition for coherence is the hyperbolic relaxation after ergodic excitation[59] which follows from a coherent nonlinear coupling among the collection of molecules within living matter.

The coherence of biophotons from the swarm of *Daphnia*[58] has been analysed in terms of an interference pattern of the photon field from individual *Daphnia*, which can be described by a destructive interference model as a consequence of Dicke's theory.

Some steps have been taken recently by Li[34–36] in order to show the possibility of coherent radiation of DNA molecules caused by the cooperative motions of DNA bases on the basis of Dicke's radiation theory which represents a general model of the real properties of all radiation sources. Also, considerations have been made on the interactions between electromagnetic waves and mechanical base oscillations which can be assigned to polaritons as in solid state physics, e.g. the joint vibrations between photons and phonons in the DNA molecular skeleton. The cooperative radiation phenomena and the possibility of communication in living systems by the use of some extreme optical effects of a quantum nature may be displayed *only* by a photon emission with very weak intensity.

More recently Gu[23,24] has performed the fundamental calculations from quantum theory in which he came to the following conclusions: a) the emission intensities are proportional to N^2 rather than N where N is the number of identical three-level molecules, which occupy a region small compared to the wavelength of all the relevant radiation modes (Dicke's model); b) the biophoton field exhibits, in different cases, partial coherence, complete coherence or nonclassical coherence, alternatively; c) the relaxation dynamics of the systems after excitation is characterized by a hyperbolic decay, which has been confirmed by many experiments[12–13,45], and by a superradiant emission. Furthermore, it has been shown that there is a critical point of non-equilibrium phase transition between an "ordered" and a "chaotic" regime. Around the critical point, biophotons display a diversity of aspects, which maintain the optical flexibility of living systems.

11.4 Perspectives

The experimental results collected so far and the theoretical considerations show that biophotons are carrying bioinformation in connection with physiological processes of biological systems, but they do not lead to a *definitive* answer to the question of sources and functions of biophoton emission. In order to make important breakthroughs in biophoton research, the contributions from a variety of disciplines such as physiology, cell biology, biochemistry, molecular genetics and physics are needed as well as a better exchange of information between these disciplines.

Bibliography

[1] F.B. Abeles, Plant chemiluminescence: An overview. *Ann. Rev. Plant Physiol.* **37** (1987) 49-72.

[2] V.V. Artem'ey, A.S. Goldobin and L.N. Gus'kev, Recording the optical emission of a nerve. *Biophysics* **12** (1967) 1278-1280.

[3] R. Barsacchi, P. Camici, U. Bottigli, P.A. Salvadoro, G. Pelosi, M. Maiorino and F. Ursinni, Correlation between hydroperoxide-induced chemiluminescence of the heart and its function. *Biochim. Biophys. Acta.* **762** (1983) 241-247.

[4] P. Bennoun, Evidence for a respiratory chain in the chloroplast. *Proc. Natl. Acad. Sci. USA* **79** (1982) 4352-4356.

[5] V.V. Blokha, G.V. Kossova, A.D. Sizov, V.A. Fedin, Yu.P. Kozlov, O.R. Kol's and B.N. Taruso, Detection of the ultraweak glow of muscles on stimulation. *Biophysics* **13** (1968) 1084-1085.

[6] A. Boveris, E. Cadenas, R. Reiter, M. Filipkowski, Y. Nakase and B. Chance, Organ chemiluminescence: Noninvasive assay for oxidative radical reactions. *Proc. Natl. Acad. Sci. USA* **77** (1980) 347-351.

[7] A. Boveris, A.I. Vasavsky, S.G. Da Silva and R.A. Sanchez, Chemiluminescence of soybean seeds: Spectral analysis, temperature dependence and effect of inhibitors. *Photochem. Photobil.* **38** (1983) 99-104.

[8] E. Cadenas, I.D. Arad, A. Boveris, A.B. Fischer and B. Chance, Partial spectral analysis of the hydroperoxide-induced chemiluminescence of the perfused lung. *FEBS Lett* **111** (1980) 413-418.

[9] E. Cadenas, Low level chemiluminescence of biological systems, in *Bioluminescence and chemiluminescence*, eds. J. Schölmerich, R. Andreesen, A. Kapp, M. Ernst and W.G. Woods (John Wiley & Sons, Chichester, Ney York, Brisbane, Toronto, Singapore, 1986) 33-40.

[10] E. Cadenas, Biological chemiluminescence. *NATO ASI Ser. Ser. A* **146** (React. Oxygen Species Chem. Biol. Med.) (1988) 117-141.

[11] A.K. Campbell, *Chmiluminescence, principles and applications in biology and medicine.* (Ellis Horwood Ltd., Chichester, England, 1988).

[12] B.W. Chwirot, Ultraweak photon emission and anther meiotic cycle in *Larix europaea* (Experimental investigation of Nagl and Popp's electromagnetic model of differentiation). *Experientia* **44** (1988) 594-599.

[13] B.W. Chwirot, Ultraweak luminescence studies of microsporogenesis in larch, in *Recent advances in biophoton research and its applications*, eds. F.A. Popp, K.H. Li and Q. Gu (World Scientific, Singapore, New Jersey, London, Hong Kong 1992) pp. 259-282.

[14] G. Cilento, Photobiochemistry without light. *Experientia* **44** (1988) 572-576.

[15] L. Colli, and U. Facchini, Light emission by germinating plants. *Nuovo Cimento* **12** (1954) 150-153.

[16] L. Colli, N. Facchini, G. Guidotti, R. Dugnani Lonati, M. Orsenigo and O. Sommariva, Further measurements on the bioluminescence of the seedlings. *Experientia* **11** (1955) 479-481.

[17] R. De Fazi, Action of ultra-violet rays on alcoholic fermentation and on yeast. *Atti Congr. naz. Chim. Ind. Appl.* **8** (1924) 449-450.

[18] B. Devaraj, R.Q. Scott, P. Roschger and H. Inaba, Ultraweak light emission from rat liver nuclei. *Photochem. Photobiol.* **54** (1991) 289-293

[19] P.R. Fischer, P. Karampetsos, Z. Wilczynska and L.T. Rosenberg, Oxidative metabolism and heat shock enhanced chemiluminescence in *Dictyostelium discoideum*. *J. Cell Sci.* **99** (1991) 741-750.

[20] E. del Giudice, S. Doglia and M. Milani, Ordered structures as a result of the propagation of coherent electric waves in living systems, in *Interaction between electromagnetic fields and cells*, eds. A. Chiabrera, C. Nocolini and H.P. Schwan (Plenum Press, New York and London, 1985) pp. 157-169.

[21] F. Grasso, C. Grillo, F. Musumeci, A. Triglia, G. Rodolico, F. Cammisuli, C. Rinzivillo, G. Fragati, A. Santuccio and M. Rodolic, Photon emission from normal and tumour human tissues, *Experientia* **48** (1992) 10-13.

[22] Q. Gu, Quantum theory of biophoton emission, in *Recent advances in biophoton research and its applications*, eds. F.A. Popp, K.H. Li and Q. Gu (World Scientific, Singapore, New Jersey, London, Hong Kong 1992) pp. 59-112.

[23] Q. Gu and F.A. Popp, Nonlinear response of biophoton emission to external perturbations, *Experientia* **48** (1992) 1069-1081.

[24] A.G. Gurwitsch, The basic laws of mitogenetic excitation. *Arch. Sci. biol. St. Petersb.* **31** (1930) 149-159.

[25] A.G. Gurwitsch and L. D. Gurwitsch, *Die Mitogenetische Strahlung* (Fischer-Verlag, Jena, 1950).

[26] M. Havaux and R. Lannoye, Temperature dependence of delayed chlorophyll fluorescence in intact leaves of higher plants. A rapid method for detecting the phase transition of thylakoid membrane lipids. *Photosynth. Res.* **4** (1983) 257-263.

[27] E Hideg, R.Q. Scott and H. Inaba, High resolution emission spectra of one second delayed fluorescence from chloroplasts. *FEBS Letters* **250** (1989) 275-279.

[28] T. Ichimura, M. Hiramatsu and N. Hirai, Two-dimensional image of ultra-weak emission from intact soybean roots. *Photochem. Photobiol.* **50** (1989) 283-286.

[29] J.E. Illigworth, Internal ultraviolet luminescence – internal carcinomas. *Medical Hypotheses* **27** (1988) 313-316.

[30] H. Inaba, Super-high sensitivity systems for detection and spectral analysis of ultraweak photon emission from biological cells and tissues. *Experientia* **44** (1988) 550-559.

[31] H. Klima and P. Roschger, Photon emission from blood cells and its possible role in immune system regulation, in *Photon emission from biological systems*, eds. B. Jezowska-Trebiatowska, B. Kochel, J. Slawinski and W. Stek (World Scientific, Singapore, 1987) pp. 153-169.

[32] S.V. Konev, T.I. Lyskova and G.D. Niesenbaum, Very weak bioluminescence of cells in the ultraviolet region of the spectrum and its biological role. *Biophysics* **11** (1966) 410-413.

[33] J. Lavorel and J.B.M. Lutz, Methodological principles of measurement of light emitted by photosynthetic systems, in *Light emission by plants and bacteria*, eds. Govindjee, J. Amesz and D.C. Fork (Academic Press, Inc. 1986) pp. 57-98.

[34] K.H. Li, F.A. Popp, W. Nagl and H. Klima, Indication of optical coherence in biological systems and its possible significance, in *Coherent excitations in biological systems*, eds. H. Fröhlich and F. Kremer (Springer, Berlin, 1983) pp. 117-122.

[35] K.H. Li, Coherence in physics and biology, in *Recent advances in biophoton research and its applications*, eds. F.A. Popp, K.H. Li and Q. Gu (World Scientific, Singapore, New Jersey, London, Hong Kong 1992) pp. 113-156.

[36] K.H. Li, Coherent radiation from DNA molecules, in *Recent advances in biophoton research and its applications*, eds. F.A. Popp, K.H. Li and Q. Gu (World Scientific, Singapore, New Jersey, London, Hong Kong 1992) pp. 157-196.

[37] *Longman dictionary of scientific usage* (Longman Group Ltd. 1979).

[38] J.W. Longworth, Yearly review: photon counting. *Photonchem. Photonbiol.* **26** (1977) 665-668.

[39] W.P. Mei, Ultraschwache Photonenemission bei synchronisierten Hefezellen (*Saccharomyces cerevisiae*) in Abhängigkeit vom Zellteilungszyklus. *PhD thesis* (University Hannover, Germany, 1991).

[40] W.P. Mei, Ultraweak photon emission from synchronized yeast (*Saccharomyces cerevisiae*) as a function of the cell division cycle, in *Recent advances in biophoton research and its applications*, eds. F.A. Popp, K.H. Li and Q. Gu (World Scientific, Singapore, New Jersey, London, Hong Kong 1992) pp. 243-258.

[41] W.P. Mei, Effects of UV-radiation on cellular systems. *Photodermatology, Photoimmunology and Photomedicine* (1993), **9**, 182 (Abstract).

[42] C. Mieg, W.P. Mei and F.A. Popp, Technical notes to biophoton emission, in *Recent advances in biophoton research and its applications*, eds. F.A. Popp, K.H. Li and Q. Gu (World Scientific, Singapore, New Jersey, London, Hong Kong 1992) pp. 197-206.

[43] M.E. Murphy and H. Sies, Visible-range low-level chemiluminescence in biological systems. *Methods in Enzymology* **186** (1990) 595-610.

[44] F. Musumeci, A. Triglia and F. Grasso, Experimental evidence on ultraweak photon emission from normal and tumour human tissues, in *Recent advances in biophoton research and its applications*, eds. F.A. Popp, K.H. Li and Q. Gu (World Scientific, Singapore, New Jersey, London, Hong Kong 1992) pp. 307-324

[45] F. Musumeci, M. Goldlewski, F.A. Popp and M.W. Ho, Time behaviour of delayed luminescence in *acetabularia acetabulum*, in *Recent advances in biophoton research and its applications*, eds. F.A. Popp, K.H. Li and Q. Gu (World Scientific, Singapore, New Jersey, London, Hong Kong 1992) pp. 327-344.

[46] W. Nagl and F.A. Popp, A physical (electromagnetic) model of differentiation: basic considerations. *Cytobios.* **37** (1983) 45-62.

[47] M. Nakano, Low-level chemiluminescence during lipid peroxidations and enzymatic reactions. *J. Biolumi. Chemilumi.* **4** (1989) 231-240.

[48] H.J. Niggli, Biophoton Re-emission studies in carcinogenic mouse melanoma cells, in *Recent advances in biophoton research and its applications*, eds. F.A. Popp, K.H. Li and Q. Gu (World Scientific, Singapore, New Jersey, London, Hong Kong 1992) pp. 231-242.

[49] H.J. Niggli, Ultraweak photons emitted by cells: biophotons. *J. Photochem. Photobiol. B: Biol.* **10** (1992) 144-146.

[50] H.J. Niggli, Artificial sunlight irradiation induced ultra-weak photon emission in human skin fibroblasts. *Photochem Photobiol. B: Biol.* **18** (1993) 281-285.

[51] I. Panagopoulos, J.F. Bornman and L.O. Björn, Effects of ultraviolet radiation and visible light on growth, fluorescence induction, ultraweak luminescence and peroxidase activity in sugar beet plants. *J. Photochem. Photobiol. B: Biol.* **8** (1990) 73-87.

[52] V.V. Perelygin and B.N. Tarusov, Flash of very weak radiation on damage to living tissues. *Biophyscs* **11** (1966) 616-618.

[53] F.A. Popp, Photon storage in biological systems, in *Electromagnetic bio-information*, eds. F.A. Popp, G. Becker, H.L. König and W. Peschka (Urban & Schwarenberg, München, 1979) pp. 123-149, and 2nd. ed. (1989) pp. 144-167.

[54] F.A. Popp, On the coherence of ultraweak photon emission from living tissues, in *Disequilibrium and self-organisation*, ed. C.W. Kilmister (D.Reidel P.C., Dordrecht 1986) pp. 207-230.

[55] F.A. Popp, K.H. Li and W. Nagl, A thermodynamic approach to temperature response of biological systems as demonstrated by low level luminescence of cucumber seedlings. *Z. Pflanzenphysiol.* **114** (1984) 1-13.

[56] F.A. Popp, W. Nagl, K.H. Li, W. Scholz, O. Weingärtner and R.Wolf, Biophoton emission: New evidence for coherence and DNA as source, *Cell Biophys.* **6** (1984) 33-52.

[57] F.A. Popp, K.H. Li, W.P. Mei, M. Galle and R. Neurhr, Physical aspects of biophotons. *Experientia* **44** (1988) 576-586.

[58] F.A. Popp, Some remarks on biological consequences of a coherent biophoton field, in *Recent advances in biophoton research and its applications*, eds. F.A. Popp, K.H. Li and Q. Gu (World Scientific, Singapore, New Jersey, London, Hong Kong 1992) pp. 357-374.

[59] F.A. Popp and K.H. Li, Hyperbolic relaxation as a sufficient condition of a fully coherent ergodic field, in *Recent advances in biophoton research and its applications*, eds. F.A. Popp, K.H. Li and Q. Gu (World Scientific, Singapore, New Jersey, London, Hong Kong 1992) pp. 47-58

[60] T.I. Quickenden and S.S. Que Hee, Weak luminescence from the yeast *Saccharomyces cerevisiae* and the existence of mitogenetic radiation. *Biochem. Biophys. Res. Commun.* **60** (1974) 764-770.

[61] T.I. Quickenden and S.S. Que Hee, On the existence of mitogenetic radiation. *Speculations Sci. Technol.* **4** (1981) 453-464.

[62] M. Rattemeyer, F.A. Popp and W. Nagl, Evidence of photon emission from DNA in living systems. *Naturwissenschaften* **68** (1981) 572-573.

[63] B. Röder, Biologische Bedeutung aktivierter Sauerstoffspezies. *Biol. Rundsch.* **25** (1987) 273-284.

[64] D.H.J. Schamhart and R. van Wijk, Photon emission and the degree of differentiation, in *Photon emission from biological systems*, eds. B. Jezowska-Trebiatowska, B. Kochel, J. Slawinski and W. Stek (World Scientific, Singapore, 1987) pp. 137-152.

[65] B. Schauf, L.M. Repas and R. Kaufmann, Localization of ultraweak photon emission in plants. *Photochem. Photobiol.* **55** (1992) 287-291.

[66] J. Schole, Theory of metabolic regulation including hormonal effects on the molecular level. *J. Theor. Biol.* **96** (1982) 579-615.

[67] J. Schole and W. Lutz, *Regulationskrankheiten: Versuch einer fachuebergreifenden Analyse* (Ferdinand Enke Verlag Stuttgart, 1988).

[68] W. Scholz, U. Staszkiewics, F.A. Popp and W. Nagl, Light-stimulated ultraweak photon re-emission of human amnion cells and wish cells. *Cell Biophys.* **13** (1988) 55-63.

[69] R.Q. Scott, S. Masiko, M. Kobayashi, K. Hishinuma, T. Ichimura, B. Yoda and H Inaba, Two-dimensional detection of ultraweak bioluminescence using a single-photon image acquisition system. *J. Opt. Soc. Am. A4* **13** (1987) 51-55.

[70] R.Q. Scott, M. Usa and H. Inaba, Ultraweak emission imagery of mitosing soybeans. *Appl. Phys. B.* **48** (1989) 183-185.

[71] R.Q. Scott, P. Roschger, B. Devaraj and H. Inaba, Monitoring a mammalian nuclear membrane phase transition by intrinsic ultraweak light emission. *FEBS Letters* **285** (1991) 97-98.

[72] L.L. Shlyakhtina and A.A. Gurwitsch, Radiations from the mouse liver at normal temperatures and on cooling. *Biophysics* **17** (1972) 1146-1150.

[73] J. Slawinski and F.A. Popp, Temperature hysteresis of low level luminescence from plants and its thermodynamical analysis. *J. Plant Physiol.* **130** (1987) 111-123.

[74] J. Slawinski, Luminescence research and its relation to ultraweak cell radiation. *Experientia* **44** (1988) 559-572.

[75] J. Slawinski, A. Ezzahir, M. Goldlewski, T. Kwiecinska, Z. Rajfur, D. Sitko and D. Wierzuchowska, Stress-induced photon emission from perturbed organisms. *Experientia* **48** (1992) 1041-1058.

[76] B.L. Strehler and W. Arnold, Light production by green plants. *J. gen. Physiol.* **34** (1951) 809-820.

[77] N. Suzuki and H. Inaba, Biophoton phenomena and active oxygen species. *Photomedicine and Photobiogy* **11** (1989) 15-26.

[78] W.B. Terzaghi, D.C. Fork, J.A. Berry and C.B. Field, Low and high temperature limits to PSII. A survey using trans-parinaric acid, delayed light emission, and F_0 chlorophyll fluorescence. *Plant Physiol.* **91** (1989) 1494-1500.

[79] R.N. Tilbury and T.I. Quickenden, Spectral and time dependence studies of the ultraweak bioluminescence emitted by the bacterium *E. coli. Photochem. Photobiol.* **47** (1988) 145-150.

[80] R.N. Tilbury, The effect of stress factors on the spontaneous photon emission from microorganisms. *Experientia* **48** (1992) 1030-1058.

[81] O. Tiphlova and T. Karu, Stimulation of *E. coli* division by low-intesity monochromatic visible light. *Photochem. Photobiol.* **48** (1988) 467-471.

[82] Y. Tsuchiya, E. Inuzuka, T. Kurono and M. Hosoda, Photon-counting image acquisition system and its applications. *J. Imaging Technology* **11** (1985) 215-220.

[83] G.K. Turner, Charpter 3: Measurement of light from chemical or biochemical reactions, in *Bioluminescence and chemiluminescence: Instruments and applications* **vol. I**, ed. K. van Dyke (CRC Press, Inc. Boca Raton, Florida, 1984) pp. 43-78.

[84] R.M. Tyrrell and S.M. Keyse, New trends in photobiology (invited review): The interaction of UVA radiation with cultured cells. *J. Photochem. Photonbiol. B: Biol.* **4** (1990) 349-361.

[85] R. van Wijk and D.H.J. Schamhart, Regulatory aspects of low intensity photon emission. *Experientia* **44** (1988) 586-594.

[86] R. van Wijk and J.M. van Aken, Photon emission in tumor biology. *Experientia* **48** (1992) 1092-1101.

[87] R. van Wijk and H. van Aken, Spontaneous and light-induced photon emission by rat hepatocytes and by hepatoma cells, in *Recent advances in biophoton research and its applications*, eds. F.A. Popp, K.H. Li and Q. Gu (World Scientific, Singapore, New Jersey, London, Hong Kong 1992) pp. 207-230.

[88] R. van Wijk, H. van Aken, W.P. Mei and F.A. Popp, Light-induced photon emission by mammalian cells. *J. Photonchem. Photobiol. B: Biol.* **18** (1993) 75-79.

[89] J.E. Wampler, Measurements and physical characteristics of luminescence, in *Bioluminescence in action*, ed. P.J. Herring (Academic Press, London, New York, San Francisco, 1978) pp. 1-50.

[90] F. Zölzer and J. Kiefer, Zelluläre Wirkungen der ultravioletten Komponente des Sonnenlichts. *Naturwissenschaften* **76** (1989) 489-495.

Chapter 12

Nonsubstantial Biocommunication in Terms of Dicke's Theory

Fritz-Albert Popp, Jiin-Ju Chang, Qiao Gu, and Mae-Wan Ho

12.1 Introduction

Bioluminescence in many species such as fireflies, dinoflagellates or deep-see fish is a well-known phenomenon[1]. At present, it is also generally accepted that some bioluminescent animals display synchronous flickering, as, for instance, the fireflies in the mangrove trees in Thailand[2,3]. Recently, some mathematical models have been developed in order to explain the phenomenon of synchronization[4–8]. However, there has been no attempt as yet to combine experimental and theoretical investigations in order to obtain a consistent model of this puzzling phenomenon, so that the results fit the theoretical predictions, and, *vice versa*, the theory should give predictions which can be subject to experimental investigations.

We started measurements on the bioluminescence of *Gonyaulax polyedra*, and after discovering the synchronous flashing in this species, we investigated the phenomenon under well-defined experimental conditions in order to examine whether a real "communication" between single cells takes place, or whether this phenomenon is due to a simple physical effect such as rescattering after light absorption.

Later on, we extended the same measurements and evaluations to other species of dinoflagellates, i.e. *Protoperidinium elegans*, *Pyrocystis lunula*, and mixtures of them, including mixtures with *G. polyedra*. In this paper, we compare the results with the density-dependent photon emission from *Daphnia magna* published elsewhere[9–11] and with the density-dependent "delayed luminescence" of tumor cells and normal cells[12–14]. We show how all these nonlinear optical phenomena can be understood in terms of Dicke's theory on coherence in spontaneous radiation processes[14], providing a new basis of the description of non-substantial biocommunication[14].

12.2 Materials and Methods

Gonyaulax polydra were cultivated as previously described[15]. For measurements, an analogical as well as a digital method have been used, i.e., a residual light amplifier (microchannel image intensifier tube AEG XX 1400) and a detector system of two photomultipliers working as photon counters, respectively.

In front of the intensifier tube with an incorporated microchannel plate and a S25 multialkali photocathode (sensitive in the range from 380 to about 900 nm) two quartz cuvettes (inside dimensions $10 \times 10 \times 26mm^3$, thickness of the glass 1mm) are centred in such a way that the distance between the front of the tube and the mid-point between the cuvettes is 5mm and a distance of 7.5mm is kept between the two cuvettes (Fig. 12.1). Both cuvettes are filled with *Gonyaulax polydra* originating from the same culture and divided into two equal parts of about 8000 cells/ml each. The luminescence of both samples in the cuvettes is then recorded by the intensifier tube with the same sensitivity. The signals of the tube are then amplified by a Proxitronic CCD-camera (type HLA UV, sensitivity 10^{-4} lux) and connected to a monitor where an image of photon emission can be registered within a measurement time interval of 50 ms. The measurements are performed in two versions, e.g. one, where the cuvettes are separated by a thin copper plate, covering the inner sides of the cuvettes such that the *Gonyaulax polyedra* cannot "see" each other, and a second arrangement, where this plate is removed (Fig. 12.1(a)). The signals are then recorded for both situations and compared with each other.

In a second series of measurements the same arrangement of *Gonyaulax polyedra* in the cuvettes with the same distance apart is placed in front of two photomultipliers in a dark chamber. However, instead of the chopper plate separating the two cuvettes, a solid wall with a shutter that can be opened or closed is used and thus the visual contact between the cuvettes can either be connected or disconnected (Fig. 12.1(b)). Each of the multipliers (EMI 9558 QA, selected type) has a cathode of 44 mm diameter and is sensitive within the range of 200 to 800 nm. The incident photons induce electric pulses, which, after travelling through an amplifier, arrive at a discriminator. This device provides a logical pulse whenever the incoming signal exceeds an adjustable threshold. The equipment thus functions as a photon counter with a quantum efficiency determined by the cathode. In order to reduce the background current of the cathode, the photomultiplier is cooled down to a temperature of -30°C, where no considerable further decrease of the dark count rate can be achieved by further cooling.

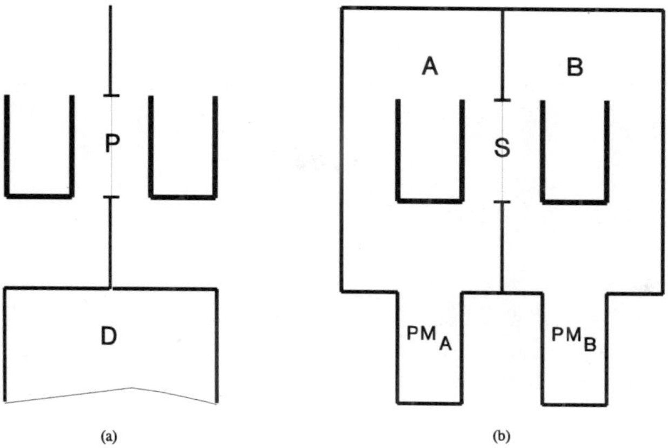

Figure 12.1: *The equipments for biocommunication experiments. These are either a residual light amplifier D or a dark chamber with two photomultipliers PM_A or PM_B. In every case, two quartz cuvettes with the samples are placed in front of the detectors, where the samples can be separated by a plate (a) or by a shutter (b). The measurements are performed in such a way that after registering the photon intensity for some time in separated samples the plate is removed (a) or the shutter is opened (b) and the measurement continues in the same way. This alternating procedures of separated and connected samples are repeated several times. The difference between the results of separated and connected samples is taken as a criterion for the analysis of biocommunication.*

The sensitivity of this equipment reaches about 10^{-17}W, where the baseline fluctuation is less than 3%/day. We can detect a current density of 5 photons/(s·cm^2) at a significance level of 99.9% within 5 hours. A more detailed description of the basic technical elements are given elsewhere[16,17].

A group of 10 successive measurements of 1024 values in each series is carried out. The shutter is closed and opened consecutively in a sequence of measurements, starting without contact, then changing the state of the shutter after 1024 values in each series. The alternation is controlled by an external electric signal in such a way that apart from the switching there is no other change or influence within the dark chamber. A series of ten measurements are performed with a preset time interval Δt of 1s for every measurement value, followed by a series with $\Delta t = 0.1$ s.

After finishing the group of measurements at a distance of 7.5 mm between the cuvettes, a further group of the same measurements is performed, where the cuvettes are placed at a distance of either 7.5 or 15mm from each other. The schedule of these measurements is displayed in Table 12.1, while Table 12.2 shows a schedule of the measurements made by including two other species *P. elegans* and *P. lunula*.

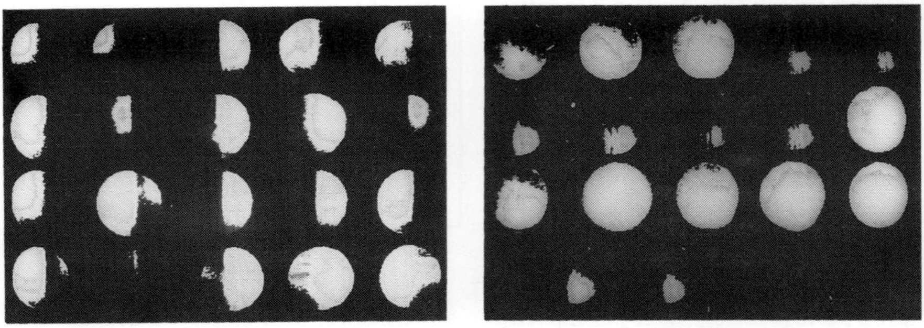

a. b.

Figure 12.2: *Biocommunication experiments with residual light amplifier. (a) When the plate is inserted between the samples of Gonyaulax polyedra, the light flashes of the two separated samples are confined either to the left or to the right side of the residual light amplifier. A series of such flashes is registered in sequence starting from the upper left and ending at the lower right. (b) When the plate is removed, the flashes on both sides appear mainly synchronous, as, the samples "become aware" of each other.*

12.3 Results

Figs. 12.2(a) and 12.2(b) show typical cases of photon emission registered by the intensifier tube when the cuvettes are separated by the copper plate (Fig. 12.2(a)) and when they are in contact (Fig. 12.2(b)). It is clear that the independent pulses in separated cuvettes become highly synchronized when the *Gonyaulax polydra* of the two cuvettes can 'see' each another.

In order to examine this effect quantitatively, the same measurements are performed with multipliers, as described above. Figs. 12.3(a,b) and 12.4(a,b) display typical cases of photon emission $i_A(t)$ and $i_B(t)$ from *Gonyaulax polyedra* of the two channels A and B without and with contact for both $\Delta t = 0.1$s and $\Delta t = 1$s.

Table 12.1: Schedule of Measurements

measurement No.	measurement-time interval Δt (s)	number of measurement-values	shutter closed/ open	distance d (mm)
1	1	1024	closed	7.5
2	1	1024	open	7.5
.	.	.	.	.
.	.	.	.	.
.	.	.	.	.
9	1	1024	closed	7.5
10	1	1024	open	7.5
11	0.1	1024	closed	7.5
12	0.1	1024	open	7.5
.	.	.	.	.
.	.	.	.	.
.	.	.	.	.
19	0.1	1024	closed	7.5
20	0.1	1024	open	7.5
21	1	1024	closed	15
22	1	1024	open	15
.	.	.	.	.
.	.	.	.	.
.	.	.	.	.
39	0.1	1024	closed	15
40	0.1	1024	open	15

Table 12.2: Further Measurements on *G.polyedra*, *P.elegans* and *P.lunula*

species	distances (mm)	time intervals (s)	number of measurements (à 1024)
G.polyedra/empty cuvette	7.5	0.1; 1	8
P.lunula/P.lunula	7.5; 15	0.1; 1	32
P.elegans/P.elegans	7.5; 15	0.1; 1	32
G.polyedra/P.lunula	7.5; 15	0.1; 1	32
G.polyedra/P.elegans	7.5; 15	0.1; 1	32
P.elegans/P.lunula	7.5; 15	0.1; 1	32
P.elegans/P.lunula	7.5	0.1; 1	16
P.lunula/P.elegans	7.5	0.1; 1	16
P.elegans/empty cuvette	7.5	0.1; 1	14
P.lunula/empty cuvette	7.5	0.1; 1	14
*G.polyedra**		0.1; 1	8
*G.polyedra**		0.1; 1	8
G.polyedra/G.polyedra	7.5	0.1; 1	16

* only one cuvette in the middle of the shutter

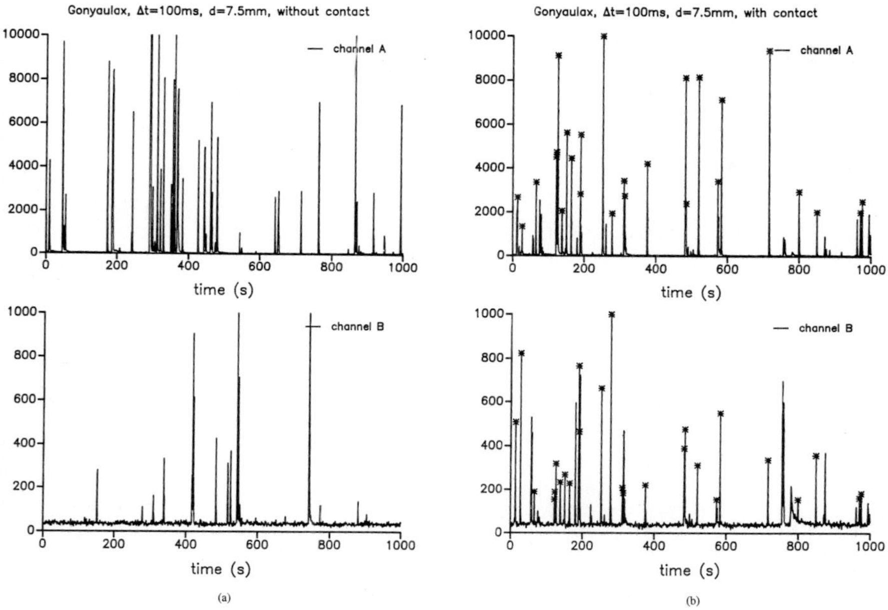

Figure 12.3: *Biocommunication experiments with photomultipliers. The preset time interval is 100ms. (a) When the shutter is closed, the flashes which are registered by the photomultiplier* PM_A *are completely uncorrelated to those of* PM_B*. (b) After opening the shutter, the samples "see" each other. As a consequence, the flashes of "channel A" (registered by* PM_A*) and "channel B" (registered by* PM_B*) become strongly synchronous.*

Table 12.3: Experimental Results of the Measurements according to Table 12.1

	time interval (s)	without contact (closed shutter)=1 with contact (open shutter)=2	distance of cuvettes in mm	mean value of i_A	standard deviation of σ_A	mean value of i_B	standard deviation of σ_B
1	1.0	1	7.5	1683.68	5545.95	289.67	251.02
2	1.0	2	7.5	1251.25	3699.33	306.23	265.40
3	1.0	1	7.5	2257.27	7277.08	243.09	203.45
4	1.0	2	7.5	1915.05	6745.67	310.01	278.09
5	1.0	1	7.5	3505.68	9702.68	233.55	254.22
6	1.0	2	7.5	1947.42	6451.35	302.17	255.07
7	1.0	1	7.5	1966.59	7006.44	242.92	165.21
8	1.0	2	7.5	2201.85	6605.17	336.75	306.70
9	1.0	1	7.5	2031.93	7477.55	279.29	217.63
10	1.0	2	7.5	1697.58	5734.09	350.64	258.27
11	0.1	1	7.5	87.58	436.83	39.36	81.71
12	0.1	2	7.5	220.31	1015.08	48.07	76.19
13	0.1	1	7.5	254.80	1247.28	41.01	73.57
14	0.1	2	7.5	241.56	916.97	51.73	102.76
15	0.1	1	7.5	368.56	1554.22	46.75	105.74
16	0.1	2	7.5	214.89	905.14	57.46	86.32
17	0.1	1	7.5	185.29	1146.41	63.62	155.64
18	0.1	2	7.5	230.97	1082.85	55.13	102.29
19	0.1	1	7.5	287.23	1843.4	43.49	93.45
20	0.1	2	7.5	183.95	832.55	55.56	96.38
21	1.0	1	15	3385.09	9888.25	511.79	403.45
22	1.0	2	15	3844.01	10723.09	625.82	502.98
23	1.0	1	15	3636.70	10016.41	499.04	379.49
24	1.0	2	15	3303.02	9181.63	436.57	430.92
25	1.0	1	15	4844.24	20365.59	403.82	2471.93
26	1.0	2	15	3472.99	9284.83	330.02	457.34
27	1.0	1	15	2916.40	9379.55	219.88	304.46
28	1.0	2	15	2443.12	7319.49	268.05	362.03
29	1.0	1	15	2651.28	7879.52	194.92	259.25
30	1.0	2	15	2941.06	8711.61	297.57	432.54
31	0.1	1	15	265.87	1592.75	17.23	4.68
32	0.1	2	15	193.98	1298.29	36.26	156.48
33	0.1	1	15	273.69	1647.14	17.26	4.18
34	0.1	2	15	397.40	2211.97	29.62	62.61
35	0.1	1	15	251.16	1648.87	20.33	44.81
36	0.1	2	15	352.35	2403.83	22.12	31.44
37	0.1	1	15	170.77	1154.09	21.56	62.98
38	0.1	2	15	251.40	1399.13	26.78	82.29
39	0.1	1	15	463.23	3099.30	23.92	94.55
40	0.1	2	15	299.40	1775.93	25.04	47.64

	mean value of $\left(\frac{i_A}{i_B}\right)$	standard deviation $\sigma\left(\frac{i_A}{i_B}\right)$	mean value of $\left(\frac{i_B}{i_A}\right)$	standard deviation $\sigma\left(\frac{i_B}{i_A}\right)$	mean value of $(i_A \cdot i_B)$	mean deviation of $\sigma(i_A \cdot i_B)$
1	6.33	20.93	0.75	0.75	529236.27	2621690.99
2	2.79	4.06	0.54	0.16	894374.40	4732816.95
3	10.21	33.45	0.47	0.40	603777.32	3478490.85
4	3.47	5.92	0.48	0.16	1764220.07	9476202.98
5	16.03	44.24	0.43	0.49	778148.69	2153259.24
6	3.65	5.47	0.45	0.16	1795602.31	10326624.73
7	8.72	31.3	0.51	0.42	452381.11	1583778.59
8	3.79	5.69	0.45	0.16	2117634.72	10395686.99
9	7.67	27.85	0.50	0.45	710146.18	5534875.81
10	3.06	4.83	0.48	0.13	1507512.91	8247696.39
11	3.11	16.12	1.21	2.58	3044.72	14742.08
12	2.17	4.18	0.95	0.46	52384.72	371046.78
13	7.76	40.25	1.55	3.25	8921.97	41792.15
14	2.53	4.82	0.86	0.45	56651.76	399840.31
15	11.38	50.94	1.43	3.33	14635.14	83481.81
16	1.96	3.86	1.06	0.47	56346.47	405837.38
17	4.7	31.18	1.94	5.10	9207.46	54478.29
18	2.15	4.88	1.02	0.44	60442.47	500958.70
19	7.91	45.76	1.14	1.98	11310.56	76826.47
20	1.64	3.28	1.04	0.40	50713.13	329978.78
21	6.99	20.47	0.50	0.46	1707034.61	5037408.94
22	4.47	7.45	0.40	0.15	4417587.97	19581489.83
23	7.85	21.6	0.39	0.35	1880773.80	6428251.67
24	6.29	8.72	0.27	0.14	2831487.44	12284410.93
25	20.17	55.29	0.20	0.35	44388107.70	1071151936.70
26	7.98	8.89	0.18	0.11	2897867.76	12999350.99
27	16.14	53.38	0.20	0.28	728094.60	4733545.55
28	6.68	7.54	0.20	0.09	1739865.32	8985364.94
29	15.67	47.27	0.20	0.30	473462.76	1354891.16
30	7.16	9.25	0.22	0.11	2243113.95	10157719.82
31	16.8	103.44	0.21	0.08	4564.85	27986.94
32	4.88	3.64	0.25	0.14	62047.08	701878.31
33	16.68	100.85	0.21	0.07	4704.78	28933.96
34	6.87	7.19	0.19	0.08	98537.40	941800.92
35	16.85	118.74	0.24	0.50	4344.48	25583.40
36	6.38	8.9	0.22	0.07	81835.10	930833.12
37	11.36	87.02	0.27	0.79	3086.37	17825.56
38	5.64	6.38	0.23	0.08	50152.66	449473.52
39	27.56	180.12	0.29	1.33	8724.10	58632.97
40	5.97	5.87	0.21	0.08	65872.78	709446.29

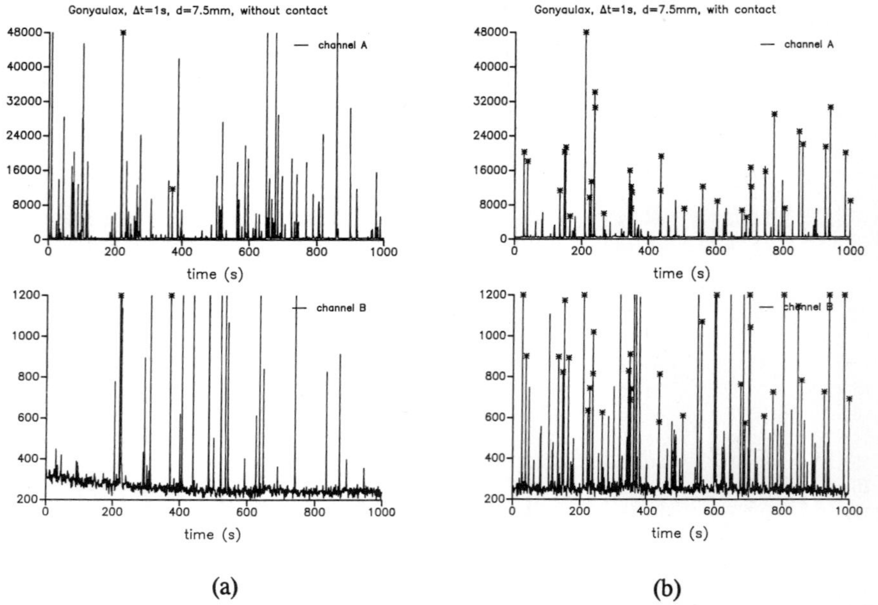

Figure 12.4: *Biocommunication experiments with photomultipliers. The same measurements as in Fig. 3 with a preset measuring time interval of 1s (a) shutter closed, (b) shutter open.*

Table 12.3 summarizes all the parameters and calculations of the measurements. The single most highly significant and consistent result is the considerable decrease of the standard deviation of both the ratios, $\left(\frac{i_A}{i_B}\right)$ and $\left(\frac{i_B}{i_A}\right)$ when the samples in the two cuvettes are switched into contact (see Figs. 12.5(a) and 12.5(b)). We shall concentrate on this point below.

12.4 Evidence of Communication

Before we discuss the implications of the highly significant decrease in $\sigma\left(\frac{i_A}{i_B}\right)$ and $\sigma\left(\frac{i_B}{i_A}\right)$ when contact is switched on between previously separated samples, we shall show that the result cannot be explained in terms of a linear additive effect. A linear additive effect is obtained when there is no essential communication between the samples, and it would necessarily follow a relation of the form:

$$\bar{i}_{A'} = X_A i_A + Y_{AB} i_B \tag{12.1}$$

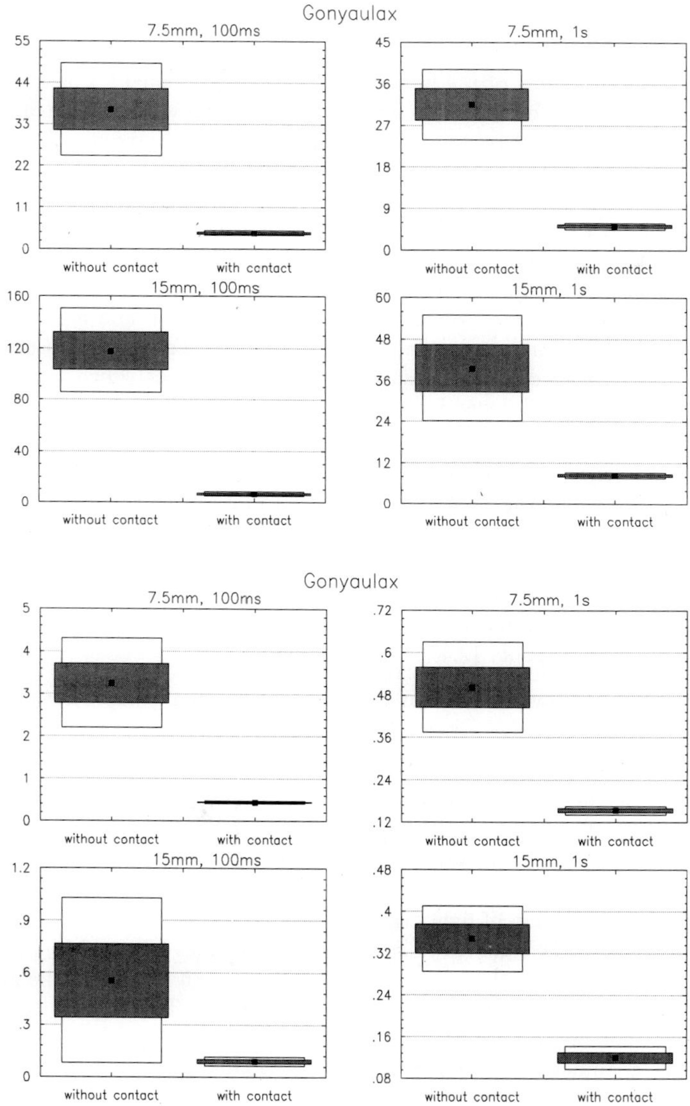

Figure 12.5: *The decrease of the variance of the ratio* $\left(\frac{i_A}{i_B}\right)$ *as well as of* $\left(\frac{i_B}{i_A}\right)$ *when the shutter is open. This holds for both the measurement time intervals (100 ms, left side of the figures and 1 s, right side of the figures) and for both a distance of 7.5 mm (upper figures) and 15 mm (lower figures) between the cuvettes.*

$$\bar{\sigma}_{A'}^2 = X_A^2\sigma_A^2 + Y_{AB}^2\sigma_B^2 , \tag{12.2}$$

where $\bar{i}$ is the mean value of the intensity of a definite channel (A or B). Thus, $\bar{i}_{A'}$ is that of channel A with contact, i_A and i_B are the mean values of the intensities of channels A and B without contact, respectively. $\bar{\sigma}^2$ are the corresponding variances. X and Y are constant values assigned to the linear scattering processes of photons within the same chamber or between two different chambers, respectively. Both X and Y depend on the geometry of the system under consideration: X accounts for the reflection of light on the closed-shutter wall and Y includes the travelling of light from the one side of the chamber through the open shutter to the other side. Both coefficients include the effects of rescattering and absorption and also the different sensitivities of the multipliers in the channels A and B. From the classical point of view it is necessary to describe X and Y as constant values which are inherent in the system but not at all dependent on the intensity fluctuations of the light itself, if the results are based on sufficiently high numbers of measurements.

On the basic of linear additive effect described by Eqs. (12.1) and (12.2), we expect that X and Y should take positive definite values, where X cannot exceed the value 1, as opening the shutter would not lead to an increase in the intensity of the biological system itself. Since, before opening the shutter, the r-th fraction of the radiation could have been reflected by the shutter wall, it is possible that $X = 1 - r$ is smaller than 1. The values of X and Y are obtained from the results of Table 12.3, and are listed in Table 12.4.

Negative values of X have been eliminated so that only the possible relevant solutions of Eqs. (12.1) and (12.2) have been taken into account. The resulting mean values of X and Y are listed in Table 12.4b (see also Fig. 12.6).

They show that there is only a small reflectance of the shutter walls, of at most 10% of the intensities ($X_A = X_B = 0.9 = 1 - r$), where r accounts for the reflected fraction on the closed shutter wall). The value of Y_{AB} is then due to the transition of light from chamber B to chamber A, if the shutter is open, and Y_{BA} concerns the corresponding fraction of light travelling from chamber A to B also in case that the shutter is open. The significant difference of the values Y_{AB} and Y_{BA} reflects the asymmetric geometry of the positions of the cuvettes with respect to the multipliers as well as the different sensitivities of the multipliers. The higher value of Y_{AB} compared to that of Y_{BA} corresponds to the higher mean values of the measured intensities in chamber A compared to those in chamber B. The values show that at most 10% of the effects can be assigned to the physical fact that after opening the shutter, light may travel from one chamber to the other one through the open shutter. Consequently, this effect, by itself cannot be responsible for the strong increase of synchronous flickering of the *G. polyedra*, when one switches from closed to open shutter. However, by investigating the significant differences of the variances of the ratios $\left(\frac{i_A}{i_B}\right)$ and $\left(\frac{i_B}{i_A}\right)$ with and without contact, respectively, one can use Eqs.

Table 12.4: A) Evaluation of X and Y according to Eqs. (12.1) and (12.2)

	δT	d	X_A^*	Y_{AB}^*	X_B^*	Y_{BA}^*	D_A	D_B
1	1.0	7.5	0.667	0.444	0.920	0.024	1.3	0.1
2	1.0	7.5	0.927	-0.728	1.047	0.025	1.4	0.3
3	1.0	7.5	0.664	-1.622	1.120	0.012	0.9	-0.4
4	1.0	7.5	0.942	1.437	1.101	0.035	2.1	1.0
5	1.0	7.5	0.767	0.500	1.202	0.007	1.6	-0.1
6	0.1	7.5	2.322	0.430	1.041	0.081	0.2	-0.1
7	0.1	7.5	0.731	1.349	0.857	0.065	0.3	0.1
8	0.1	7.5	0.582	0.005	0.955	0.035	0.2	-0.2
9	0.1	7.5	0.937	0.903	0.749	0.040	0.2	-0.1
10	0.1	7.5	0.447	1.278	1.149	0.019	0.2	-0.1
11	1.0	15	1.084	0.339	1.035	0.028	1.6	0.2
12	1.0	15	0.917	-0.061	0.610	0.036	1.7	0.8
13	1.0	15	0.487	2.756	0.262	0.046	-0.1	-0.4
14	1.0	15	0.780	0.766	0.870	0.026	0.5	0.1
15	1.0	15	1.106	0.050	0.897	0.046	0.6	0.2
16	0.1	15	0.815	-1.319	0.589	0.098	13.6	13.6
17	0.1	15	1.343	1.730	1.115	0.038	17.1	16.9
18	0.1	15	1.458	-0.678	0.978	0.009	0.2	-0.3
19	0.1	15	1.207	2.101	0.793	0.057	0.1	0.0
20	0.1	15	0.571	1.453	0.776	0.014	0.1	-0.2

$$D_A = \left(\frac{i_A}{\sigma_A}\right)^2 + \left(\frac{i_B}{\sigma_B}\right)^2 - \left(\frac{i'_A}{\sigma'_A}\right)^2$$

$$D_B = \left(\frac{i_A}{\sigma_A}\right)^2 + \left(\frac{i_B}{\sigma_B}\right)^2 - \left(\frac{i'_B}{\sigma'_B}\right)^2$$

Table 12.4: B)

$\bar{X}_A$	$\bar{X}_B$	$\bar{Y}_{AB}$	$\bar{Y}_{BA}$
0.94 ± 0.1	0.9 ± 0.05	0.56 ± 0.25	0.04 ± 0.02

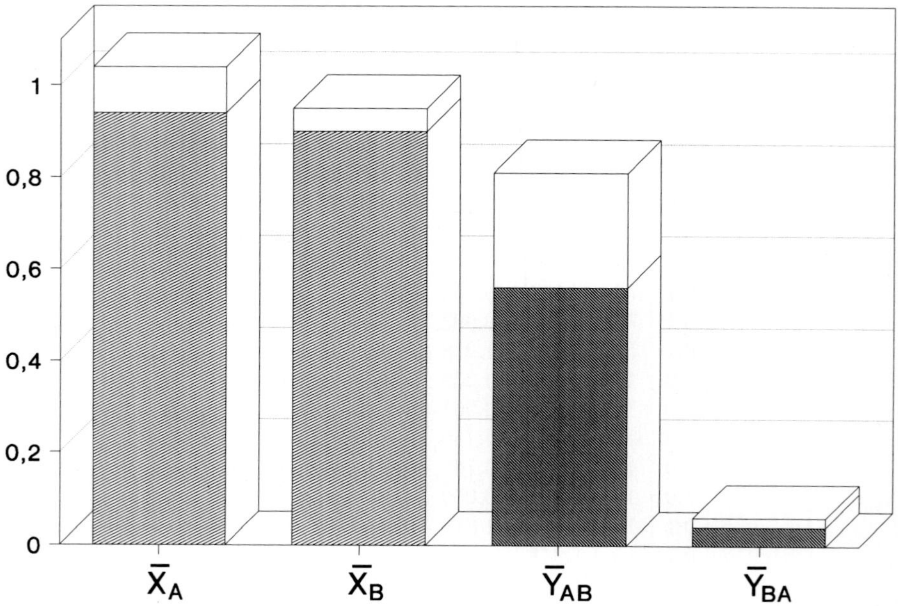

Figure 12.6: *The calculated linearized coupling coefficients between the chambers. The rather high value Y_{AB} compared to the low value Y_{BA} implies that the flashes from chamber B may be registered directly in chamber A while (in view of the asymmetry of the arrangement) the flashes of chamber A do not penetrate directly to the photomultipler* PM_B

(12.1) and (12.2) again in order to calculate the expected (theoretical) magnitudes of these variances, which can then be compared with the actual observed values.

Straightforward calculation leads to the following relations, if Eqs. (12.1) and (12.2) hold,

$$Q_{(T)AB} = \frac{\sigma\left(\frac{\dot{i}_A}{\dot{i}_B}\right)'}{\sigma\left(\frac{\dot{i}_A}{\dot{i}_B}\right)} = |X_A X_B - Y_{AB} Y_{BA}| \left(\frac{\overline{\dot{i}_B}}{\overline{\dot{i}_{B'}}}\right)^2 \tag{12.3}$$

$$Q_{(T)BA} = \frac{\sigma\left(\frac{\dot{i}_B}{\dot{i}_A}\right)'}{\sigma\left(\frac{\dot{i}_B}{\dot{i}_A}\right)} = |X_A X_B - Y_{AB} Y_{BA}| \left(\frac{\overline{\dot{i}_A}}{\overline{\dot{i}_{A'}}}\right)^2 \tag{12.4}$$

Table 12.4a displays all the relevant values which enable us to examine the validity of relations (12.3,12.4). The statistical analysis presented in Table 12.4b (see also Fig. 12.7) shows, however that there is a significant difference between the l.h.s and the r.h.s. of Eqs. (12.3, 12.4) i.e., between the theoretical expected values and the observed values, rejecting decisively the hypothesis that a linear additive effect suffices to explain our observations. Rather, some biological response (or coherent nonlinear interaction) has to be taken into account, in order to understand the significantly lower values of the standard deviation of the ratios $\left(\frac{\dot{i}_A}{\dot{i}_B}\right)$ and $\left(\frac{\dot{i}_B}{\dot{i}_A}\right)$ when contact is present compared to when it is not.

It should be noted that the results suggest that the systems become unified as soon as they "see each other", such a unified system will display correlation of parts so that the ratio of the intensities of the parts will remain constant, i.e.,

$$\frac{\dot{i}_A}{\dot{i}_B} = \text{const.}\,, \tag{12.5}$$

which results in $\sigma\left(\frac{\dot{i}_A}{\dot{i}_B}\right) \to 0$.

Table 12.6 summarize the results in all the other experiments (see Table 12.2). It shows clearly that the nonlinear response suggesting communication depends on
(1) the measurement time interval,
(2) the species under investigation, and
(3) the distance between the cuvettes.

These will be the subject for future investigations.

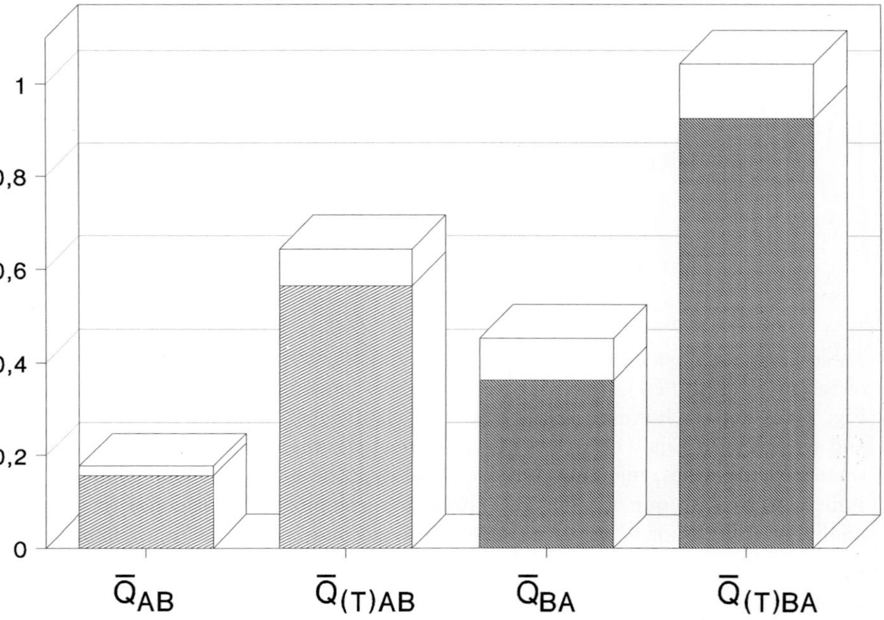

Figure 12.7: *The actual and theoretical values of the ratio of variances. Provided that the linear coupling coefficients are responsible for the decrease of the variances* $\left(\frac{i_A}{i_B}\right)$ *and* $\left(\frac{i_B}{i_A}\right)$, *the theoretical values* Q_{T} *of the ratios of the variances without contact and with contact are significantly almost one order of magnitude higher than the actually measured values for both* $\left(\frac{i_A}{i_B}\right)$ *and* $\left(\frac{i_B}{i_A}\right)$. *This indicates that the effect of synchronous flashing cannot be explained by linear optics.*

Table 12.5: A)

	δT	d	Q_{AB}: $\frac{\sigma\left(\frac{i_A}{i_B}\right)'}{\sigma\left(\frac{i_A}{i_B}\right)}$	Q_{BA}: $\frac{\sigma\left(\frac{i_B}{i_A}\right)'}{\sigma\left(\frac{i_B}{i_A}\right)}$	S_{AA}: $\frac{\left(\overline{i_A}\right)^2}{\left(\overline{i'_A}\right)^2}$	S_{BB}: $\frac{\left(\overline{i_B}\right)^2}{\left(\overline{i'_B}\right)^2}$	XY: $X_A X_B - Y_{AB} Y_{BA}$	$Q_{(T)AB}$: $\lvert XY \rvert \cdot S_{BB}$	$Q_{(T)BA}$: $\lvert XY \rvert \cdot S_{AA}$
1	1.0	7.5	0.19	0.21	1.81	0.89	0.60	0.54	1.09
2	1.0	7.5	0.18	0.40	1.39	0.61	0.99	0.61	1.37
3	1.0	7.5	0.12	0.33	3.24	0.60	0.76	0.46	2.47
4	1.0	7.5	0.18	0.38	0.80	0.52	0.99	0.51	0.79
5	1.0	7.5	0.17	0.29	1.43	0.63	0.92	0.58	1.32
6	0.1	7.5	0.26	0.18	0.16	0.67	2.38	1.60	0.38
7	0.1	7.5	0.12	0.14	1.11	0.63	0.54	0.34	0.60
8	0.1	7.5	0.08	0.14	2.94	0.66	0.56	0.37	1.63
9	0.1	7.5	0.16	0.09	0.64	1.33	0.67	0.89	0.43
10	0.1	7.5	0.07	0.20	2.44	0.61	0.49	0.30	1.19
11	1.0	15	0.36	0.33	0.78	0.67	1.11	0.74	0.86
12	1.0	15	0.40	0.40	1.21	1.31	0.56	0.73	0.68
13	1.0	15	0.16	0.31	1.95	1.50	0.00	0.00	0.00
14	1.0	15	0.14	0.32	1.42	0.67	0.66	0.44	0.94
15	1.0	15	0.20	0.37	0.81	0.43	0.99	0.42	0.80
16	0.1	15	0.04	1.75	1.88	0.23	0.61	0.14	1.14
17	0.1	15	0.07	1.14	0.47	0.34	1.43	0.49	0.68
18	0.1	15	0.07	0.14	0.51	0.84	1.43	1.21	0.73
19	0.1	15	0.07	0.10	0.46	0.65	0.84	0.54	0.39
20	0.1	15	0.03	0.06	2.39	0.91	0.42	0.39	1.01

Table 12.5: B)

Table 12.5: B)

$\bar{Q}_{AB}$	$\bar{Q}_{(T)AB}$	$\bar{Q}_{BA}$	$\bar{Q}_{(T)BA}$	p-level of T-test and F-test
0.154 ± 0.022	0.565 ± 0.079	0.364 ± 0.089	0.925 ± 0.119	< 0.01

Table 12.6: Q_{AB}-,$Q_{(T)AB}$-, Q_{BA}- and $Q_{(T)BA}$-values for the different measurements

species	distance (mm)	time-interval (s)	$Q_{AB}/Q_{(T)AB}$	p	$Q_{BA}/Q_{(T)BA}$	p
G.polyedra	7.5	1	0.17/0.54	0.00	0.32/1.41	0.00
	15	1	0.25/0.47	0.11	0.34/0.66	0.00
	7.5	0.1	0.14/0.70	0.00	0.15/0.85	0.00
	15	0.1	0.06/0.55	0.00	0.64/0.79	0.10
P.lunula	7.5	1	0.31/0.81	0.04	0.04/0.49	0.00
	15	1	0.36/1.16	0.01	0.63/0.67	0.56
	7.5	0.1	1.28/12.8	0.00	0.96/5.22..	0.00
	15	0.1	0.76/1.01	0.05	0.78/0.80	0.36
P.elegans	7.5	1	0.007/0.37	0.00	0.003/0.36	0.00
	15	1	0.058/0.38..	0.05	0.027/0.15..	0.00
	7.5	0.1	0.253/2.59	0.00	0.692/0.75	0.39
	15	0.1	1.488/0.52	0.60	0.713/0.67	0.39
P.elegans/ P.lunula	7.5	1	0.027/0.48	0.00	0.017/0.08	0.02
	15	1	0.032/1.68..	0.00	0.023/0.25	0.03
	7.5	0.1	1.036/7.21	0.01	0.637/1.64..	0.01
	15	0.1	0.044/0.15	0.03	0.136/0.12	0.85
G.polyedra/ P.elegans	7.5	1	0.012/0.32	0.00	0.007/0.44	0.00
	15	1	0.008/0.31	0.00	0.007/0.21	0.00
	7.5	0.1	0.067/0.06	0.43	0.076/0.81..	0.00
	15	0.1	0.451/9.74	0.00	0.349/0.29..	0.87
G.polyedra/ P.lunula	7.5	1	0.013/0.12	0.00	0.019/0.20	0.00
	15	1	0.023/0.24	0.00	0.023/0.32	0.00
	7.5	0.1	0.039/0.25	0.17	0.148/2.27	0.01
	15	0.1	0.303/0.72	0.05	0.057/0.19	0.01

Direct evidence of nonsubstantial communication of this system has been shown recently by J.J. Chang *et.al.*[18]. They used a yellow filter between the flickering samples and two blue filters between the samples and the multipliers. The degree of synchronisation increased rather than disappeared. Consequently, a contribution of scattered light between the channels A and B for this kind of communication is truthfully excluded.

12.5 Discussion

In order to understand the mechanism behind this kind of biocommunication, we compare the results of dinoflagellates reported here with measurements of the pho-

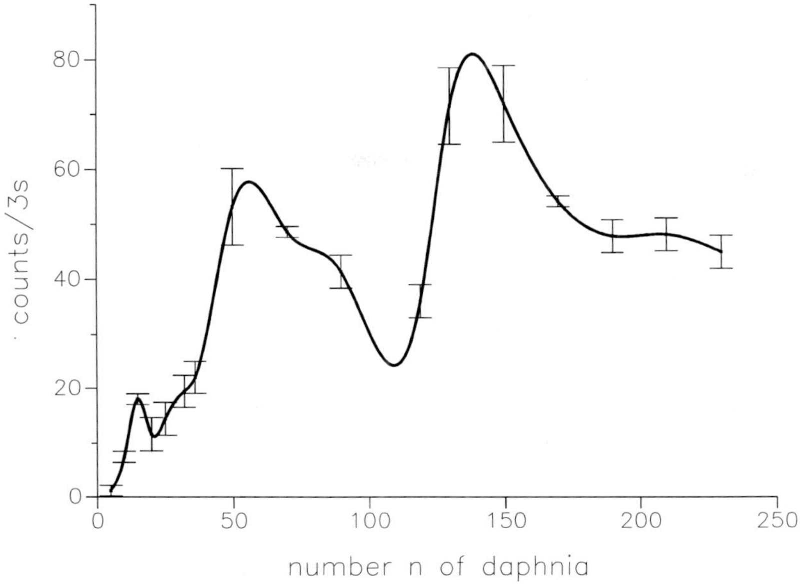

Figure 12.8: *Photon-emission as a function of number of daphnia. The photon intensity of daphnia displays an "interference-like" dependence on the number of daphnia (see Galle 1991, 1993), as indicative of nonsubstantial interaction between the individuals (see text).*

tonemission reported elsewhere from *Daphnia magna*[9–11] and cell cultures[12,13] (see Fig. 12.8 and 12.9).

The common denominator of all these results is a nonlinear increase of the photon emission as soon as the system undergoes a perturbation that can be recognized by the whole population cells. On the other hand, the photon intensity takes on a minimum value when the coherence of the photon field reaches the highest degree[19]. We suggest a common explanation for all these experimental results in terms of destructive interference according to Dicke's theory[14]. The principle is displayed in Fig. 12.10. The advantage of this kind of communication is discussed below.

The mechanism of approaching destructive interference can work as an identification or recognition signal. Every biological system will display a complex wave pattern which is most probably species-specific and which may extend into nonoptical regions of the electromagnetic spectrum. The capacity for mutual destruction of the emitted wave patterns is then a sufficient condition for belonging to the same species. A perfect destruction between the two systems implies that they are identical with each other, with higher or lower deviations from perfect destruction indicating that the systems are more or less different. The deviations may work as modulations

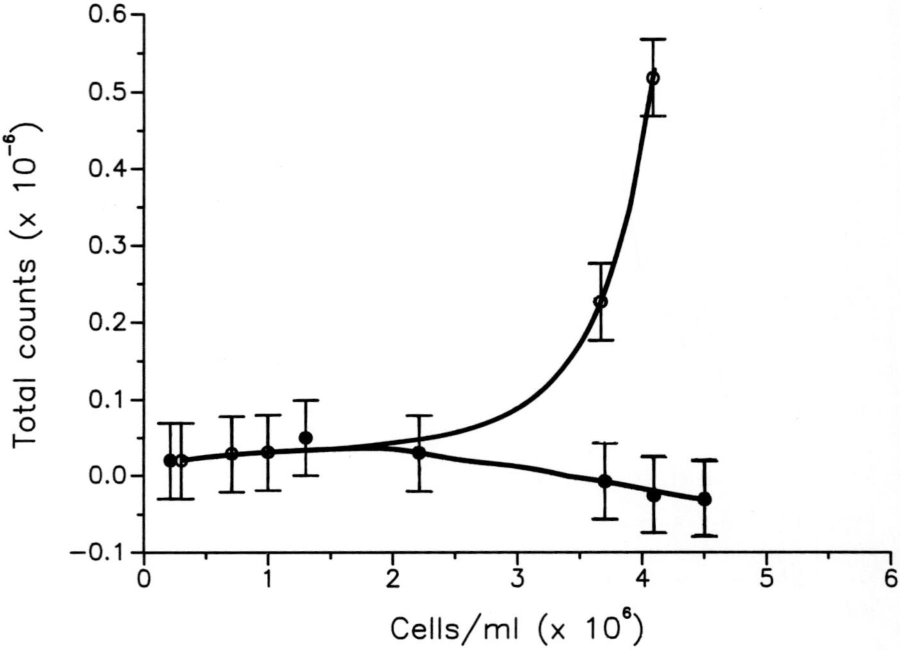

Figure 12.9: *Delayed luminescence as a function of cell number. The "delayed luminescence" of cells shows qualitatively different density dependence for normal cells (lower curve) compared to tumour cells (upper curve) (see Schamhart and van Wijk, 1987; Scholz et al., 1988). This behaviour indicates different intercellular interactions between the cells (see Popp, 1992).*

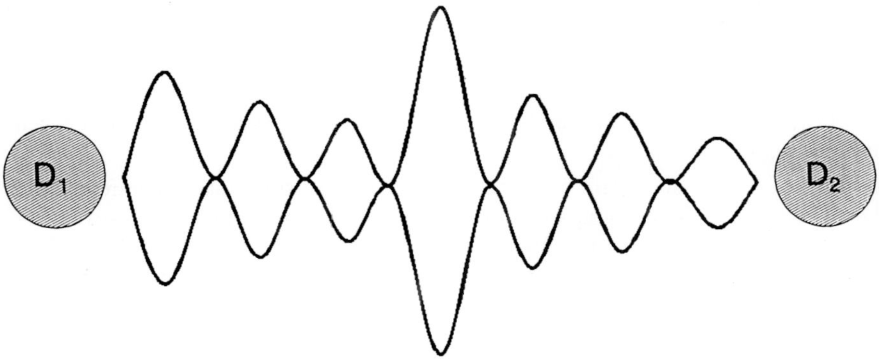

Figure 12.10: *Nonsubstantial biocommunication as "destructive interference" according to Dicke's theory. This means that organisms send wave-like electromagnetic patterns which others, in response, try to cancel by destructive interference. This leads to a miminum of free coherent electromagnetic energy between the biological systems.*

transferring actual information.

Another way to describe this phenomenon is in terms of mutual absorption of coherently emitted photons, a reciprocal effect which minimizes the intensities observed externally. This reciprocal, mutual compensation occuring in parts of systems that are working together as a coherent whole is emerging as a general principle of biological organization. One of the clearest examples is in the phenomenon on free energy compensation between enzyme and their specific substrates in the process of catalysis, which minimizes the free energy changes involved[20]. There is also evidence that enzyme and substrate may specifically attract each other at a distance by means of "coherent excitations"[21] of specific frequencies.

As the zone between the communicating systems remains completely field-free in the ideal case, an infinitesimally small perturbation can work as a signal of theoretically infinite signal/noise ratio.

In the ideal case of destructive interference, the 'nonsubstantial' communication between the individual living units of the whole organism cannot be observed from the outside, as it involves the disappearance of the field! This means that only an individual of the same species, which can engage in specific destructive interference, can receive the complete information.

The range of information is unlimited as it increases with the coherence volume of the electromagnetic waves which provide the physical basis of this kind of communication. In practice, this means that entive populations or societies can be engaged in specific information transfer[21].

The velocity of this communication by means of destructive interference is based on the velocity of light. While the phase velocity determines the awareness of communication due to the dependence of the interference pattern on the phase, the group velocity remains the responsible factor for the transfer of actual information via this communication channel.

Destructive interference provides an elementary organization principle of living matter, irrespective of whether cells, organs, organisms or complicated living beings are concerned. Of course, the zone which allows destructive interference of the species-specific field pattern depends on the organization of the living units, just as the organization of the units is in turn dependent on the geometry of the zones where destructive interference can be achieved. Destructive interference provides at the same time an attractive force between the communicating systems. Consequently, this principle is crucial for processes of growth and differentiation in the development of organisms.

Destructive interference also provides a basis of evolutionary development, where, as an optimization-principle, the f_ν=constant-rule may play the role of extending the coherent-state-formation towards longer and longer wavelengths. Living matter governed by this rule may provide a subradiant state in which photons are trapped, resulting in a destructive interference extending over the entire spectrum.

Some of the consequences of communication by destructive interference can be illustrated by analysis of the experiment results in dinoflagellates, daphnia and cell cultures.

The synchronous flickering of dinoflagellates can be described in terms of the breakdown of destructive interference as the result of a perturbation. Actually, synchronous flickering can also be obtained when nonspecific perturbations of separated systems occur, for example, mechanical vibrations, change of the pH-value etc. On the other hand, contact between samples has the tendency to reduce the average value of photon intensity, thus indicating that the degree of destructive interference between them is increased.

This interpretation is in line with the results on "delayed luminescence" of normal cells and tumor cells. While normal tissue show an increasing degree of optical coherence with increasing cell density, thus giving rise to the decrease of observed photon intensity, the opposite is the case in tumor tissues. Since the coherence of the photon field of tumor cells breaks down with increasing cell density, in other words, the capacity of photon storage becomes diminished, the photon emission increases.

The experiment in daphnia shows the sensitive dependence of destructive interference on the distance between (almost identical) animals. During the minima of photon emission, the population is beginning to build up a 'super-organism', such as that which occurs in group formation. During the maxima of photon emission, by contrast, destructive interference breaks down. The consequences are higher variances and stronger fluctuations of photon emission as well as an increased indisposition of the animals.

A theoretical work on destructive interference, based on Dicke's theory, is in preparation, whereby we can expect to explain some of the phenomena of biological coherence, including the hypothesis presented here, that strong bioluminescence may be triggered by an ultraweak biophoton emission.

Bibliography

[1] E.N. Harvey, *Bioluminescence.* Academic Press. New York 1952.

[2] J. Buck and E. Buck, *Synchronous fireflies.* Scientific American, **234** (1976), pp. 74-85.

[3] J. Buck, Synchronous rhythmic flashing of fireflies. II, *Quart. Rev. Biol.*, **63** (1988), pp. 265-289.

[4] A.T. Winfree, Biological rhythms and the behavior of populations of coupled oscillators, *J. Theoret. Biol.*, **16** (1967), pp. 15-42.

[5] C.S. Peskin, *Mathematical aspects of heart physiology.* Courant Institute of Mathematical Sciences, New York University, New York, 1975, pp. 268-278.

[6] S.H. Strogatz and R.E. Mirollo, Collective synchronisation in lattices of nonlinear oscillators with randomness. *J. Phys. A*, **21** (1988), pp. L 699-L 705.

[7] S.H. Strogatz, C.M. Marcus, R.M. Westervelt, and R.E. Mirollo, Collective dynamics of coupled oscillators with random pinning. *Phys. D.*, **36** (1989), pp. 23-50.

[8] R.E. Mirollo and S.H. Strogatz, Synchronization of pulse-coupled biological oscillators. *SIAM J. Appl. Math.*, **50** (1990), pp. 1645-1662.

[9] M. Galle, R. Neurohr, G. Altmann, F.A. Popp, and W. Nagl, Biophoton Emission from *Daphnia magna*: A possible factor in the self-regulation of swarming. *Experientia* **47** (1991), pp. 457-460.

[10] M. Galle, Untersuchungen zum dichte- und zeitabhängigen Verhalten der ultraschwachen Photonenemission von parthenogenetischen Weibchen des Wasserflohs *Daphnia magna.* Dissertation. Universität Saarbrücken 1993.

[11] M. Galle, Population density-dependence of biophoton emission from *Daphnia magna.* In: *Recent Advances in Biophoton Research and its Applications*, pp. 345-356. Eds. F.A. Popp, K.H. Li and Q. Gu. World Scientific, Singapore, New Jersey, London, Hong Kong 1992.

[12] W. Scholz, U. Staszkiewicz, F.A. Popp, and W. Nagl, Light-Stimulated ultraweak photon reemission of human amnion cells and Wish cells. *Cell Biophysics* **13** (1988), pp. 55-63.

[13] D.H.J. Schamhart and R. van Wijk, Photon emission and the degree of differentiation. In: *Photon Emission from Biological Systems*, pp. 137-152. Eds. B. Jezowska-Trzebiatowska, B. Kochel, J. Slawinski and W. Strek. World Scientific, Singapore 1987.

[14] F.A. Popp: Some remarks on biological consequences of a coherent biophoton field. In: *Recent Advances in Biophoton Research and its Applications*, pp. 357-376. Eds. F.A. Popp, K.H. Li and Q. Gu. World Scientific, Singapore, New Jersey, London, Hong Kong 1992.

[15] W. Volknandt and R. Hardeland, Circadian rhythmicity of protein synthesis in the Dinoflagellate *Gonyaulax polyedra.* Biochemical and radioautographic investigation, *Com bio chem physiol.* **77B** (1984), pp. 493-500.

[16] B. Ruth, Experimenteller Nachweis ultraschwacher Photonenemission aus biologischen Systemen. Disseration, Marburg 1977.

[17] B. Ruth and F.A. Popp: Experimentelle Untersuchungen zur ultraschwachen Photonenemission biologischer Systeme. *Z. Naturforsch*, **31C** (1976), pp. 741-745.

[18] J.J. Chang, F.A. Popp, and W.D. Yu: Research on Cell Communication of Pyrocistis elegans by Means of Photon Emission. *Chinese Science Bulletin 1994*, in press.

[19] F.A. Popp and K.H. Li: Hyperbolic Relaxation as a Sufficient Condition of a Fully Coherent Ergodic Field. *International Journal of Theoretical Physics* **32** (1993), pp. 1573-1583.

[20] R. Lumry and R.B. Gregory. In: *Fluctuating Enzymes* (G.R. Welch, eds.), John Wiley & Sons, New York, 1986.

[21] F. Fröhlich. In: *Fluctuating Enzymes* (G.R. Welch, ed.), John Wiley & Sons, New York, 1986.

[22] M.W. Ho. *The Rainbow and the Worm, The Physics of Organisms*, World Scientific. Singapore, 1993.

Chapter 13

Estimates of Brain Activity Using Magnetic Field Tomography and Large Scale Communication within the Brain

A. A. Ioannides

13.1 Introduction

Biomagnetism is the science dealing with the non-invasive measurement and interpretation of the minute magnetic fields originating in biological organisms. The first biomagnetic signal was recorded from the heart in 1963 by Baule and McFee[1]. David Cohen at MIT obtained the first magnetic signals associated with brain activity in the late 60's and early 70's[2,3]. These pioneering studies were followed by a period of steady progress, particularly in instrumentation. Point-like models for the generators were surprisingly effective at times, despite their rather simplistic nature, from a neurophysiological point of view. A number of reviews are available describing these early days and the more recent developments. The most thorough and up-to-date review of biomagnetism and its applications to non-invasive studies of the human brain, using the widely used point-like model for the generators, is the recent[4] one from the prolific group at Helsinki University of Technology, which is highly recommended.

This chapter deals with the distribution and evolution of electrical activity in the body, especially within the brain. It draws on continuous, three-dimensional estimates of electrical currents in the body extracted from biomagnetic signals. The resulting images are fundamentally different from the point like descriptors used in earlier studies. The key features of the measurements and the resulting estimates of activity in the body are:

1. The completely non-invasive and non-contact nature of the measurements which do not affect in any way the processes under investigation (except of course through the constraints of the experimental protocol).

2. The analogical description of activity extracted from the data which can be displayed as animated 'images' with a static anatomical background.

3. The spatial accuracy for focal cortical activity, which is well within one centimetre.

4. The temporal resolution, which can be a millisecond or better.

For historical reasons, our early biomagnetic works were not in-depth studies, but glimpses into brain function from widely different perspectives, involving evoked responses from normal subjects, as well as investigations of normal or pathological spontaneous activity. Measurements from the brain and heart will be discussed more or less independently from each other. Particular emphasis will be placed on activity within isolated cortical regions, where the techniques we have developed so far are most reliable. I will discuss how these findings may fit into the global view of brain function, and briefly the relationship between structure and function; I shall not address the largely unexplored relationship between rhythmic brain activity and the global, body-wide electrical activity[5].

The chapter is as far as possible, self-contained, with emphasis on concepts, and sufficient details and references to direct the interested reader to more technical descriptions. The next section outlines the key non-invasive methods for imaging structure and function, with emphasis on biomagnetism. In the same section, magnetic and electric measurements will be compared. Section 13.3 discusses the inverse problem and the related task of modelling the generators and the biological medium. Point and distributed source models are compared and the problem of non-uniqueness is discussed, with emphasis on the way three dimensional estimates of activity can be extracted from MEG signals, millisecond by millisecond. The study of processes can not be divorced from the knowledge of the structures within which the processes take place. I shall begin the presentation of results in section 13.4 by discussing the cardiac cycle, where relationship between structure and function is well understood. Section 13.5 contrasts the regular rhythmic activity of the heart with the variability of brain responses, and sets out the case for and against averaging. The historical link between averaging and point-like models for the generators is identified, and the reasons why averaging was necessary in MEG and EEG are differentiated. Section 13.6 focuses on early applications where the traditional evoked potential techniques have been applied to magnetic measurements. For the sake of concreteness, the distributed solutions will be restricted to auditory evoked responses. The subject of section 13.7 is oscillations and spontaneous activity; examples from both normal and pathological conditions will be discussed. Section 13.8 pulls some of the early strands together linking, in a rather speculative fashion, a number of recent findings and theories. In the concluding section 13.9, I will discuss the opportunities opening up by the new wave of functional brain imaging techniques.

13.2 Brain imaging

Survival - the moment to moment maintenance of the living state - depends on electrical processes encompassing a wide spatial and temporal range, from the very small and swift (e.g. the electrical events associated with the opening of neuronal ion channels) to the slow and large (e.g. the the near dc fields around limbs and developing embryos). These processes are interwoven in space and time, forming organised sequences of complex electrical events. In some cases the sequences are hierarchically and precisely ordered (e.g. the action potential of neurons or the sinus of the heart), with different stages associated with well understood physiological processes at anatomically well-defined places. A rather rigid sequence of electrical events is involved when the electrical activity subserves mechanical function. In that case the function itself can be studied both by electromagnetic and mechanical means, for example, echo-cardiography, where the cardiac movement is monitored. The normal electrical activity of the central nervous system does not couple to muscular activity directly; it is highly variable, embellished with uniquely rich and elaborate cascades of electrical events. The absence of muscular activity limits the tools available for studying brain function: either a direct measure of electrical activity (local or global) is required, or an indirect measure must be defined which is usually based on the increase or decrease of metabolic requirement of neuronal populations.

The most direct way of studying the brain involves recordings of the actual electrical activity *in situ*. Such invasive studies either on animals or on humans, whenever presurgical evaluation procedures provide the opportunity, have greatly advanced our understanding of brain processes. However it is difficult to relate isolated measurements from single or a few neurons to global activity in the intact animal, where synergistic action of many millions of neurons is involved. The understanding of how electrical processes in the brain relate to emotions and thought is a long way off, and is not significantly advanced by invasive experiments alone.

Within the last few decades, it has become possible to image the intact human brain and body in a number of non-invasive ways[6]. Different techniques work best at different spatial and temporal resolution windows[7]. For our purposes we distinguish non-invasive techniques into those imaging structure, those capable of revealing nodes of activity and a third category which can monitor the fluctuations of function itself.

- The main techniques to image structure are magnetic resonance imaging (MRI) and Computer assisted Tomography (CAT).

- Positron Emission Tomography (PET)[8,9,10] and functional MRI (fMRI)[11,12] are the two most prominent examples of techniques which can delineate regions where activity preferentially increases or decreases when different tasks are attempted. Both methods depend on the accumulation of byproducts of local neuronal activity; in the case of fMRI these metabolic byproducts have different contrast properties while for PET they are radioactively labelled. The heamodynamic processes are sufficiently fast to produce measurable concentration changes in a matter of seconds but not fast enough to respond to changes in the distribution at the millisecond rate, which corresponds to the temporal scale applicable to typical brain processes.

- Techniques which can monitor activity in milliseconds, as it spreads from one brain centre to the next, are exclusively bioelectromagnetic. These are Electroencephalography (EEG), which maps the distribution of the electrical potential on the scalp, and Magnetoelectroencephalography (MEG), which records the magnetic field just outside the head.

Understanding brain function requires information from all of these techniques. This chapter concentrates on the analysis of MEG signals, which is, at present the best way to image fast activity on the cortical mantle.

13.2.1 Bioelectromagnetism

Channels and types of signals Bioelectromagnetic signals are named according to the method used and the organ from which they emanate. The first part of the name, and the first letter in its abbreviation, distinguishes between electrical potential (E) or magnetic flux (M) measurements. The second part of the name and the second letter in the abbreviation defines the organ under investigation, e.g. Cardiac (C) for heart measurements and Encephalon (E) for the brain. Hence, the two main types of magnetic measurements we are concerned with are termed Magnetoencephalography (MEG) and Magnetocardiography (MCG); the corresponding and more familiar electrical measurements are Electroencephalography (EEG) and Electrocardiography (ECG).

For biomagnetism, the most basic element in the sensing device is a coil of superconducting wire, and the channel output is proportional to the flux threading one or more such coils. The sensor and its housing is placed just outside the body. The arrangement is completely non-invasive: the signal is generated by electrical activity within the body as part of the normal biological processes associated with the task under investigation. The signal is minute, and its amplification into a macroscopic voltage is achieved by coupling the coil(set) to the input of SQUID (Super Quantum Interference Device)[13]. Discrimination against noisy background is achieved partly by the arrangement of coils (gradiometer) and partly by the use of expensive magnetically and Eddy-current shielded rooms. In recent years, software noise elimination techniques are increasingly used to obviate the need for the expensive and cumbersome shielded room. The technology remains expensive however, since the coil-SQUID arrangement can only operate at very low temperatures, just a few degrees above absolute zero. The sensors are immersed in liquid helium and housed within a dewar.

For electrical measurements, the most basic sensing device is an electrode which makes a good electrical contact with the scalp. The EEG signal consists of the amplified differences in the electrical potential between an electrode and some reference potential, which can be either that of another electrode or some linear combination (e.g. average) of many electrodes. The term channel is used to describe a single biomagnetic or electric sensor. The output from a single channel is a time-series of instantaneous values at successive time intervals, or timeslices.

The instrumentation for a single electrical channel is simpler and far less expensive than that for a single magnetic channel. As we will see, however, the output from a single channel tells us little, unless we already know a lot on how the generators are arranged in space. In the case of the brain, where the activity is distributed over many different centers which operate in a highly variable way, it is essential that a distribution of the electric and/or magnetic field is obtained.

There are two ways of getting such a distribution. One can use many channels and measure the signal at many different points simultaneously, or one can design experiments where the same activity is repeated exactly and then use one or a few channel probes to record the signal many times with time markers defining the beginning of each sequence. Such time markers can be defined either by the onset of a stimulus or by the detection of a characteristic event (e.g. a sharp wave in interictal epileptic activity). The signals from the brain is highly variable, and the use of successive measurements is only possible if many repetitions are made *at each point* so that the (small) component in the signal which is timelocked to, say, an external stimulus, is enhanced by averaging. Until 1989, when only a single or few sensor MEG probes were available, there was no alternative but to use the average signal. The introduction of the 37 channel family of probes has removed this need and it is now possible to record a spatiotemporally coherent signal over a substantial area of the head. Most of the MEG signals used in this work have been obtained with either the SIEMENS AG 37 channel probe, the KRENIKON or with the BTi 37 channel probe, MAGNES. The introduction of the helmet-like systems by CTF and Neuromag has finally allowed the magnetic field around the head to be mapped with no loss of coherence. A concise description of the multichannel instruments available today can be found in[4].

EEG and MEG Since the early days of biomagnetism, it was assumed that MEG is better than EEG in localising tangential generators, at least on the cortex. The publication of a comparison between the two methods claiming that in practice there is little difference in the localisations ability of MEG and EEG[14] has naturally created a controversy. A number of reports have highlighted flaws in the original experiment and analysis[15,16,17], and the intrinsic limitation of the point-like model which was central in both the experimental arrangement and the analysis[18]. The question has not really been resolved, because the experiments performed so far have not been designed to maximise the capabilities of each technique, and the analysis of what little common data there exist between the two techniques has not been carried out with the best available methods.

In recent years, rapid advances have been made in both MEG and EEG in hardware and software. The combination of each or both of these techniques with other brain imaging methods is opening up new possibilities which make the comparison of MEG with EEG in terms of simple analysis of low quality signals of little relevance.

The sum total of electrical activity generated by a single or a few neurons is far too small to be measured by either MEG or EEG. The signal which can be picked up by the the sensors is many millions of times stronger than what a single neuron can produce. The very presence of strong EEG and MEG signals suggest that the generators must be widespread distributions of neuronal populations acting coherently. When we consider just the macroscopic properties of such generators, the total current density vector $\mathbf{j}(\mathbf{r})$ can be expressed as,

$$\mathbf{j}(\mathbf{r}) = \mathbf{j}^p(\mathbf{r}) + \mathbf{j}^v(\mathbf{r}) \tag{13.1}$$

The primary current, $\mathbf{j}^p(\mathbf{r})$, is the biological source, depending on both the intracellular current and the electrical properties in the immediate vicinity of neurons. The volume current, $\mathbf{j}^v(\mathbf{r})$, is the Ohmic current induced in the medium by the macroscopic electric field. For the sake of clarity, consider a focal, macroscopic generator in the body. This generator need not be point-like, it can be extended (e.g. along a gyrus or fissure). The MEG and the EEG signal generated by the activation of such a generator can be written as the sum of two terms: a local contribution which can be expressed in terms of properties of the *focal* generator (strength and shape) and its immediate vicinity, i.e. $\mathbf{j}^p(\mathbf{r})$, and a *medium* contribution which depends on the properties and geometry of the conducting medium and its boundaries. The task of the analysis of MEG and/or EEG signals is to determine the focal generators of the signals, their location, shape and strength and how these parameters vary with time. The task is neatly divided into two parts, the modelling of the properties of the medium and its boundaries, which determine the medium contributions to the signal, and the main task, i.e. the extraction of the distribution and time dependence of the focal generators.

The presence of the skull affects differently the MEG and the EEG signal. Magnetically the skull is transparent, i.e. the magnetic field propagates with almost no modification through the skull, and hence the local contribution can be described with no reference to the conductivity value(s) of the intervening layers of tissue and bone[19]. Hence, in the case of MEG the primary currents account for the local contribution to the signal and they dominate, except when the sensors are close to regions of abrupt change in conductivity, e.g. the eye sockets or the lower part of the head. *The MEG signal away from sharp discontinuities is dominated by local contributions from* **nearby** *generators.* In contrast, the electrical resistivity of the skull is high and it modifies drastically the EEG signal. In the case of EEG, the local contributions are scaled by the conductivity value, they are diminished and distorted as they cross the highly resistive skull. The effective distance between a generator and an EEG electrode is dominated by the sharp drop in conductivity at the skull, and hence the differential between nearby and distant generators is lost: *for EEG, all generators are effectively distant.* Consequently, the local contribution, as much as can be separated out, is only a small fraction of the total EEG signal, even if they are generated very close to the measurement site. Hence, the distinction between local and medium contributions to the signal is more appropriate for MEG than EEG.

The complex conductivity profile of the head is often approximated by simpler, more manageable models. The most popular model is as a set of concentric spheres with piecewise uniform conductivity within each spherical shell. In the case of MEG, and as long as the measurements are taken outside the outermost sphere, the signal is not influenced by either the radii of the different spherical shells, or their conductivity values; the only model parameter that influences the signal is the coordinate of the centre of the conducting sphere. In contrast, the EEG signal depends crucially on both the conductivity value(s) as well as the radius(ii) of the spherical shell(s).

13.3 The Inverse Problem and Models of the Generators

13.3.1 The Inverse Problem and Non-uniqueness

It is important to realise that the MEG and EEG signal respond linearly to changes in the strength of a given electrical source. Consider a fixed pattern of electrical current producing an EEG (MEG) signal of strength S_{EEG} (S_{MEG}). If the identical pattern of electrical activity recurs with strength p times the original strength, then the resulting EEG (MEG) signal will have a strength p x S_{EEG} (p x S_{MEG}). Of course threshold effects in the brain may induce additional strength dependent activity in other places coupled to the original site.

Thus, if the source configuration has a fixed geometry, the relative strength and orientation of the primary current density at each point in space is constant, and only the overall strength is allowed to vary, then a single MEG or EEG sensor would provide a precise record of the temporal variation of source strength, irrespective of how complex the generator and the biological medium might be. In short, the bioelectromagnetic inverse problem for fixed sources is unique within a normalisation constant, even if we know nothing about the biological medium within which the generators are embedded.

The corresponding inverse problem relating to spatial variations of source activity is much more complex. It is impossible to deduce uniquely the spatial distribution of activity within a confined volume from measurements of the surface potential and/or the magnetic field outside the body, as Helmotz originally pointed out, nearly 150 years ago[20]. In fact, it is possible to reproduce exactly all these observables from a distribution of sources arranged along any closed surface within the body. Since an infinite set of such surfaces can be defined, an infinite set of solutions can be obtained. This is one way of expressing the well known non-uniqueness problem.

13.3.2 Models of the Generators and Silent Sources

It appears that any attempt to deduce the distribution of sources within, say, the brain from either EEG or MEG measurements, or even combined measurements, is doomed to failure on account of the non-uniqueness problem. Despite this obvious mathematical hurdle, it is perfectly feasible to obtain reliable estimates for the electrical activity within the brain from electrographic signals, provided sufficient *a priori* information is injected in the analysis, and the level of detail demanded in the reconstruction is not finer than the sensitivity of the sensors. For MEG, substantially less *a priori* information is required than for EEG, particularly for superficial generators. One way of introducing *a priori* information is through a model. A model often involves hidden assumptions which, even if they are appropriate in the original use of the model, have a habit of creeping unobtrusively into other applications where they might not be so appropriate. This is the case for the most widely used model in both EEG and MEG, the point-like generator or current dipole.

There is a beautiful complementarity between the EEG and MEG sensitivities, which is based on deep physical properties of the electromagnetic field. This is reflected in what each technique is most sensitive to, but also in the types of sources which produce no signal; these sources are termed silent sources. For MEG a monopole, i.e. a point source or sink of current produces no magnetic signal. Also for the case of the conducting sphere (or spherical shells with piecewise uniform conductivity) a radial current dipole (see below) is silent. A radial source is a source which passes through the centre of the conducting sphere, when projected from its point of application either forward or backward along the direction of the current flow. A close loop of current produces no surface electrical potential difference, and hence no EEG signal.

The current dipole model In simple terms, we can think of the units of signal generation as monopoles for EEG and close loops for MEG. Current conservation makes a monopole impossible, so the next available structure, as far as EEG is concerned, is a pair of monopoles (a source and a sink) very close together. It is important to realise that such a pair of monopoles is still a silent source as far as MEG is concerned. The term current dipole has come to symbolise a slightly different construct: a source sink pair connected by an infinitesimal line element of current. A current dipole embedded in the biological medium will induce a flow of current to complete the circuit. Hence we can think of a current dipole as a pair of source sink monopoles for the purposes of EEG and as an elementary flow of primary current for MEG. Note that as far as MEG is concerned, the monopoles are silent, and as far EEG is concerned, the shape of the primary current loop connecting the source and sink is of no consequence.

The most widely used constraint is to assume that the signal is produced by one or more current dipoles; these have the same mathematical form as the idealised point generators described above, but are descriptors of distributions - they are therefore equivalent current dipoles (ECD). The task in this case is to identify the location and direction of the ECD which in some sense best describes the data. As long as the activity emanates from one or two well separated focal generators the identification of the ECDs is feasible. The process becomes cumbersome and unreliable when the number of dipoles increases. Recently, elaborate methods have been developed[21] using a scanning procedure for multiple directional emitters[22]. The problem however remains when a fully distributed source generates the signal.

There is nothing wrong with the current dipole model in general: it is an attempt to explain all with too little. Provided too little is sufficient to explain all, it produces results with minimal effort; unfortunately in all but the simplest of cases, we have no reason to believe that such a simple model could reflect biological reality.

An ECD approximation makes more sense for EEG than MEG; for EEG the shape part of the distribution is silent, and only the focal point-like parts contribute. In the case of MEG where the point-like contributions are silent the ECD makes little sense, unless we know in advance that the underlying activity is highly focal.

13.3.3 Probabilistic Distributed Source Solutions

For the rest of the article we will restrict the discussion to biomagnetic signal analysis, and hence focus on fully distributed solutions which are appropriate for biomagnetic signals. The first software system capable of extracting 3D 'images' of brain function from MEG signals was developed at the Open University UK[23], and the first results reported in 1989[24,25]. The system produces an estimate of the *expectation value of the primary current density*, $\langle \mathbf{j}^p(\mathbf{r}) \rangle$. This expectation value is constructed so that it satisfies large scale physiological reality (e.g. the generators are inside the head) and the precise details of the experimental arrangement (e.g. the solutions reflect the sensitivity of the sensors). Mechanisms are also provided for adding detailed *a priori* information and for allowing for noise in the signal. In summary the quantity $\langle \mathbf{j}^p(\mathbf{r}) \rangle$ is expressed as a linear sum of s (s=number of sensors) terms:

$$\langle \mathbf{j}^p(\mathbf{r}) \rangle = \sum_{k=1}^{s} A_k \phi_k(\mathbf{r}) w(\mathbf{r}) \tag{13.2}$$

The vector value function $\phi_k(\mathbf{r})$, is the lead field of the kth detector; it is defined so that the output of the kth sensor m_k is given by

$$m_k = \int_Q \phi_k(\mathbf{r}) \cdot \mathbf{j}^p(\mathbf{r}) d^3 r \tag{13.3}$$

Thus the lead field, $\phi_k(\mathbf{r})$, expresses the sensitivity profile of the kth sensor; it is confined within a precisely defined source space **Q**. The form of $\phi_k(\mathbf{r})$ is determined by the sensor properties and the electrical properties of the medium. The source space, **Q**, defines the region of space where primary currents are allowed - it does not restrict the region over which the total current density, $\mathbf{j}(\mathbf{r})$, is non-zero. This neatly insulates the assumptions about the primary current density and the properties of the medium. The latter are incorporated in as much detail as necessary through the computation of the lead fields; the assumptions about the primary currents can be introduced either in an all-or-none fashion through the definition of the source space, **Q**, or in a probabilistic fashion through the weighting factor, $w(\mathbf{r})$.

The quantity $w(\mathbf{r})$ is a probability weight for the current density, incorporating any *a priori* information about source location. Fig. 13.1 shows an example of the sensitivity profile for a typical MEG sensor and the widely used spherically symmetric conductivity profile, and how the profile changes when a simple form for $w(\mathbf{r})$ is introduced. Finally the A_k's are expansion coefficients to be determined from the data, with the precise nature of the fit determined by the value of a regularisation parameter[26,27]. The choice for this value specifies how spatial resolution and stability are compromised[23].

The probability weight $w(\mathbf{r})$ can be used to introduce both global and a more local *a priori* information. The inverse problem is separated into two parts: a part which is data independent and deals with readjusting the biases introduced by experimental contingencies and a second part where the refined sensitivity acquired from the first part allows us to home in to accurate descriptions of activity as defined by the instantaneous distribution of signals at each timeslice in the data. In the first part, constraints can be introduced[23] from other techniques. For example, MRI slices can be used to restrict the source space within the grey matter and the probability weight can be enhanced in the regions which are known (e.g. from animal experiments) to be activated in the given experimental protocol. The combination of information

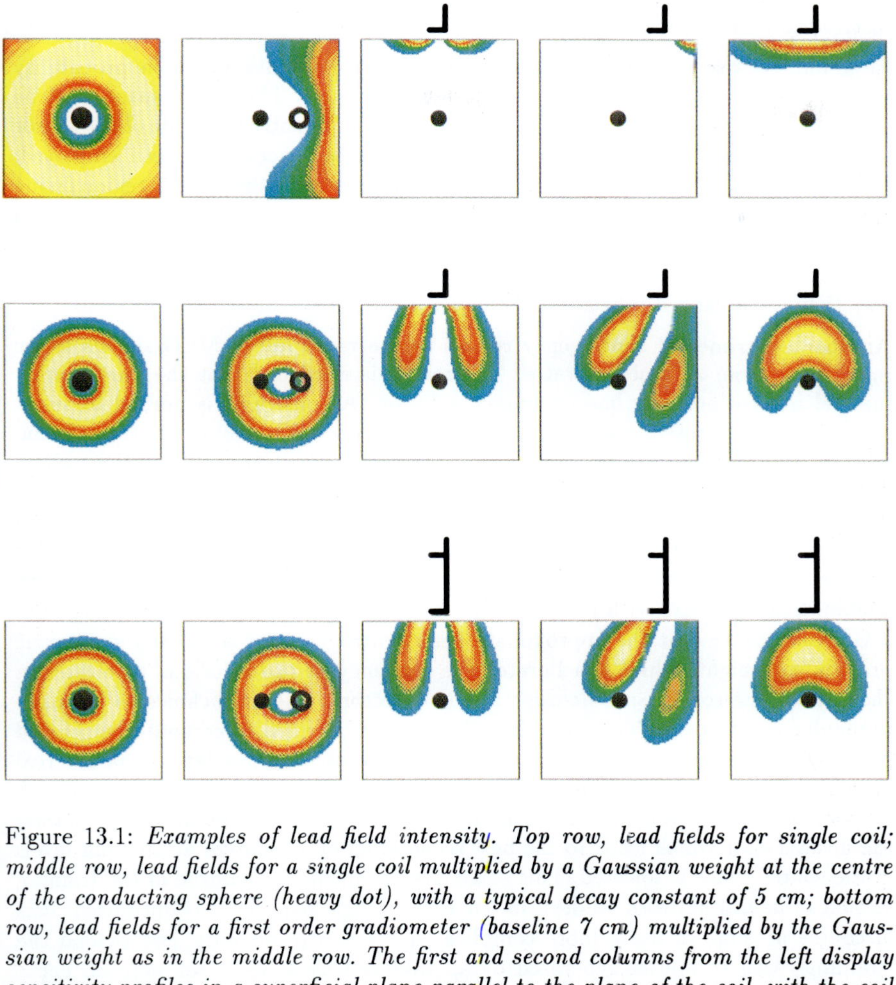

Figure 13.1: *Examples of lead field intensity. Top row, lead fields for single coil; middle row, lead fields for a single coil multiplied by a Gaussian weight at the centre of the conducting sphere (heavy dot), with a typical decay constant of 5 cm; bottom row, lead fields for a first order gradiometer (baseline 7 cm) multiplied by the Gaussian weight as in the middle row. The first and second columns from the left display sensitivity profiles in a superficial plane parallel to the plane of the coil, with the coil (open circle) directly over the centre of the conducting sphere, or displaced to the right; the corresponding profiles in the perpendicular plane are displayed in columns three and four; the last column of frames display the sensitivity profiles in a plane perpendicular to the plane of the coil(s), but 3 cm in front of that passing through the centre of the conducting sphere.*

from type two methods, e.g. PET and/or fMRI is particularly appealing, but the relationship between changes in metabolic activity and changes in the current density distribution are not yet well understood.

In all our work so far, we have used the source space definition and the probability weight as a means of introducing large scale constraints. The source space must be consistent with large scale physiology and the overall sensitivity profile of the sensor arrangement; both two-and-three-dimensional source spaces have been employed; either a disk or part of the surface of the sphere representing the cortical region facing the sensor arrangement or a cylinder and hemisphere. The choice in each case is such that the source space boundary follows closely the cortical surface below the sensors. The extent (depth) of the source space is determined by the sensor properties, the way they are arranged in space and the noise level in the measurements.

Another experimental contingency relates to the rapid fall of MEG signals as we move away from a focal generator. The reader is reminded that the pure mathematical inverse problem has no unique solution. A constraint is needed to select one of the many possible solutions. If the usual choice is made, where the solution with minimal strength is selected[28], then the activity gravitates to the surface where the sensitivity is highest. The introduction of a probability weight can be thought of as a redefinition of the inverse problem so that non-uniqueness is absorbed into the choice of $w(\mathbf{r})$. Complicated minimisation procedures can be developed at this point to select an optimal $w(\mathbf{r})$ within the source space. We have found that a very simple choice for $w(\mathbf{r})$ with just one adjustable parameter is surprisingly successful: a Gaussian centred at the approximate head centre, creating a radially symmetric probability weight, which ameliorates the tendency of the solutions to cluster in the part of the source space closest to the detectors. The only choice to be made is the decay rate of the Gaussian. This is selected from fits to computer generated data *before* any measurements are analysed. Noise at the same level as in the real data is added and the appropriate regularisation parameter is used throughout. A number of model inversions are carried out with different Gaussian decay factors; the decay factor which leads to the best reconstruction of sources at different levels of the source space is selected. This stage is a rudimentary analogue of the training session in neural network studies. In our case a single parameter is optimised. It is of course possible to use a more complicated *a priori* probability, and extend the 'training' with computer generated data, produced by a wider set sources. For a given shape of the source space and model for the medium, the decay factor for $w(\mathbf{r})$ and the regularisation parameter are the only two adjustable parameters in the method; both are fixed before the real data are handled. Fig. 13.1 shows how the effective expansion functions are modified when a Gaussian weight is applied and the source space is confined to the area of the head.

The first solution, $\mathbf{j}_0^p(\mathbf{r})$ is iterated by repeating the inversion, but with a new probability weight which is the product $\mathbf{j}_0^p(\mathbf{r}) * w(\mathbf{r})$. This has the effect of sharpening up the image. This iterative scheme converges within one iteration, leading to distributed solutions with fine detail. The major disadvantage of the iteration scheme is that it makes heavy computational demands, which we have satisfied by adopting a transputer based implementation of our algorithms[29].

The lead fields, ϕ_k, are not the most convenient expansion functions, but they reflect precisely the sensor sensitivities and they exclude by construction 'silent components'. They include both the local and medium contributions to the signal: the medium effects are fully described and they are not confined to the source space. The source space confines only the focal contributions. Irrespective of how complicated the medium is, the ϕ_k need only be computed once, since they are independent of the signal. The major advantage of this choice, and the one which is only partially exploited, is that it allows, in a consistent way, the incorporation of additional information through the probability weight $w(\mathbf{r})$.

The full method has been tested with computer generated data, MEG signals from evoked responses where the focus of activity is known from other invasive techniques, and with MEG signals generated by physical dipoles implanted in the head of a living human subject. Estimates were obtained from signals generated by a single implanted dipole and by a pair of simultaneously active implanted dipoles[30]. In all these studies the method was able to recover highly localised activity almost as well as a current dipole fit can[23,30,31,32,33], and could also cope with activity spread over different regions, a situation where the ECD model failed; the ECD was, however, reliable whenever an extremely good fit to the data was obtained. These results provided much needed explanation for recent ECD successes, where stringent selection criteria produced ECD solutions of clinical usefulness[34,35,36].

13.3.4 Magnetic Field Tomography (MFT)

In a strict mathematical sense, any solution of the bioelectromagnetic inverse problem can only be interpreted as a probability estimate. However, given the sharp MEG sensitivity to nearby sources (with practically no distortion from the intervening tissue and skull) and that the strength of the generators can not exceed physiologically feasible limits, the ambiguity is all but eliminated for estimates of strong superficial generators directly below the sensor array. Provided the source space does not extend into regions with little sensitivity the main uncertainty appears to be the relative strength of the sources at different levels. Thus, if sources at the same level are compared, then their relative strength is maintained in the reconstructions, while the strength of sources at different levels is modified. The choice for the decay constant for the Gaussian probability is optimised so that superficial and deep focal generators are correctly localise. This choice enhances the images of deep sources above the level they would appear if a uniform *a priori* probability is used. The enhancement is not however complete; in the reconstructions, the strength of a deep source is considerably reduced compared to the strength of a similar superficial source. Intuitively, this is not surprising given that we start with very small sensitivity for deep sources; compensating fully for localisation need not necessarily compensate fully for strength. The tempting modification to artificially strengthen the deep sources has been resisted, because other model uncertainties (e.g. sphere centre) make the reconstruction of deep sources problematic. We therefore choose to "see" deep activity at a given timeslice, only if it is considerably stronger than the weakest superficial activity the system can recover. At each depth, the limit beyond which a source will not be recovered depends on the regularisation parameter, which in turn is set according to the noise level in the data. The overall uncertainty in relative strength at different depths does not change with time, and the time dependence of the solutions provides a good estimate of how the relative power between superficial and deep activity changes with time.

The probabilistic estimates of the distribution of the primary current density are not easy to visualise. Although the solutions are continuous functions, they are usually displayed along cuts (tomes) through the source space. These cuts are normally parallel to the plane or surface of sensors, so that the relative sensitivity within each slice is constant. Such displays are referred to as Magnetic Field Tomography (MFT) images. Specially designed software has been developed to help with the display and analysis[37] of MFT solutions. Sequences of MFT images are recorded on video to provide a record of the dynamics of the activity in the brain[38,39].

Even if no background anatomy is to be displayed, one needs to show a three-dimensional vector field which changes in time. For theoretical as well as practical reasons it is better to display the intensity, $I(\mathbf{r},t) = |\mathbf{j}^p(\mathbf{r},t)|^2$. The practical advantage is that we are dealing with a (positive) scalar than a vector. Pixel by pixel statistics, commonly used in other brain imaging methods, become possible, but they have not yet been applied to MFT. The consequences of the uncertainty in the relative strength at different depths and model errors (e.g.uniform conducting sphere approximation which makes the head centre silent), must be fully understood before the MFT estimates are treated as true images. No temporal constraints have been introduced either, e.g. as is commonly done in studies using multiple point source models[40]. The use of anatomical constraints will undoubtly improve resolution[41], and although the mechanism for introducing such constraints was present from the beginning[23], it has not yet been exploited fully. The MFT analyses performed so far have used minimal *a priori* assumptions on the spatial domain and no assumptions in the temporal domain (other than what is implicitly assumed by the filtering applied to the signal). At this early stage, it is safer to avoid forcing assumptions in the solutions, but use instead the comparison between spatial accuracy and anatomy, and the way the solutions at different timeslices vary as a way for testing the self-consistency of the solutions.

The analogical nature of MFT estimates, and particularly the fact that the intensity is always positive, allows post-inversion processing, which is simply not available if all signal details are collapsed into an ECD point description. The intensity can be integrated in space or time, leading respectively to estimates of the temporal variation in intensity from a given region, and estimates of the three dimensional distribution of intensity over a finite time interval. We can think of MFT intensity estimates of current intensity as PET-like images expanded along the time direction. It is tempting to interpret the PET image as the integral over time of the MFT intensity sequence. This is only approximately true for the reasons outlined above and because the metabolic rate is only approximately proportional to the square of the primary current density, while MEG and PET have different resolution limits and thresholds[42]. These uncertainties do not vary with depth, and hence PET or fMRI measurements may yield constraints to resolve the MEG uncertainty in relative strength with depth.

13.4 MFT Estimates of Cardiac Activity

The normal activity of the heart at rest is associated with a well-defined cycle of electrical events. The cycle begins with the spread of an electrical wave from the Sinoatrial node (SA node), situated on the posterior wall of the right atrium. This activity begins spontaneously, about once a second in the intrinsically electrically unstable cell membranes of the SA node. From the SA node a wave of depolarisation spreads across the two atria initiating atrial systole. The wave arrives within a few milliseconds at the atrioventricular node (AV node) situated in the lower posterior part of the artrial septum. The electrical activity is delayed in the AV node, for about 100 ms, before it is conducted through the bundle of His across into the ventricles. The matrix of large diameter Purkinje fibres ensures that within a few tens of milliseconds the polarisation wave spreads through the left and right ventricles, stimulating the ventricular contraction. The geometric location of the nodes, the cell arrangements within them and the resulting electrophysiological properties, the length and diameter of conducting fibres and the overall geometry of the atria and ventricles determine the sequence of electrical events which in turn drive the mechanical processes of systole and diastole.

The electrical events in the heart produce measurable electromagnetic signals at the surface and just outside the body. The 37 channel MAGNES probe of BTi, placed over the thorax directly above the heart, has been used to record MCG signals from a normal subject. In Fig 13.2, the traces from two MCG channels are displayed, and correlated with the standard P, QRS, and T segments, of the more familiar ECG record; the active regions at key stages and the direction of propagation of the excitation wave are also shown. Fig 13.3 shows the same two MCG channels and the results of MFT analysis. The source space, where the primary currents are confined, is a cylinder which has the same symmetry axis with the MAGNES probe. The top surface of the cylinder is 15 cm below the sensor array into the body. The cylinder has a radius of 8 cm and depth of 8 cm. In the lower part of Fig 13.3 the activity at distinct timeslices is shown in cylinder plot style. In a cylinder plot[37,39], slices are cut

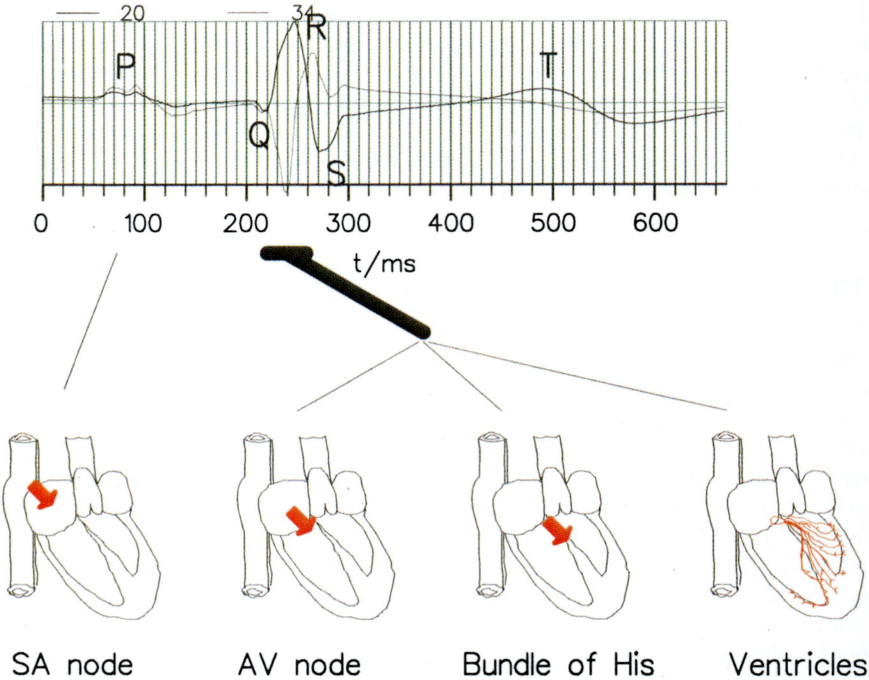

Figure 13.2: *The cardiac cycle. Top trace: the output of two MCG channels of the MAGNES 37 channel probe. In the diagrams of the heart, red colour is used to highlight the activity at key stages; the arrow shows the direction of propagation as the activity is initiated at the SA node, AV node and through the bundle of His. The data in this and the next figure have been provided by David Kynor of Bti.*

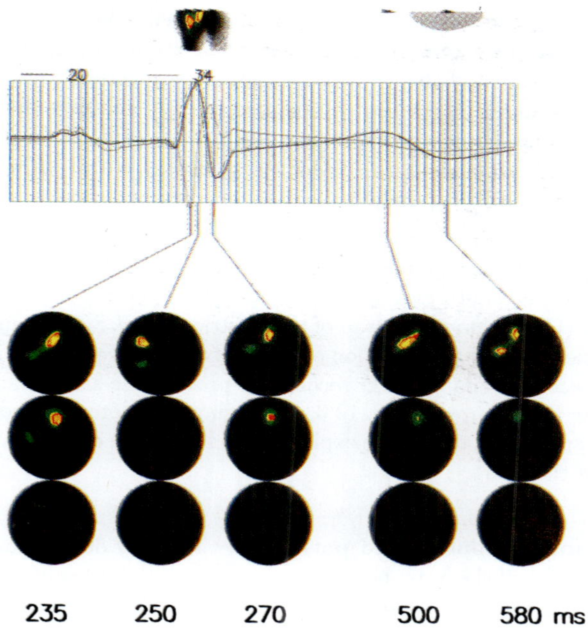

Figure 13.3: *MFT analysis of MCG signals. The MCG traces are shown again (middle) together with the MFT estimates (above and below). Above, a depth-time plot provides a summary of the spatiotemporal variation, while below, cylinder plots are used to show snapshots of activity at the QRS and T segment of the cycle. The mid-point of the cylinder (centre of middle slice) corresponds to the position of the bundle of His, with the head in top-right direction.*

at regular intervals through the cylinder. The activity across each slice is computed and displayed as contour plot. Each display is rotated so that it is flat on a page and displays at successive depths are placed in order, with the most superficial at the top and the deepest slice at the bottom. In this way a column of contour plots describes the instantaneous distribution of activity. By placing MFT solutions at successive time intervals next to each other, the variation in space and time can be displayed. In the top part of Fig 13.3 a depth-time-plot is used, to show how the activity changes as a function of time. In a depth-time-plot[37,39] the activity is integrated across cuts through the cylinder and the integrals of activity are displayed as contour plots of average intensity across depth and time. The cylinder plots are appropriate for displaying the activity over a small number of time slices showing how activity changes in distinct steps, while depth-time-plots are convenient ways of summarising the changes in activity over long time intervals.

For a normal heart, the sequence of main events is well defined, and at each point of the cardiac cycle we know where the activity is strong on the heart. However, because of the movement of the heart, we do not know precisely where these regions of strong activity are in relation to the sensors, even if we know the heart outline at one point in time. Despite these complications, the problem is almost reduced to finding the variation of source(s) with fixed geometry, which, as we have seen in section 13.3, is much easier to handle.

Depth plots are useful even if no anatomical background is provided, because they show variations over long intervals with low spatial resolution. In the case of the heart, we see that the depth plot delineates clearly the periods of atrial (P segment) and ventricular (QRS segment) spread of the electrical wave. Such a representation is probably sufficient if only the duration of each segment is required. It is no accident that the more easily available ECG record can provide the same information. The cylinder plots in the last row of Fig. 13.3, although much more detailed, are not very informative. If spatial detail is made explicit, then it must be displayed together with the anatomy of the heart.

Cardiac variability, control and pathology The tidy description given above for electrical activity of the heart belittles the potential usefulness of MFT estimates. Fine details are latent during the normal heart cycle at rest. To begin with, cells within the AV node and the bundle of His and some Purkinje cells have an intrinsic but slower rhythm than the dominant SA node rhythm. These other rhythms do not manifest themselves because the faster SA node rhythm always arrives first. During exercise, a different, but equally normal heart cycle is established, always compatible with the component properties and overall heart geometry and function. In pathological conditions, changes in either the properties of the nodes or the conduction system lead to unstable activity. The central nervous system lies silent but active behind the scene. If all influences from the brain and brainstem are removed, the intrinsic heart rate increases by 50% showing that during rest, inhibitory central control is in place. As the complexity and variability of the cardiac activity increases so does the potential advantage of MFT estimates, provided structure and function can be accurately coregistered and displayed.

13.5 Brain Signals: Variability and Averaging

The brain is a single organ which subserves many functions. At the most basic level, usually below conscious control, the brain ensures homeostasis, i.e. it maintains a stable internal environment. In this primary task, the brain receives and processes impulses from the sensory organs. The body posture and movement is also under the direct control of the brain, not just in controlling and directing muscular activity, but also in prohibiting, pro-enacting the consequences of these activities, in storing complex sequences of motor output and weaving them together to initiate and guide purposeful movement. Finally the brain constructs reality, self and consciousness.

It is important to emphasise right at the beginning that the signals we have studied were generated when the subject was asked to perform in artificially constructed scenaria, where experimental contingencies allow only biologically irrelevant stimuli, where the flow and novelty linking each moment to the next was replaced by the meaningless stability of sequences of identical stimuli. When we look at brain activity under these artificial conditions, and particularly when we use averaging to extract a few percentages of signal, we have almost certainly eliminated the 'communication' between regions which binds perception and thought, linking activity in isolated centres into meaningful awareness. Yet this reductionist approach is of immense value, as long as its fundamental limitation is recognised. The rest of this section sets out the case for and against averaging, with emphasis on averaging signals from modern multichannel MEG probes. Almost all EEG and/or MEG studies dealing with the localisation of the generators rely on average signals. The next section traces the history which has linked averaging so strongly with source localisation. I will argue that averaging is still necessary for EEG, but it is no longer required for MEG signals recorded with modern multichannel probes.

13.5.1 The Case for Averaging

Almost all analysis of evoked brain responses in EEG and MEG relies on averaging a large number of responses timelocked to identical stimuli. The 'need' for averaging is evident when the average signal and traces from single epochs are compared. Even if the channel with the strongest response in the average record is displayed, the average and single epoch traces appear very different. The variability in brain responses is in sharp contrast to the regularity seen in the heart activity, where if individual cycles were displayed with the average, one would hardly notice a difference in the traces. However, if we know where the activity emanates from, we can construct simple signal combinations and use them as templates. Figure 13.4 shows the simplest such examples, where by taking the difference between the signal in two adjacent channels, the activity in the auditory cortex, lying between the two channels is made apparent, despite some jitter in latency and magnitude. The MEG data are from a CMV experiment (see section 13.6), using the GO/NOGO avoidance paradigm, recorded with the BTi MAGNES system[43].

Away from the main peak, the signal picked up by averaging is typically 10 to 100 times smaller than the strength of the signal in single epochs. Averaging allows this small component to be extracted from the stronger background activity. The case for averaging rest on the assumption that the small component thus extracted relates to the way the brain reacts to the stimulus, and it can therefore tell us something about how the brain functions.

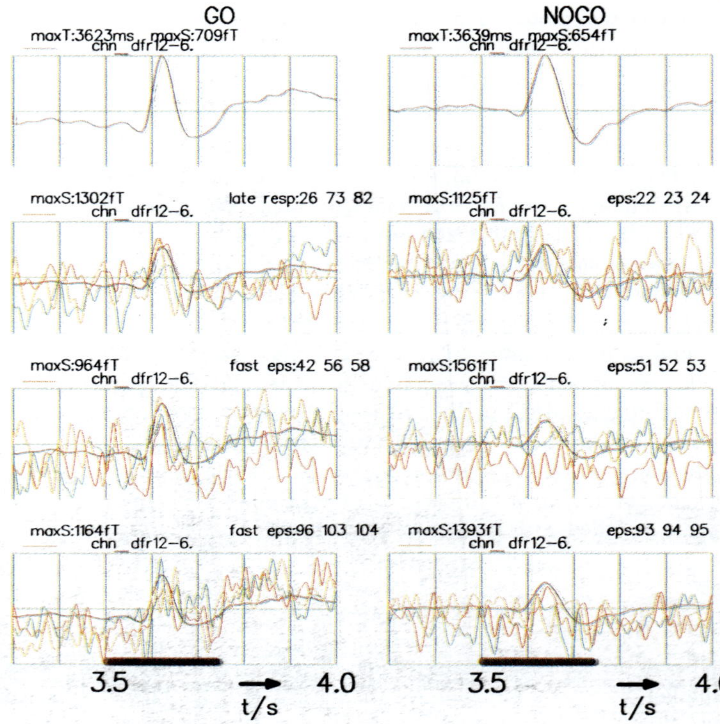

Figure 13.4: *Single trial variability in brain responses. Each trace corresponds to the difference of two MEG channels, selected so that the activity in the auditory cortex is highlighted. In the traces on the left (GO condition) the subject had to make a movement after the tone, while on the right (NOGO condition) no movement was to be made in response to the tone. The top row shows the average response, where priming of the auditory cortex in the GO condition is evident (offset before the imperative stimulus). The average signal difference is shown as a black heavy trace in the other rows, and displayed together with the same channel differences for single epoch responses. On the left the single epoch GO responses show less latency jitter and better phase match with the main peak of the average, than the corresponding responses in the NOGO condition on the right. The heavy horizontal line in the bottom two figures marks the duration of the imperative tone. The epoch numbers and the maximum signal value are printed above each set of traces.*

13.5.2 The Case Against Averaging

The limitations of averaging have been discussed by many authors, and it has lead to alternative ways of analysing the data. Usually, these other methods involve time series analysis, which are not easy to relate to the topography of the generators in the brain. The emphasis in this chapter is very much on localisation; I shall therefore explore how the drawbacks of averaging can be avoided without sacrificing the ability to extract information about the spatial distribution of generators. The starting point is the realisation that the causes of localisation uncertainty are different for EEG and MEG, and that once averaging is abandoned, the historical link with the successes of the ECD analysis is broken.

The high resistivity of the skull and the presence of alternative easy routes of low resistivity are major sources of uncertainty for EEG localisation. Recall that for EEG the focal contribution, even for nearby sources is swamped by other signals. Consider a generator directly below an EEG electrode; the relative contribution to the EEG signal through the skull is similar to that generated by the alternative routes through skull openings, like the eye sockets. Since at any one timeslice, many widely distributed generators are active simultaneously, a given evoked response of interest, say from a primary sensory area, will be swamped by 'biological noise' even at electrode sites close by. Averaging is necessary if the irrelevant activity is to be eliminated. In fact, for EEG, the extraction of reliable information about the generators usually involves, not only averaging the responses of identical stimuli, but also averaging across subjects, to eliminate the troublesome and unknown variation in skull conductivity. The remaining irreducible spread is considerable, and even for the most favourable case of a radial, cortical dipole this spread is several centimeters[44]. The computation of the local Laplacian[45] provides a convenient way of sharpening the EEG distribution; it highlights regions of strong radial currents through the skull[46]. A way of removing the effect of the skull has been developed by Gevins and collaborators[47]: The skull conductivity is computed from estimates of the skull thickness, extracted from MRI slices; under the assumption that no sources exists beyond the surface of the brain, the potential on the other side of the skull can be computed from the measurements using finite element methods. The method produces reliable results if many electrodes are used (typically 64 or 128) and the signal has a high signal to noise ratio, hence again the need for averaging.

As we have seen, very little can be said about the location of the generators from the MEG signal at a single point in space. If spatial detail is required then a map of the MEG signal is needed surrounding the cortical area of interest. The limitations of earlier generations of MEG hardware made correspondingly small demands on the analysis software. In particular, the need to move the single sensor or a few sensor probes from one location to another so that the magnetic field could be mapped over a fairly large area outside the head left no option but to use the average signal derived from many repetitions of a stimulus. The need for averaging thus arises for quite different reasons for MEG and EEG.

Many EEG and MEG studies have shown that the ECD model provides an exceptionally good description of the early peaks of the average signal, evoked by simple stimuli. The characteristic patterns generated by point-like sources have become a major preoccupation, and the success of an experiment has often been linked to whether or not such a pattern can be seen in the data.

The distribution of generators is not immediately apparent on inspection of the traces recorded by modern multichannel MEG probes. At some of the peaks in the average signal distribution, the characteristic and "much loved" signatures of dipolar activity are evident. For much of the time in the average traces, and nearly at all times in single epochs, the distribution is not dipolar. The interepoch variability is large and it is often referred to as "brain noise", but the pattern in each epoch has a structure which does not appear to be due to random uncorrelated activity. Even a simple signal transformation[48] can highlight single epoch features, as Fig. 13.4 demonstrates.

The MFT solutions from individual epochs show most of the activity to be fairly well localised but with variability in both latency and precise shape. When the signal is averaged, any variability in the traces is likely to lead to large cancellation, because of the rapid fluctuations of the signal about zero.

Integrals of intensity distributions provide a new and immensely more powerful capability, with little loss of information, because no cancellation is involved. This is in sharp contrast to signal averaging where only a small fraction of the original signal survives. MFT analysis of modern multichannel MEG data has provided glimpses about what is actually lost. Comparisons of MFT solutions from spontaneous unaveraged and averaged signals and evoked and steady-state responses supports the notion, expressed by many investigators in the past, that brain function cannot be separated into background "brain noise" and an evoked response. Such artificial separation, implicit in averaging, destroys most of what is truly brain function, and leaves only the intrusion of sensory input into the cortical circuitry, hence the success of the ECD model with average signals. The average signal is the record of activity which, for one reason or another, exhibits large scale coherence with time markers which can be picked in advance (eg time locked response to stimulus onset/offset etc) or retrospectively (eg by a spatiotemporal template technique).

The most fundamental new capability of multichannel MEG probes with 30 or more sensors is that they capture the magnetic field (flux gradient) with no loss of spatiotemporal coherence. This allows single epoch analysis, or at least consideration of signal assemblies other than the simple average[49]; it is therefore possible to compare both the mean response and its variability from exactly the same generators under different conditions.

13.6 MFT Analysis of Evoked Responses

A number of recent MFT studies have contrasted the evolution of activity in the two hemispheres, when either identical stimuli are processed, or identical tasks are executed (e.g. finger movement). In one set of experiments[50], identical simple auditory stimuli (600 ms long, 1 kHz tones) were presented monaurally to the left and right ear and the magnetic field from the ipsilateral and contralateral hemisphere was recorded with the SIEMENS AG 37 channel KRENIKON system. For each of the four conditions two experiments were performed, one with eyes open and one with eyes closed. The subject in this experiment was asked to stay alert but not to pay attention to any particular feature of the stimulus. The 8 experiments were performed at Erlangen university. The processing of the auditory stimulus during the first 40 ms after stimulus onset, begins with a weak fast interplay between deep (probably thalamic) and cortical activity[51]. The strongest cortical activity is seen between 70 and 90 ms after stimulus onset, close to the posterior end of the Sylvian fissure. The activity is stronger and earlier by a few ms in the contralateral hemisphere. On the right the activity spreads in the anterior direction by one to two cm and then into the depth of the temporal lobe, probably in the hippocampus. At about 150 ms after stimulus onset, the cortical activity reappears in the anterior end, at the same cortical site. On the left hemisphere a similar sequence of activity is seen, except that the superficial activity is strong and focal, rooted around a small area throughout the period. After about 160 ms the activity does not differ much between left and right ear delivery, but differences appear to be more sensitive to whether the eyes are open or closed. The period from 240 ms after stimulus onset is characterised by a resurgence of activity throughout the deep levels (probably both thalamic and hippocampal regions) as well as sites earlier excited on the cortical mantle. Preliminary, single epoch MFT analysis suggests that this reflects strong reactivation of all these sites, particularly on the right hemisphere, in some but not all of the epochs. This resurgence of activity probably relates to the positive electrical potential at the vertex, around 300 ms after stimulus onset (P300), and elicited strongly by deviant or unexpected stimuli.

In another set of experiments[31,43] the GO/NOGO avoidance paradigm[52] was used, with 3.5 seconds between the warning and imperative auditory stimuli. In the GO condition, the subject has to move the index finger, as soon as the imperative tone arrives, to avoid punishment. Movement with the left and right index finger (in the GO condition) was compared. In both experiments the MFT estimates, obtained from the average MEG signals for each condition, revealed substantial interhemispheric differences. in the NOGO condition the subject must withhold movement to avoid punishment. The auditory evoked response is very different in the GO and NOGO condition, for the same subject, and the difference is particularly noticeable for the response to the imperative tone. A difference is evident even in single epochs as Figure 13.4 demonstrates. MFT analysis shows that activity in the auditory cortex around the time of the imperative stimulus in the GO condition is not evoked by the stimulus: the activity is built up and for at least 500 ms, *before* the onset of the stimulus, and at about 100 ms after the stimulus onset, at the time when normally the maximum evoked response is seen, the activity, already built up in the auditory cortex drops sharply. The responses in cortical areas relating to movement grow sharply from this point in time. Thus the auditory cortex is used as a trigger for the initiation of activity in regions of the brain associated with movement execution. The 'trigger' is primed so that the arrival of the minimum necessary information sets the movement related system to action. In addition to the expected activity in the auditory cortex and sensorimotor strip, strong activity was observed in the supplementary motor area (SMA) and medial parietal area (MPA). Strong activity in the MPA was reported recently, 220-285 ms after eye blinks[53]. This activity was interpreted as a mechanism ensuring a continuous image of the environment despite the 0.1 second loss of visual input during each blink. In the CMV experiments, strong post-movement MPA activity is observed when no strong pre-movement SMA activity is present, while it is absent when the SMA is strongly activated before S2. Interpreting MPA activity as associated with the resumption of normal (sensory) function after a well learned, almost automatic task is completed, is consistent with both observations.

13.7 Oscillations, Synchrony and Spontaneous Activity

Distinct oscillations, in addition to the prominent 8-12 Hz alpha activity is evident in both the EEG and MEG signals. Low frequency activity in the delta (0-3 Hz), and theta (3-6 Hz) range is often associated with pathology while fast activity in the beta range 15-25 Hz, and gamma band 35-55 Hz, are associated respectively with motor and attentional or cognitive processes. It is not always possible to capture these oscillations in in the average signal, although recently some notable successes have been reported, including some intriguing measurements of the gamma band activity around the 40 Hz range[51,54].

The suggestion that at least a component of the 40 Hz signal is generated by a thalamo-cortical loop becoming resonant at 40 Hz is of theoretical as well as of practical interest because it addresses the fundamental problem of unifying perception, what is often referred to as the binding problem. The localisation of function is by now well established. Even within distinct modalities, the neuronal response appears to be channeled to specific areas and then segregated into component features which are separately analysed in parallel. This process is particularly evident in the visual cortex, where specific areas are selectively activated by specific visual features, e.g. colour or movement[55]. No area appears to exist where all the "results" of the segregated analysis converge, and hence the binding problem, i.e. how these separate views are merged into a unified percept. The excitement with the 40 Hz story is that the binding may be achieved in time, by the synchronous or at least phase-related activity of neurons in different areas, dealing with the properties of an entity to be bound into one percept.

MFT analysis of steady-state auditory evoked signals, recorded with the BTi, 14-channel Gemini probe at the MEG laboratory of New York University Medical Center (NYUMC) has revealed an ordered sequence of superficial and deep activity when the the average of 1,000 epochs, filtered in the 35-45 range was used. A small number of subjects have been investigated, including a normal young subjects (below 25 years), old (over 70) Alzheimer's patients and matched old controls. A fascinating change in the distribution of superficial and deep activity has been observed: in young subjects the 40 Hz coherence is observed at both the deep and superficial levels, with the superficial activity dominating. In old controls the 40 Hz coherence is nicely balanced between deep and superficial regions, while in Alzheimer's patients the 40 Hz coherence in the depth is still evident, but the coherence at the cortical level is sharply reduced in accordance with the severity of the illness. Recently, intriguing changes in the 40 Hz rhythm have been observed in different states of awareness[57].

The comparison of MFT solutions of single epochs and average MEG signals is a formidable task, partly because of the computational burden, but mainly because of the huge number of 3-dimensional images that must be compared with each other. Recently, we have applied a simple signal transformation which highlights nearby sources[48] to compare focal 40 Hz activity in specific cortical regions. Single epoch analysis of early 40 Hz activity in the primary auditory cortex showed that a very small fraction of epochs (5 to 10%) accounted for much of the 40 Hz component just after stimulus onset in the average signal. Some features of the (unfiltered) average signal appear to originate in epochs which match the early 40 Hz response of the average, while different features of the average appear to come from epochs which show no strong early 40 Hz activity[58]. The 40 Hz activity in single epochs occurs throughout the period, before and after the onset of the stimulus, and it is far less precisely organised than what the MEG analysis of the average signal implies[51,54]. The orderly behaviour in the average is probably due a small fraction of epochs where the onset of the stimulus is accompanied by a reset of the 40 Hz burst. In tasks demanding attention an increase in the 40 Hz component of the average signal is observed[59]; the origin of this increase is likely to be an an increases of the fraction of epochs with 40 Hz reset rather than stronger activity in individual epochs.

We have observed activity with strong 40 Hz components in both spontaneous and evoked responses. In a recent study[60] MFT solutions from interictal epileptic MEG data averaged with a spatiotemporal correlation template technique showed an orderly (40 Hz) build up leading to the peak of the average sharp wave; however, the MFT solutions from a single sharp wave showed irregular interplays between superficial and deep activity; no build up to the peak of the sharp wave is seen, but a relatively long silent period follows the sharp wave. In another study, MFT analysis of MEG signals from a head-injured patient[33], obtained with the BTi 7-channel system at the University of Texas Medical Branch in Galveston was studied. In this experiment measurements from the left side of the head were taken under ipsilateral and contralateral auditory stimulation (noise to the other ear). The errors due to false geometry assumptions are identical in each condition since the same source space definition is used; differences in the solutions for ipsilateral and contralateral stimulation should therefore reflect either differences in the pathways involved or the effect of injury. The response from this patient shows marked differences in both position and latency compared to the responses from normal subjects. Particularly relevant to our discussion is the comparison of ipsilateral and contralateral MFT solutions (from averaged signals): the ipsilateral profile was very different to the contralateral one, but the two coincided if the ipsilateral response was shifted forward by 25 ms. This can be interpreted as a failure of the cortical circuitry, in many of the epochs, to organise the standard evoked response within the normal time, requiring at least one more (40 Hz) cycle. Clearly analysis of single epoch data from patients with similar injuries is required to elucidate the relationship between injury and fast brain activity.

13.8 Pulling the Early Threads Together

Thanks to advances in many aspects of neuroscience research, not least in biomedical imaging, a new consensus is beginning to emerge about how the brain works. This consensus emerges slowly from new insights and apparently contradictory views of brain function. There is a need for a self-consistent formulation which accommodates both the average response, as exhibited in the average of evoked responses or the PET and fMRI images, and the dance of transient events in the cortex.

The assimilation of environmental influences is achieved through a web of processes involving segregation and integration of stimulus attributes. The segregation begins at the sensory organs, which sense different kinds of stimuli and maintain separate streams as they home in from the periphery, through the thalamus to the cortex. Interestingly, in olfaction, the only sense where differences are sensed directly by specialised sensors at the periphery, the cortical input bypasses the thalamus. At the cortex, stimuli from different senses are as separate as possible, except for the somesthetic and motor areas which occupy adjacent sites on either side of the central sulcus.

In the visual system, by far the best studied system, the main input to the cortex is channeled through area V1, around the calcarine fissure in the occipital part of the head. This area is arranged so that different properties of the visual stimulus (colour, form and motion) are segregated and fed forward to other nearby areas on the cortical mantle which appear to specialise so that they can construct from the earlier segregation and analysis what we sense as colour and movement. It appears that the integrity of both the specialised areas and the primary areas (V1 and V2) are necessary for conscious perception of the final product. Each one of the specialised areas receives precise specialised input and returns diffuse output to the more primitive areas. This principle of highly precise feed-forward arrangement with diffuse feedback appears to be a common way of sensory processing on the cortical mantle[55], with integration within sensory modalities facilitated by parallel recursive signalling along cortico-cortical connections[56]. For the large integration of different sensory modalities the involvement of deep brain centres is probably necessary[57], with the thalamic complex and the surrounding sheet of the reticular nucleus acting more like a primary global sensory area, rather than simply the gate to the cortex[62].

The new consensus, a resolution of the old argument regarding localised centres versus global function can be sensed in the recent findings, but still far from completion. I have suggested on the basis of the very first, two-dimensional, distributed source analysis, that a 'wireless' mode of operation may also be involved in the process of integration[61]. Return currents, which so far have been treated as a purely mathematical inconvenience, have the capacity to carry information instantly and efficiently, across large cortical areas. Changes of the intensity and direction of volume currents generated by rhythmic impressed currents in primary sensory areas associated with a stimulus, and deep brain nuclei, could pace and correlate distant areas in the brain which must be recruited for the task. This will provide a role for the variability which separates the average and evoked electrographic signals. The commonality must be sought in large-scale spatio-temporal patterns of activity, and not just in epoch by epoch similarity signal traces of individual channels. Not every neuron, not even every area has to be activated in each epoch, particularly if the same, biologically irrelevant stimulus is presented many, many times.

These considerations beg the discussion of the interplay of structure and function; is the cortical shape dictated simply by the requirement to pack more neurons into a small volume, or are the physical properties of the geometric arrangement necessary to subserve function. For example, the areas subserving different senses are, in general, separated as far as possible from each other, a fact which will allow much freedom in the form of the large scale volume current flows, so bifurcations in cortical regions specialising in each sensory modality interfere as little as possible with each other. The two specialised areas which are very close to each other are the sensory and motor strip. MFT studies with the NYU group have revealed oscillations, around 25 Hz, of primary currents, occurring intermittently before voluntary finger movement. At each quarter cycle the oscillation crosses the central sulcus. The possible relationship of these oscillations to population coding of the direction of movement[64] is intriguing. What is the precise functional role of these oscillations is not yet known, but, whatever it is, the flow of current must match in position and direction the geometry and somatotopy on either side of the central sulcus.

Looking at the larger, body-wide scene, structure and function integration must be maintained at all levels. Our earlier discussion of the heart cycle was an oversimplification; its goal was to contrast the precise coupling of structure and mechanical function associated with the heart with the more variable programs of the brain. However, if one moves a layer deeper in the analysis, a surprisingly more complex picture emerges[63]. The heart and brain dynamics are not independent, one affects the other in a profound way. Even a cursory inspection of MEG signals shows a weak correlation between the cardiac cycle and evoked or epileptic activity, but this must be the topic of future works.

13.9 Conclusion

The dawn of Magnetoencephalography (MEG) as a clinically useful tool, rather than an expensive research toy, must be identified with the introduction of the BTi and SIEMENS AG multichannel probes (both with 37 channels), and the development of theoretical models, practical algorithms and associated software to analyse effectively the resulting sets of data. Interest in MEG is based on two factors: the almost instantaneous response of the signal to generator activity and the rapid rate of data acquisition, and the belief that information about the spatial distribution of the generators can be extracted from the signal. The most important new capability of multichannel MEG probes with 30 or more sensors is that they capture the magnetic field (flux gradient) with no loss of spatiotemporal coherence. This allows single epoch analysis, or at least consideration of signal assemblies other than the simple average; it is therefore possible to compare both the mean response and its variability from exactly the same generators under different conditions. The potential for functional imaging is realised with analysis tools and software capable of extracting details in the entire spatiotemporal window made accessible by the measurements, and not just at convenient peaks of the average signal. MFT analysis of single epoch data has shown that a distribution of activity, fairly well localised but not point-like is evident in this signal. The shape and variation of these MFT solutions are consistent with the anatomy and physiology.

If the brain is likened to a city, then techniques such as MRI and CAT scans provide a blue print of the buildings, roads and subways, while techniques like fMRI and PET provide measures of how traffic and people accumulates over a few days. MEG and to a lesser extend EEG provide the snapshots of moment to moment city life, showing how the instantaneous distribution changes in timescales which correspond to the time it takes to cross major roads or travel from one subway station to the next. The images display fairly well the surface distribution (cortical activity), but the deep distribution (subcortical activity) leads to a fuzzy description because the trains are not directly visible. Given the map and the measures of accumulated activity the entire pattern can be inferred from the data.

It is possible to increase the MEG resolution substantially through the introduction of detailed anatomical constraints, but at the moment the most promising way forward appears to be the pooling of information from MRI, fMRI, PET and MEG and EEG. Although the precise way of achieving this integration is somewhat unclear, a non-invasive and truly functional imaging capability of the brain and body is within our reach. The impact will is likely to be strongest in the mind sciences[65], where often hard quantifiable correlates of pathology are rare.

Bibliography

[1] G.M. Baule and R. McFee , Detection of the magnetic field of the heart. *Am. Heart J.,* **66** (1963) 95-96.

[2] D. Cohen, Magnetoencephalography: Evidence of magnetic fields produced by alpha-rhythm currents. *Science* **161** (1968) 784-786.

[3] D. Cohen, Magnetoencephalography: Detection of the brain's electrical activity with superconducting magnetometer. *Science* **175** (1968) 664-666.

[4] M. Hämäläinen, R. Hari, R.J. Ilmoniemi, J. Knuutila and O.V. Lounasmaa , Magnetoencephalography - theory, instrumentation, and applications to the noninvasive studies of the working human Brain. *Rev. Mod. Phys.* **65** (1993) 413-497.

[5] M. Keidel, W.D. Keidel, W.S. Tirch and S.J. Pöppl , Possible central oscillators synchronizing brain waves and muscle oscillations in man?, in *Chronobiology & Chronomedicine, Basic research and applications,* G. Hildebrandt, R. Moog and F. Raschke (Eds). Peter Lang, Frankfurt (1987), pp. 202-208.

[6] R.P. Grease, Biomedicine in the Age of Imaging, *Science* **261** (1993) 554-561.

[7] A.A. Ioannides, Comparison of MEG with other functional imaging techniques. *Clin. Physics and Physiol. Meas.*, **12** (1991) A23-A28.

[8] J.C. Mazziotta, S.C. Huang, M.E. Phelps, R.E. Carson, N.S. MacDonald and K. Mahoney, A non-invasive positron computed tomography technique using oxygen-15 labeled water for the evaluation of neurobehavioral task batteries, *J. Cereb. Blood Flow Metab* **5** (1985) 70-78.

[9] P.T. Fox and M.A. Mintun, Noninvasive functional brain mapping by change-distribution analysis of averaged PET images of $H_2^{15}O$ tissue activity, *J. Nucl. Med.* **30** (1989) 141-149.

[10] J.G. Colebatch, M.-P. Deiber, R.E. Passingham, K.J. Friston and R.S.J. Frackowiak, Regional cerebral blood flow during voluntary arm and hand movements in human subjects. *J. Neurophysiol.*, **65** (1991) 1392-1401.

[11] J.W. Belliveau, D.N. Kennedy, R.C. McKinstry, B.R. Buchbinder, R.M. Weisskoff, M.S. Cohen, J.M. Vevea, T.J. Brady and B.R. Rosen, Functional mapping of the human visual cortex by magnetic resonance imaging, *Science* **254** (1991) 716-719.

[12] R.W. Thatcher *et al.*, Eds., *Functional Neuroimaging: Technical Foundations* (Academic Press, Orlando, 1993).

[13] O.V. Lounasmaa, *Experimental principles and methods below 1K* (Academic Press, London, 1974).

[14] D. Cohen, B.N. Cuffin, K. Yunokuchi, R. Maniewski, C. Purcell, G.R. Cosgrove, J. Ives, J.G. Kennedy and D.L. Schomer, MEG versus EEG localization test using implanted sources in the human brain, *Ann. Neurol.* **28** (1990) 811-817.

[15] R. Hari, M. Hämäläinen, R.J. Ilmoniemi and O.V. Lounasmaa, MEG versus EEG localization test (Letter to the Editor), *Ann. Neurol.* **30** (1991) 222-224.

[16] S. Williamson, MEG versus EEG localization test (Letter to the Editor), *Ann. Neurol.* **30** (1991) 222.

[17] G. Annogiannakis, J.-M. Badier, G. Barret, S. Erné, R. Fenici, P. Fenwick, F. Grandori, R. Hari, R. Ilmoniemi, F. Mauguière, D. Lehmann, F. Perrin, M. Peters, G.-L. Romani, and P.M. Rossini, A consensus statements on the relative merits of EEG and MEG, Electroencephalogr. *Clin. Neurophysiol.* **82** (1992) 317-319.

[18] J.P. Wikswo,Jr, A. Gevins and S.J. Williamson, The future of the EEG and MEG, Electroencephalogr. *Clin. Neurophysiol.* 87 (1993) 1-9.

[19] S.J. Williamson, and L. Kaufman, Biomagnetism, *J. Magn. Magn. Mat.* **22** (1981) 129-201.

[20] H. von Helmoltz, "Ueber einige Gesetze der Vertheilung elektrischer Ströme in körperlichen Leitern, mit Anwendung auf die thierisch-elektrischen Versuche". *Ann. Phys. Chem.* **89** (1853) 211-233, 353-377.

[21] J.C. Mosher, P.S. Lewis, R. Leahy, and M. Singh, Multiple Dipole Modelling of Spatio-Temporal MEG Data, in *Digital Image Synthesis and Inverse Optics,* A. F. Gmitro, P. S. Idell and I. J. Lahaie (Eds.). Proc SPIE 1351, (1990), pp 364-375.

[22] R.O. Schmidt, Multiple emitter location and signal parameter estimation. *IEEE Trans. Antennas and Propagation,* **AP-34** (1986) 276-280.

[23] A.A. Ioannides, J.P.R. Bolton and C.J.S. Clarke, Continuous probabilistic solutions to the biomagnetic inverse problem. *Inverse Problems,***6** (1990) 523-542.

[24] C.J.S. Clarke, A.A. Ioannides and J.P.R. Bolton, Localised and distributed source solutions for the biomagnetic inverse problem I, in *Advances in Biomagnetism,* S.J. Williamson, M. Hoke, G. Stroink and M. Kotani (Eds.). Plenum Press, New York 1989, pp 587-590.

[25] A.A. Ioannides, J.P.R. Bolton, R. Hasson and C.J.S. Clarke, Localised and distributed source solutions for the biomagnetic inverse problem II, in *Advances in Biomagnetism,* S.J. Williamson, M. Hoke, G. Stroink and M. Kotani (Eds.). Plenum Press, New York 1989, pp 591-594.

[26] C.J.S. Clarke and B.S. Janday, Probabilistic methods in a biomagnetic inverse problem. *Inverse Problems,* **5** (1989) 483-500.

[27] C.J.S. Clarke, Probabilistic methods in a biomagnetic inverse problem. *Inverse Problems,* **5** (1989) 999-1012.

[28] M.S. Hämäläinen, and R.J. Ilmoniemi, Interpreting Measured Magnetic Fields of the Brain: *Estimates of the Current Distributions.* Preprint TKK–F–A559, Helsinki University of Technology (1984).

[29] M.J. Liu, R. Hasson, and A.A. Ioannides, "A transputer-based system for Magnetic Field Tomography", in *Transputer Applications and Systems '93*, R. Grebe, J. Hektor, S.C. Hilton, M. Jane and P.H. Welch (eds.). IOS Press, Amsterdam 1993, Vol. 2, pp. 1290-1297.

[30] A.A. Ioannides, R. Muratore, M. Balish and S. Sato, In vivo validation of distributed source solutions for the biomagnetic inverse problem. *Brain Topography* **5** (1993) 263-273.

[31] P.B.C. Fenwick, A.A. Ioannides, G.W. Fenton, J. Lumsden, P. Grummich, H. Kober, A. Daun and J. Vieth, Estimates of Brain Activity using Magnetic Field Tomography in a GO/NOGO avoidance paradigm, *Brain Topography* **5** (1993) 275-282.

[32] A.A. Ioannides, E. Hellstrand and K. Abraham-Fuchs, Point and distributed current density analysis of interictal epileptic activity recorded by Magnetoencephalography. *Physiol. Meas.* **14** (1993) 121-130.

[33] A.A. Ioannides, K.D. Singh, R. Hasson, S.B. Baumann, R.L. Rogers, F.C. Guinto and A.C. Papanicolaou, Comparison of Current Dipole and Magnetic Field Tomography Analyses of the Cortical Response to Auditory Stimuli, *Brain Topography* **6** (1993) 27-34.

[34] E. Hellstrand, K. Abraham-Fuchs, B. Jernberg, L. Kihlström, E. Knutson, C. Lindquist, S. Schneider and A. Wirth, MEG localization of interictal epileptic focal activity and concomitant stereotactic radiosurgery. A non-invasive approach for patients with focal epilepsy, *Physiol. Meas.* **14** (1993) 131-136.

[35] C. Gallen, B. Schwartz, C. Pantev, S. Hampson, D. Sobel, E. Hirsschkoff, K. Rieke, S. Otis and F. Bloom, Detection and localization of Delta frequency activity in human stroke, in *Biomagnetism: Clinical aspects,* M. Hoke, S.N Erné and G. L. Romani (Eds.). Elsevier Science, Amsterdam (1992), pp. 301-305.

[36] J. Vieth, H. Kober, G. Sack, P. Grummich, S. Friedrich, A. Mõger, E. Weise, A. Daun and H. Pongratz, The efficacy of the discrete and the quantified continuous dipole-density-plot (DDP) in multichannel MEG, in *Biomagnetism: Clinical aspects,* M. Hoke, S.N Erné and G. L. Romani (Eds.). Elsevier Science, Amsterdam 1992, pp. 321-325.

[37] K.D. Singh, A.A. Ioannides, R. Hasson, U. Ribary, F. Lado and R. Llinas, Extraction of dynamic patterns from distributed current solutions of brain activity, in *Biomagnetism: Clinical aspects*, M. Hoke, S.N Erné and G. L. Romani (Eds.). Elsevier Science, Amsterdam 1992, pp. 767-771.

[38] A.A. Ioannides, R. Hasson and J.P.R. Bolton, *Spatial and temporal evolution of brain activity.* Open University video, July 1989.

[39] A.A. Ioannides, *Searchlights into the brain,* Open University video, August 1993.

[40] M. Scherg, Fundamentals of Dipole Source Potential Analysis, in *Advances in Audiology,* (eds Grandori, F., Hoke, M., Romani, G.L. (Eds.). Karger, Basel 1990, pp. 40-69 .

[41] A.M. Dale and M.I. Sereno, Improved Localization of Cortical activity by combining EEG and MEG with MRI cortical surface reconstruction: a linear approach. *Journ. of Cognit. Neurosc.* **5** (1993) 162-176.

[42] A.A. Ioannides, K.M. Stephan, P.B.C. Fenwick, J. Lumsden, G.W. Fenton, M.J. Liu, J. Vieth, K.C. Squires, D. Lawson, R. Myers, G.R. Fink, and R.S.J. Frackowiak, Analysis of MEG signals from a GO/NOGO avoidance paradigm and comparison of estimates of brain activity using PET, in *BIOMAG'93* (Elsevier, Amsterdam, in press).

[43] A.A. Ioannides, P.B.C. Fenwick, J. Lumsden, M.J. Liu, P.D. Bamidis, K.C. Squires, D. Lawson, and G.W. Fenton, *Activation of discrete brain areas during cognitive processes: Results from Magnetic Field Tomography* (1994) (In preparation).

[44] P.L. Nunez, *"Electric Fields of the Brain"*, New York-Oxford, 1981.

[45] B. Hjorth, An on-line transformation of EEG scalp potentials into orthogonal source derivations. *Electroenc. clin. Neurophysiol.,* **39** (1975) 526-530.

[46] D.M. Tucker, Spatial sampling of head electrical fields: the geodesic sensor net. *Electroenceph. clin. Neurophysiol.* **87** (1993) 154-163.

[47] A.S. Gevins, P. Brickett, B. Costales, and B. Reutter, Beyond Topographic Mapping: Towards Functional-Anatomical Imaging with 124-channel EEGs and 3-D MRIs. *Brain Topography* **3** (1990) 53-64.

[48] A.A. Ioannides, Graphical solutions and representations for the biomagnetic inverse problem, in ***Advances in electronics and electron physics, supplement 19, inverse problems: an interdisciplinary study,*** Sabatier PC (ed). Academic press, Orlando (1987), pp. 205-216.

[49] F.N. Alavi, J.G. Taylor, and A.A. Ioannides, Estimates of current density distributions: I. Applying the principle of cross-entropy minimization to electrographic recordings. *Inverse Problem* **9** (1993) 623-639 .

[50] K.D. Singh, A.A. Ioannides, N. Gray, H. Kober, H. Pongratz, A. Daun, P. Grummich, and Vieth, Distributed Current and single current dipole analyses of ipsilateral/contralateral auditory N1 responses, in *BIOMAG'93* (Elsevier, Amsterdam 1993, in press).

[51] U. Ribary, A.A. Ioannides, K.D. Singh, R. Hasson, J.P.R. Bolton, F. Lado, A. Mogilner and R. Llinas, Magnetic Field Tomography *Proc. Natl. Acad. Sci. U.S.A.* **88** (1991) 11037-11041.

[52] D. Brown, P.B.C. Fenwick and R.C. Howard, The Contingent Negative Variation (CNV) in GO/NOGO avoidance task: relationships with personality, subjective state. International Journal of *Psychophysiology* **7** (1988) 35-45.

[53] R. Hari, R. Salmelin, S.O. Tissari, M. Kajola, and V. Virsu, Maintaining perception during eye-blinks. *Nature* (in press 1994).

[54] C. Pantev, S. Makeig, M. Hoke, R. Galambos, S. Hampson and C. Gallen, Human auditory evoked gamma-band magnetic fields *Proc. Natl. Acad. Sci. U.S.A.* **88** (1991) 8996-9000.

[55] S. Zeki, A vision of the brain (Blackwell, Oxford, 1993).

[56] G. Tononi, O. Sporns and G.M. Edelman, Reentry and the problem of integrating multiple cortical areas: simulation of dynamic integration in the visual system. *Cerebral Cortex* **2** (1992) 310-335.

[57] R. Llinas and U. Ribary, Coherent 40-Hz Oscillations characterizes dream state in Humans, *Proc. Natl. Acad. Sci. U.S.A.* **90** (1993) 2078-2081.

[58] L.C. Liu and A.A. Ioannides. Single epoch analysis of MEG signals, in *BIOMAG'93* (Elsevier, Amsterdam, in press).

[59] H. Titinen, J. Sinkkonen, K. Reinikainen, K. Alho, J. Lavikainen and R. Nätänen, Selective attention enhances the auditory 40-Hz transient response in humans. *Nature* **364** (1993) 59-60.

[60] A.A. Ioannides, E. Hellstrand, P.D. Bamidis and K. Abraham-Fuchs, Estimates of brain activity from unaveraged interictal multichannel Magnetoencephalographic signals, in *BIOMAG'93* (Elsevier, Amsterdam, in press)

[61] A.A. Ioannides, Brain function as revealed by current density analysis of magnetoencephalography signals. *Physiol. Meas.* **14** (1993) A75-A80.

[62] J.G. Taylor and F.N. Alavi, A global competitive network for attention. *Neural Networks World* **5** (1993) 447-502.

[63] P.J. Kilner, G.Z. Yang, R.H. Mohiaddin, D.N. Firmin and D.B. Longmore, Helical and Retrograde Secondary Flow patterns in the Aortic arch studied by three-directional Magnetic resonance velocity mapping. *Circulation* **88** (1993) 2235-2247.

[64] A.P. Georgopoulos, M. Taira, and A. Lukashin, Cognitive neurophysiology of the motor cortex *Science* **260** (1993) 47-52.

[65] P.B.C. Fenwick, A.A. Ioannides and J. Lumsden, The use of MEG in Psychiatry, in *Imaging of the Brain in Psychiatry and related fields* (Ed. K. Maurer) Springer-Verlag, Heidelberg 1993, pp. 185-202.

Chapter 14

Log-Normal Distribution of Physiological Parameters and the Coherence of Biological Systems

Chang-Lin Zhang and Fritz-Albert Popp

14.1 Introduction

It is well-known that physiological parameters such as blood pressure, drug tolerance, body size, survival rate of living populations and so on do not follow a Gaussian (normal) distribution but a lognormal one[1]. It has been shown, that the probability distribution of skin resistance values of healthy people fits a lognormal curve much better than those of afflicted with a wide variety of diseases patient[2,3] (see also Fig. 14.1).

This physiological lognormal distribution has been interpreted in terms of a "multiplicative Gestaltungs-principle of nature"[4,5], which, however, has never been explained at a more basic level. In this paper we present a first attempt to do so. The "multiplicative Gestaltungs principle" is traced to the formation of coherent states in living systems[6,7]. This offers a new tool for understanding certain kinds of diseases in terms of their deviations from the optimal regulation principle.

14.2 Normal and Lognormal Distribution

The difference between normal and log-normal distribution can be explained in terms of random fluctuations of (1) either the original measurement values, or (2) their logarithms, for conductivity or other parameters, around some mean value giving rise respectively either to a normal distribution or to a lognormal one.

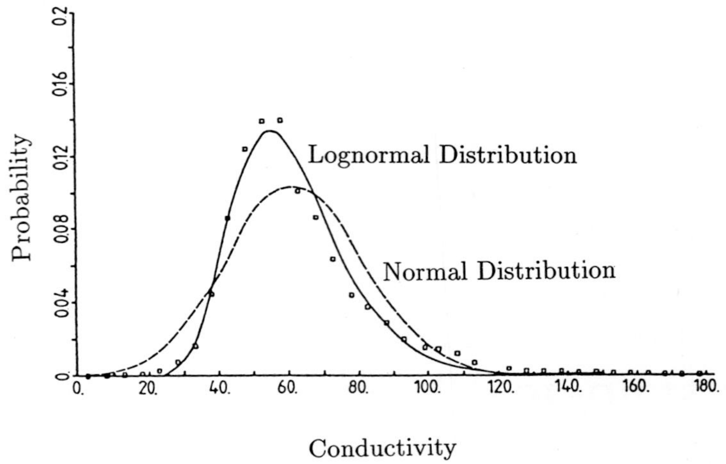

Figure 14.1: *The probability distribution of 18000 skin-conductivity values of 200 healthy people follows a lognormal curve (continous line) instead of a normal one.*

For the measurement values x_e, we can write

$$x_e(t) = \sum_{j=1}^{n} x_j p(x_j, t) \tag{14.1}$$

where x_j is a fixed set of values within the system under investigation and $p(x_j, t)$ represents the propability of the appearance of x_j at a time t. A random fluctuation of $p(x_j, t)$ leads to fluctuations of x_e according to

$$\Delta x_e(t) = \sum_{j=1}^{n} x_j \Delta p(x_j, t) \tag{14.2}$$

Fig. 14.2 shows an example where x_e is the sum of n numbers which are obtained by throwing a die n times. The values are distributed around a mean value according to a function which, after normalization approaches more and more a normal distribution, the higher the number n and the more values x_e are taken into account.

If, however, instead of following the "additive" principle in Eq. (14.1), the measurement values $x_e(t)$ depend on the product of single probabilities according to

$$x_e(t) = f(x_1, x_2, \cdots, x_n) \prod_{j=1}^{n} p(x_j, t) \tag{14.3}$$

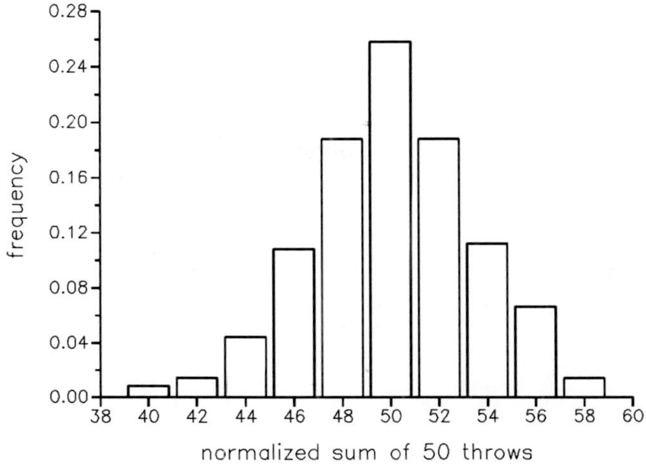

Figure 14.2: *The (normalized) sum of 50 throws of a die follows approximately a normal distribution.*

we can write

$$\ln x_e(t) = \ln f(x_1, x_2, \cdots, x_n) + \sum_{j=1}^{n} \ln p(x_j, t) \tag{14.4}$$

such that

$$\Delta \ln x_e(t) = \sum_{j=1}^{n} \frac{\Delta p(x_j, t)}{p(x_j, t)} = \sum_{j=1}^{n} \lambda_j \Delta p(x_j, t) \tag{14.5}$$

where $\lambda_j = \frac{1}{p(x_j)}$ and $p(x_j, t) = p(x_j)$.

Consequently, the logarithms of x_e according to (14.3)-(14.5) follow, in case of an explicitly time-independent function $p(x_j)$, the same mathematical dependence as x_e itself according to (2). This means that a normal distribution based on an "additive" principle turns into a lognormal one as soon as this additive principle turns into a "multiplicative" principle due to (3), under some constraints on the time-behaviour of $p(x_j, t)$, (see also Fig. 14.3).

Note that, in any case, the $\Delta p(x_j, t)$ for $j = 1, 2, \ldots, n$ are subjects of completely random fluctuations.

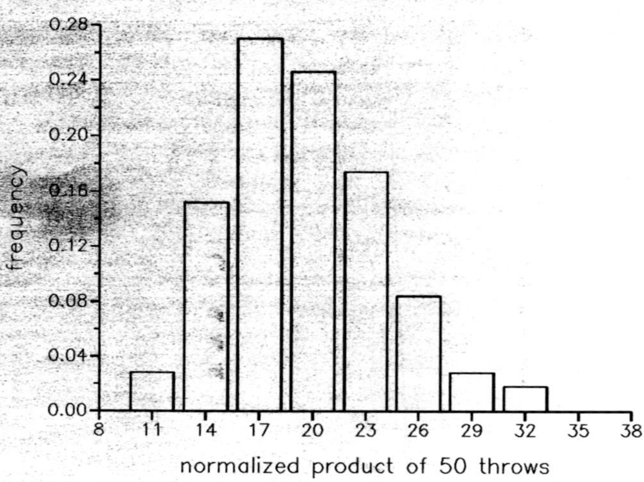

Figure 14.3: ***The (normalized) product of 50 throws of a die follows approximately a lognormal distribution.***

14.3 Organization of Possibilities

Any biological system can be considered as consisting of a network of n equal or essentially equal subunits (like molecules or cells) which are coupled together in such a way that the interactions and the distribution of the interactions over all the units determine the values of the physiological parameters at any instant. For the purpose of our discussion it is not necessary (and, of course, also not possible) to specify the nature of the subunits and their interactions.

Let us distinguish two limiting cases, i.e.

1. the units behave like perfect individuals such that

$$\underbrace{① + ① + \ldots + ① = n}_{\mathbf{n-times}} \tag{14.6}$$

   or

2. the units are strongly coupled to each other by a stiff link such that only a single one new entity results from coupling of always n units:

$$\underbrace{① + ① + \ldots + ① = 1}_{\mathbf{n-times}} \tag{14.7}$$

The circle around the numbers denotes the symbolic character of this kind of mathematics.

As the measuring values x of any physiological parameter will depend on the coupling forces and their distribution between n units, in case of Eq. (14.6), a normal probability distribution $p(x)$ is expected in view of the independence of all the links in the measuring curcuit, while Eq. (14.7) will give rise to a δ-function distribution. However, Eq. (14.6) and Eq. (14.7) are only limiting cases of couplings between n units, i.e. $\binom{n}{1}$ and $\binom{n}{n}$. The most general case can be written as

$$M = \sum_{j=0}^{n} p(n,j)\binom{n}{j} \tag{14.8}$$

$$\sum_{j=0}^{n} p(n,j) = 1 \tag{14.9}$$

where $p(n,j)$ is the probability of distributing j couplings over n units in such a way that all possible kinds of couplings ($\binom{n}{j}$, where j runs from 0 to n) are taken into account. If all possible couplings $\binom{n}{j}$ are equally likely, we have $p(n,j) = \frac{1}{n}$, and hence $p(n,j)$ do not dependent on j. The value M then takes its maximum which is $M = \frac{2^n}{n}$. This can be regarded as an optimization of the regulation of the system, as all the possible links are used. In fact, this kind of organization always leads to a lognormal distribution as soon as n, the "number of degrees of freedom", follows a normal one. Fig. 14.4 shows an example.

In order to understand the point behind this strategy, i.e. that $p(n,j)$ does not depend on j, take the case $n = 3$ as an example. Fig. 14.4 shows the maximum number of different possibilities of connecting three units. The first case corresponds to the highest individual freedom, where no coupling takes place at all. The members remain individuals, giving rise to a normal distribution. In that case we have the symbolic equation ① + ① + ① = 3.

In the second case, all members are strongly coupled together such that their individuality gets completely lost. This case is described by the symbolic relation ① + ① + ① = 1. There are further possibilities between these limiting behaviour, e.g. the cases

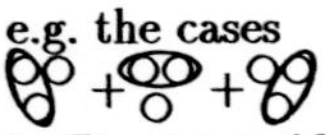

in Fig. 14.4. All at once are described by the binomial coefficients $\binom{3}{1}$, $\binom{3}{2}$ and $\binom{3}{3}$. The case $\binom{3}{0}$ has been omitted here for simplicity.

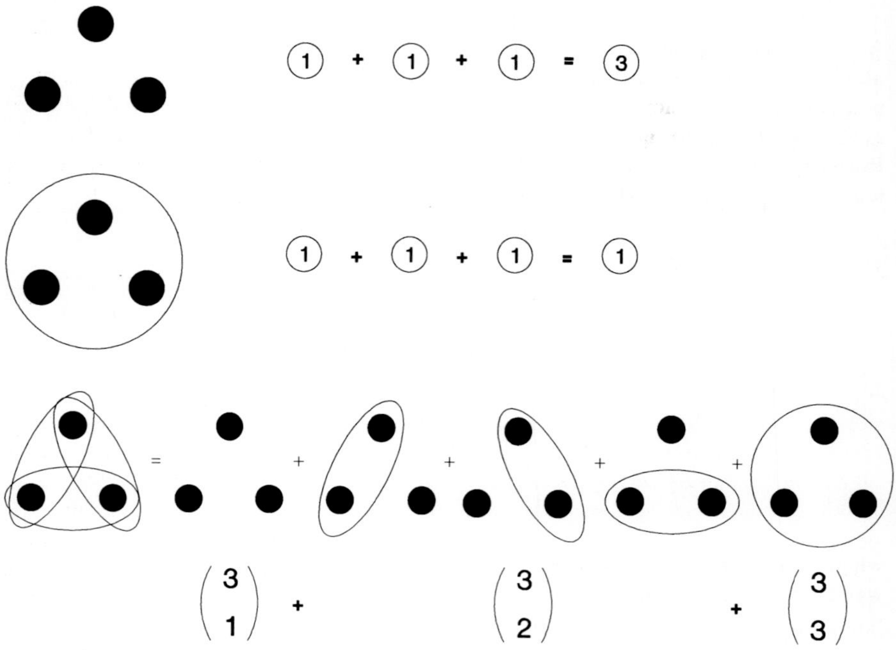

Figure 14.4: *The organization of links between three elements can be analysed in terms of Bernoulli-coefficients. No link at all provides the individuality of the elements (upper symbol). It is described by* $\binom{n}{1}$. *A stiff link between all elements reduces the number of degrees of freedom to 1 (middle symbol). It is described by* $\binom{n}{n}$. *The complete organization involves also the cases* $\binom{n}{2} \cdots \binom{n}{n-1}$. *The lower symbols show this in case of* $\binom{3}{2}$ *besides of* $\binom{3}{1}$ *and* $\binom{3}{3}$.

14.4 Lognormal Distribution and Coherence

It has been substantiated that coherent states in the strong sense of its physical definition provide the basis of biological regulation[6,7]. From Eq. (14.8), we know that the lognormal distribution of physiological parameters can be traced back to the highest efficiency of using all possible couplings between n identical elements which are considered to be responsible for the physiological state. Consequently, there should be a connection between the coherent state and the lognormal distribution, that is between the photocount statistics (PCS) of the coherent state and the Bernoulli distribution of Eq. (14.8). It is well known that the PCS of a coherent state follows a Poissonian distribution:

$$p(n,k) = \exp(-k)\frac{k^n}{n!} \tag{14.10}$$

This means that if the mean value of an ergodic and stationary coherent photon field is k, the probability of measuring n photons (more generally n bosons) follows the Poissonian distribution Eq. (14.10). Now, let us imagine that two identical units which we considered as representing the elements of the physiological state "communicate" via coherent bosons with identical mean values $k_1 = k_2 = k$. An observer of this communication measures the PCS and by registering N bosons there is the possibility that N_1 originated from source 1 and $(N - N_1)$ originated from source 2, where the number N_1 runs from 0 to N. Consequently the PCS of the connected units follows

$$\begin{aligned} p(N,2k) &= \sum_{N_1=0}^{N} p_1(N_1,k)\cdot p_2((N-N_1),k) \\ &= \sum_{N_1=0}^{N} \exp(-k)\frac{k^{N_1}}{N_1!}\exp(-k)\frac{k^{(N-N_1)}}{(N-N_1)!} \\ &= \exp(-2k)\frac{k^N}{N!}\sum_{N_1=0}^{N}\binom{N}{N_1} \end{aligned} \tag{14.11}$$

$$p(N,2k) = \exp(-2k)\frac{(2k)^N}{N!} \tag{14.12}$$

The Bernoulli distribution given by Eq. (14.11) is responsible for the connection of the two photon fields. It provides the conservation of coherence of the coupled fields and the use of all the possibilities in the interaction of the sources.

Of course, as a coherent state is completely delocalized, the highest possible number of ways to distribute n bosons over n elements is to distinguish the sequence of the elements by the phase relations of the bosons and thus to get the number $n!$ combinations for all the n elements. Consequently, apart from unimportant questions of normalization, the highest possible number of ways to distribute n bosons of a coherent field over the whole system is

$$n!p(n,k) \propto k^n \tag{14.13}$$

which is a consequence of Eq. (14.10).

This means that if the biological system represents a coherent state where the information of the bosons in terms of using the possibities is optimized, the random fluctuation of n bosons (following a normal distribution) causes a lognormal distribution of physiological parameters, where we tacitly assumed that the physiological parameter reflects just the number of ways to distribute the bosons over the system. This is reasonable as the "information" of the "physiology" can originate in that case only from the distribution of bosons over matter, that is in the pattern of excited states.

If, on the other hand, $p(n,k)$ describes the probability distribution of registering n bosons of a chaotic field with mean value k, the enumeration of the highest possible number of ways to distribute n bosons over n elements has to be based on $\binom{n}{1}$ in the series of Bernoulli-elements. This can be understood as follows: As soon as chaotic bosons excite n units (say molecules), the memory about the sequence of excitations becomes lost in view of the loss of the phase relations. Instead of $n!$ possible ways of excitations as in the case of a coherent field we have only n possibly different ways in the case of a chaotic field. This reflects just the loss of information due to the loss of phase relations. Consequently, after registering n counts, we are then unable to decide in what sequence originally distinguishable elements may have emitted the n bosons. The $p(n,k)$ of a chaotic field has then to be multiplied by n instead of $n!$ when we look for the number of ways to distribute n bosons over n elements. Consequently, apart from bunching phenomena (which are not considered here), the number of ways to distribute n chaotic bosons over a system is

$$np_{chaot.}(n,k) = n\frac{k^n}{(k+1)^{n+1}} \propto n \tag{14.14}$$

Chaotic fluctuations of the boson number n following a normal distribution will therefore cause a normal distribution of physiological parameters.

Our studies have shown that the "degree of health" correlates significantly with the agreement of the probability distribution of the skin resistance values to a lognormal distribution as well as with the disagreement to a normal distribution. In particular, cancer patients show the tendency to display a normal distribution of their skin resistance values[8].

Bibliography

[1] L. Sachs: *Statistische Auswertungsmethoden*. Springer, Berlin 1969.

[2] F.A. Popp: *Neue Horizonte in der Medizin*. Karl F. Haug Verlag. Heidelberg 1987.

[3] H. Rossmann and F.A. Popp: Statistik der Elektroakupunktur nach Voll I,II: *Ärztezeitschrift für Naturheilverfahren*. Medizinisch Literarische Verlagsgesellschaft mbH, Uelzen, 1 (1986), 51-59; 9 (1986), 623-630.

[4] H. Schuermann and M. Schirduan: Untersuchungen über Fragen der Penicillin-Dosierung. *Klin. Wschr.* **26** (1948), 526.

[5] H. Gebelein and H.J. Heite: Über die Unsymmetrie biologischer Häufigkeitsverteilungen. *Klin. Wschr.* **28** (1950).

[6] F.A. Popp, K.H. Li and Q. Gu (eds.): *Recent Advances in Biophoton Research and its Applications*. World Scientific, Singapore-London 1992.

[7] F.A. Popp: Evolution as the Expansion of Coherent States. In: *The Interrelationship between Mind and Matter* (B. Rubik, ed.), The Temple University, Marketing Graphics, Southampton, PA, 1992.

[8] F.A. Popp: Zur Theorie der Elektroakupunktur. *Erfahrungsheilkunde* **4** (1990), 240-248.

Chapter 15

Electromagnetic Sensitivity of Animals and Humans: Biological and Clinical Implications

Ulrich Warnke

15.1 Examples of the Use of the Earth's Magnetic Field by Organisms

From the beginning of biological evolution, electromagnetic interactions have been determining factors. It seems unlikely that they are not significant for organisms during a long phylogenetic history of at least 10^6 to 10^9 years.

Actually, the natural magnetic fields on the surface of the earth contain a lot of biologically relevant information. For instance, the geographical position can be characterized by the flux density, the direction and the temporal variation of the magnetic field. The time of day or the year can be determined by magnetic field influences originating from periodic movements of the sun and the moon. And the change of weather and winds is linked to the emission of electromagnetic waves in the VLF-range (the so-called spherics).

The biosphere on earth is connected to the electromagnetic fields of the universe by means of two narrow radiation bands, (1) the window in the narrow range from UV to visible light, including the near-infrared radiation with mean intensities of about 10^{-3} W/m^2, and (2) windows in the HF-range from wavelengths of about 0.1 m to 100 m, where the intensities vary from an average of about 10^{-9} W/m^2 up to 10^{-3} W/m^2 during eruptions of the sun.

Effects of the earth's field or of cancelling the earth's field as well as of weak artificial fields have been found for all living systems at all organizational levels, in bacteria, unicellular or multicellular algae, higher plants, protozoans, flatworms, insects, molluscs and vertebrates.

Magnetobacteria (Aquaspirillum magnetotacticum) make use of the total amplitude of the earth's magnetic field for orientation in the mud of the ocean. The mechanism is based on magnetite crystals which by means of their magnetic moment of about 3.6×10^{-21} Vsm, form a chain of compass needles orienting the bacteria against the thermal motion of the water molecules. The magnetic field of the earth works with an energy of about 1.4×10^{-18} J on a bacteria which is 200 times higher than that of thermal motion at 22°C[82].

Fish are navigating in the magnetic earth-field. For example, moving sharks and rays experience different strong induced electric fields. The amplitude of the fields depends on the orientation of the movement with respect to the earth's field. In addition, local water flows interfere mechanically in such a way that further electric fields are induced. Electric-field sensors in fish which work as Lorenzinian ampoules has a sensitivity as high as 10^{-7} V/m. It may work even for mechano-reception.

Among insects, the Compass-Termites (Amitermes) build their galleries in north-south direction. In other termites and some isopodes the desire for food depends on natural magnetic AC-fields (spherics) and the earth's magnetic field. The orientation and communication of bees is influenced by the earth's magnetic field. They receive information about the weather by means of electromagnetic waves in certain frequency ranges[77].

Among birds, the carrier pigeons are able to recognize variations of the earth's magnetic field which have inductions as low as nano-Tesla. Migratory birds navigate by flying into directions either perpendicular or parallel to the induction lines of the earth's magnetic field. For example, storks (Ciconia) fly perpendicular to the lines of equal induction.

There is no doubt anymore that whales perceive the earth magnetic field[76].

Humans react to atmospheric electromagnetic AC-fields of about 10 to 50 kHz in the form of central nervous perturbations, i.e. measurable subjective time-shifts of 20 s-intervals. There are, in addition, correlations between the earth's-magnetic activity and insomnia as perturbation of circadian rhythms, enzyme activities, and hormone production in the CNS, vitamine B_1-level and the iron concentration in blood, the mean skin temperature and twilight seeing.

In conclusion, there is abundant evidence of biological sensitivity to electromagnetic fields. The electromagnetic sensitivity of organisms exhibit a number of features. Weak fields are often more biologically effective than strong fields on account of amplitude or gradient "windows" for the biological effect. Similarly, frequency windows also exists so that only specific frequencies or sequences of pulses are effective. Sometimes, the effect depends on the form of the electric or magnetic pulse and on a certain complexity of the frequency spectrum. A time threshold often exists for some effects so that the field must be applied for a certain period of time. Finally, there may be specific factors which act synergistically with the electromagnetic field. For example, the effect may depend on the presence or absence of light during field exposure.

15.2 Magneto-Sensibility

Since birds, insects, fish and snails can orient with respect to the magnetic fields in air or in water, it has been suggested that a specific receptor organ is required. One may ask, however, whether such an organ is necessary. Of course, magnetic fields penetrate the body practically without attenuation. Hence, the energy of the earth's field within the body is about 10.000 times higher than that of the highest electric field in air, which amounts to 3 Megavolt/m[82].

The forces arising from quasi-static magnetic fields and the low-frequency electromagnetic fields do not need a peripheral amplifying receptor.

In all animals which can navigate by means of an endogenous compass, magnetite has been found, partially in the form of ferritin-protein[32]. In addition, the human brain contains magnetite[33]. However, whether and in what way magnetite is involved in the orientation and the magneto-sensitivity is not known, in particular as there does not appear to be neuronal contact to these magnetic "isles". In the tissues of birds, bees, fish and whales the magnetite concentration is higher than in the human brain[76]. However, most regions of our brain contain not less than about 5 Million magnetite-crystals per gram, the brain-membrane even contains a higher amounting of 100 Million crystals per gram. Since magnetite has about 10^6-times higher response an external field than normal paramagnetic or diamagnetic tissue, one has to consider the possibility of information transfer without neurons. For instance, oscillating magnetite which can be excited by ELF-fields could play a modulating role for ion-transport channels or junctions between the cells.

It is easy to show evidence of the effect on insects of relatively strong magnetic fields. We have demonstrated the following (Warnke, unpublished):

- A new bee-swarm is extraordinarily sensitive to magnetic forces and a magnet of a few mT can excite a swarm collectively as soon as the magnet approaches the wooden hive.
- Bees in cages take up a position at night which can be influenced by an artificial magnetic field of some mT.
- Dead bees and flies (and many other insects), swimming on an electrostatically neutral surface of water, can be attracted (in some cases also repelled) by an electrostatically neutral magnet of high induction.

Besides the field direction, bees can perceive also the intensity and the gradient[64] Whether the magnetite in bees is responsible for this sensitivity to magnetic fields has yet to be investigated[35]. We found in our laboratory, ferrite particles in the bristles which could be responsible for the magnetic moment.

The magnetic compass of a bird works as an inclination compass, where the axial direction of the magnetic field lines and the inclination are used for orientation[89]. After the bird crosses the magnetic equator, it reacts to the new vectorial force in a "reversed mode", finding its destination in South-Africa as to after it has started in Northern Europe. The question remains how the bird takes account of this change. The investigations of Wiltschko show evidence that the birds do not find the destination under the same conditions of the equator artificially produced in the laboratory. One assumes therefore that natural cofactors like the actual position of the sun or the constellation of the stars may be responsible. During the flight of the birds there is a permanent coupling of the magnetic earth field with the position of the sun and the constellation of the stars[7].

It has been shown that the magnetic compass of the birds functions only in the intensity range from 43 to 56 μT. However, after a habituation of three days the animals could orient due to fields of 16 up to 150 μT[65] which has been interpreted as an adaptation to the environment.

The duck-billed platypus (Ornithorhynchus anatinus) has electroreceptors in its beak which are used for the detection of prey. These receptors can respond also to DC- as well as AC-electric fields with induction voltages of about 20mV. They are connected to the trigeminus nerve. Fish with similar receptors use for the transport of electro-stimuli, the acoustic nerves instead. This shows that evolution can make use of electric and magnetic environment in quite different way. The Lorenzini ampullae of fish are able to distinguish magnetically induced electric stimuli from electrically stimulated ones[14]. Whether this can be done also by the beak receptors of the duck-billed platypus is not yet clear. Although they are specialized for detecting mechanical stimuli, they react so sensitively that even electrical forces, may provide a mechanical component and hence trigger the recept[78].

According to experiments of Schulten[67], an induction of 1mT corresponding to 10^{-7} eV suffices for perturbation of the spin orientation. Even weaker fields, as, for instance, the earth's magnetic field with only 10^{-9} eV may induce those effects[36]. Schulten et al.[69] propose a spin-compass-sensor model for magnetic orientation and navigation of birds. In contrast to the magnetite model, this model was explain the observation that birds need light in addition to magnetic field induction for orientation, and also the fact that birds cannot distinguish north- and southpole. In another paper[9], it has been shown that the products of radical reactions can be influenced by an external magnetic field. An oscillating polymer chain prevents the escape of radicals. In this concept the electron conductivity of redox systems plays as dominant a role as in photosynthesis, where photoproducts and the photoactivation of enzymes are involved[68]. Apart from that, micellae can work as cages for the radicals. Magnetic fields smaller than the earth's field may influence those systems. However, the consequences of all that for the organism are not yet clear[38].

15.3 The Magnetic Sensitivity of Certain Centers of the Brain

Since Semm et al.[70] and Welker et al.[84] found electrophysiological evidence of the magnetic sensitivity of mammalian pinealocytes, an intensive research activity started on this topic. Today, the reactivity of the pineal organ to static as well as pulsed and sinoidal magnetic fields belongs to the best-reproduceable effects. In addition, this organ reacts also to electric fields, however only at unphysiologically high amplitudes of 2-130 kV/m, which means that the electric field effect alone may not likely be of biological significance.

The pineal organ has high strategic significance for the physiology of mammalians. It controls certain circadian functions linked to hormone production, the immunological response as well as to regeneration, to deep sleep, the emotions and behaviour. It has morphological and biochemical similarities to the retina. Thus, it contains many proteins which are specific also for the retina, like rhodopsin kinase, retinol-binding protein, retinal protein S-antigen. The pineal organ has neuronal connections to the eye, e.g. with the suprachiasmatic and the paraventricular nuclei, with the neurons of the sympathicus in the spinal cord and also with the cervical ganglions. Very surprising is the finding that the magneto-sensitivity of the pineal organ is connected to the undisturbed function of the retina[56]. According to these experiments the magnetic stimuli can influence the pineal organ only by means of the retina[51,52,60]. Consequently, the retina takes part in the magnetoreceptor system as soon as it is exposed to weak light irradiation. It is obvious that magnetic fields play a role in the whole visual processing of humans. When the earth's magnetic field is compensated down to 30 nT/m, the majority of people display a significant decrease of the critical threshold of the flicker-frequency fusion. The longer the compensation takes place, the lower the resolution of rhythmic light variations[8]. The retina of the eye and the pineal organ are sensitive to the earth's field as well as to weak magnetic AC-fields[56]. The enzymatic activity of hydroxyindole-O- methyltransferase (HiOMT) depends on the variation of the earth's magnetic field. In addition, N-acetyl-transferase (NAT) shows such a dependency. Both these enzymes are connected to the melatonin content of the retina and the pineal organ, thus influencing also twilight vision[34]. This dependence of dark-adaptation of the visual system on the earth's magnetic field has been described in an earlier paper[17]. Interesting enough the enzymes NAT and HiOMT of melatonin production can be inhibited by weak magnetic fields in the night only when there is weak red light illumination at the same time.

Also, the spontaneous neuronal activity of single cells of the carrier pigeons visual and the vestibular system can be influenced by purely magnetic orientational stimuli. Again this works only if, at the same time, an exposure to external light takes place. In the experiments concerning the vestibular system, the animal has to be tilted in addition[70]. Furthermore, of decisive importance is the wavelength of the light.

Light with wavelenghts either smaller than 450 nm or higher than 500 nm can induce newts (Notophthalamus) to recognize the direction of a magnetic field, while for light of wavelenghts around 475 nm or in darkness, an orientation due to the magnetic field direction is impossible. Wehner[83] suggested that this light-dependent reaction of specialized photoneurons could be explained in terms of electron spin resonance effects.

Our own experiments indicate that the cervical ganglions of the sympathicus and the baroreceptors may exhibit magneto-sensitivity as soon as they are activated by mechanical pressure. The cervical ganglions are connected to the suprachiasmatic nuclei (SCN) of the hypothalamus by means of sympathic adrenerge fibres bridged over the intermediolateral cell column of the spinal cord. In these nuclei a circadian rhythm of the pineal melatonin (= N-acetyl-5-methoxytryptamine) secretion is generated. This rhythm is mediated by the eyes, and thus controlled by light. Actually, light perceived by the eyes, inhibits rapidly the production and secretion of melatonin by means of the inhibition of NAT. In this way, the baroreceptors are coupled to the light and to magnetic fields.

Always, when light plays a role in the organism, melanin is involved partly as an anti-radical, and partly as an energy transducer[4]. However, there is evidence that melanin is significant for magneto-sensitivity, too. Albinos react to magnetic fields, while animals with normal pigmentation do not[50]. This result has importance also for humans, simply because melanin as well as melatonin play a key role in the regulation of the transmitter substances dopamine and norepinephrine (noradrenaline). Actually, after exposure to weak light and magnetic fields, these transmitter substances are significantly lower in the retina than in the control groups[52]. However, there is still a lack of satisfactory explanation.

An effect of weak electric fields, the influence on the zeitgeber of the human circadian rhythm, has been demonstrated already by Wever[85,86] who applied 10 Hz rectangular impulses to humans. More recent results confirm the influence of magnetic fields on hormone production in the brain, like that of endorphines[88,44]. Some recent work support the original results of Wever[62,87], and seems to rule out the influence of weak magnetic fields on multi-oscillators of the brain.

There are some papers which report on the rhythm entrainment of neurons in the cerebral cortex by means of weak exogenous magnetic and electromagnetic fields. A magnetic field of 1 nT, with a carrier frequency of 3 Hz and a modulation period of 30 s, slows the brain rhythm considerably, whereas the amplitude increases at the same time up to $70\mu V$[72]. A sinoidal magnetic field of 4.5 Hz and the same induction as the earth's magnetic field induces also the reduction of the EEG-frequencies. After switching off the field, these lower frequencies continue to run for 1-2 minutes before it returns to the normal rhythm[74,23]. In summary, frequencies of 0.01 to 10 Hz at inductions from 1nT to 100 nT change the rhythms of the EEG. Parallel to these alterations one observes often a change of the heart-pulse frequency, i.e. a lowering of about 5% ($p < 0.02$), when the induction of about 100 μT works on the head with frequencies between 0 and 10 Hz[75,43].

The deepness of sleep in humans, characterized by the electric activity of the brain theta- und delta-waves, depends on the orientation with respect to the earth meridian as well as on the influence of low frequency-beat spherics[21]. Actually, the time lag between falling asleep and the first REM period is longer for men who lie north-south compared to those in east-west direction[61]. The REM-latency period is about 7% longer for people sleeping in north-south direction compared to those in east-west. Medical experts take this difference seriously. In addition, Sanker Narayan et al.[63] found that people who lie north-south produce less 5-hydroxytryptamin (5-HT, Serotonin) and less products of the catabolism of serotonin 5-HiAA in urine, if their head is exposed to very weak magnetic fields between 5 and 50 nT and pulse frequencies between 0.1 and 10 Hz. At the same time, the electric activity of the Alpha- and Beta- Rhythm is considerably damped, and the blood flow decreases. The behaviour is described as unquiet and confused. On the other hand, in case of east-west orientation, the electric activity of the head increases significantly as well as the blood flow, and people feel quiet and relaxed.

15.4 The Key Role of Melatonin

The long-term change of pineal function is connected to cancer induction, i.e. mammary carcinoma[18], prostata carcinoma[53], ovarial carcinoma[37], and melanoma[47]. It is likely that magnetically induced electric fields in the 50 to 10 kHz range modify the β-adrenergic receptors in the membranes of pinealocytes. The relaxation time after removing the field can take 3 days. There are a variety of consequences, e.g., deterioration of the circadian rhythm, increased incidence of cancer[11,59], suppression of the immunological sytem[41], change of reproduction physiology[58], alterations of the endocrine system, a permanent feeling like "jet lag"[1], and psychic depression[26,39].

The pineal organ works with its hormone, melatonin, as an "oncostatic gland"[11]. Melatonin can prevent the growth of mammary carcinoma if it is applied in physiological concentrations[12]. Some catabolic products of melatonin can also work for inhibiting different tumours[37]. Melatonin stimulates the immunological system[41] by means of opoid receptors which themselves can be influenced by magnetic fields[31]. Melatonin controls the level of gonadotropic hormones and suppresses their excretion. At the same time patients suffering from these kinds of tumors show a significantly lowered melatonin level during the night[5,37].

The reasons why the magnetically induced inhibition of melatonin production promotes tumor growth are manifold. During the night, there is a lack of oncostatic efficiency. The steroid level in the body increases as a consequence of the diminished concentrations of melatonin. The immunological response becomes weakened, as the anabolic growth hormone somatotropin will not be secreted thus reducing the regeneration activity of the immune system. Finally, the sleep hormone, vasotocin.

is not sufficiently stimulated, and the deficiency of sleep causes deteriortion of the central nervous system (Fig. 15.1).

The increased risk of tumor formation due to magnetic fields are mainly due to exposure to artificial magnetic fields during the twilight and night, where, in particular, steep spatial and temporal field gradients may impede the adaptative or habituating response of the organism.

15.5 Spherics

In the 1960's, some valuable reviews about the effects of spherics on the organism have been published[57,3]. Spherics are indicators of labile processes in the troposphere, related to the charge-recombination of convecting clouds. They reach the population much faster than the mechanical turbulences, as they are electromagnetic waves. Spherics are, in this sense, harbingers of weather changes, and may provide biological information for preparations to the changes of weather conditions. Many people show some sensitivity to weather changes. The ground oscillation of the spherics has a frequency of about 7.5 Hz with a bandwith of some kHz. It accounts for the group velocity of electromagnetic waves induced by flashes of lightning and spreading out between the earth surface and the ionosphere as resonator modes.

The honey bees return immediately as soon as the 10-20 kHz component of the spherics increases within a distance of about 200 km[77]. At the same time, the sucking efficiency changes with the movement of the weather fronts[66]. Furthermore, bees use the receptor channel for spheric-like electromagnetic waves also for communication[79]. Technical perturbations within the frequency range of natural spherics lead to considerable loss of orientation in birds, as soon as they exceed the natural field amplitudes. This takes the form of a destruction of the V-formation of cranes. We found that the V-formation of migrating birds has its origin in electrical AC-fields, originating from static charges on the feathers of the wings[78].

Not only birds but also vertebrates and humans are influenced by spherics. Independent of the amplitude, spherics pulses shift the pH-value of the tissues. This works under natural conditions, as well as in the laboratory with simulated pulses up to higher amplitudes. Just the band between 2 to 20 kHz which provides the highest energy of atmospheric electric waves displays also give the greater effect. At the same time amputation pain as well as pain due to injured brain correlates in the laboratory and in nature with the presence of spherics[57]. In Reiter's paper, one finds indications for many deseases, as, for instance, spherics-stimulated bronchial asthma, cardio-vascular disorder, insomnia, headache, glaucoma, gall-bladder convulsions, infarct and apoplexia. It has been known for a long time that there is a significant increase of thrombosis, heart infarct as well embolies in certain weather conditions[2,13]. A distinct increase of the adhesion of thrombocytes has been documented for different electromagnetic waves which are generated during the exchange

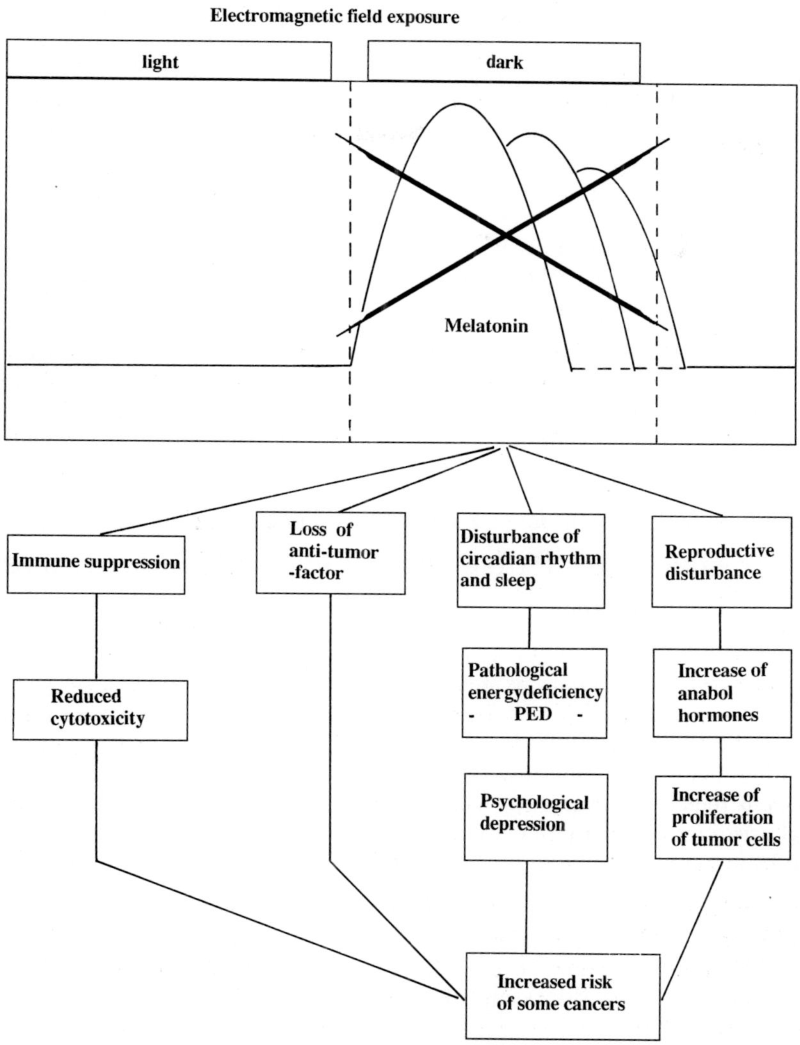

Figure 15.1: *Chronically induced inhibition of melatonin production by artifcial magnetic fields can promote tumor growth by different mechanisms.*

processes of charges in the weather front areas of the atmosphere. These spherics very likely penetrate into buildings due to their long wavelength. The average pulse train frequency is about 5 to 15 per second, covering a physiological window. Jacobi et al.[28] investigated the adhesion of thrombocytes under controlled conditions with the aid of a spherics simulator. They found a significant increase of the adhesion ($p < 0.0005$) at a carrier frequency of 10 kHz und a pulse train frequency of 10 per second. In contrast, adhesion of thrombocytes decreased at pulse train frequences of 2.5 and 20 per second, or without electric stimulation. Remedies like Dipyridamol (75 mg) and Acetylsalicylic acid (300 mg) block the spherics-induced adhesion of trombocytes. Psychically labile persons have been more affected by the change of the adhesion than stabile ones.

There are correlations between the daily working efficiency and the activity of spherics[55]. According to Jacobi[27] the head is the physiological target of spherics. If one protects the head only, the adhesion of thrombocytes goes down.

15.6 Coherent Oscillator - Triggering as a Synchronous Pulse for Neuronal Networks and Hormones

Many receptors and neurons, as well as muscles and other cell types display rhythmic variations of information carriers like cAMP, substrates such as glucose and ions like Ca^{2+} and K^{+}. The organism regulates a variety of body functions by means of coupled oscillating neuronal networks in specific centers of the brain, i.e. in the hippocampus (emotions, learning), thalamus and medulla oblongata (vital functions), pineal organ (circadian rhythm and sleeping, regeneration), baroreceptors (circulation), hypothalamus (temperature and many other functions).

These oscillations trigger the manifold physiological functions by means of the membrane potentials of the cells. The special property of the neuronal membrane oscillator is the lack of a threshold for exciting its activity. In contrast, it takes all the intermediate states between complete rest and extremely high excitation activity immediately before the next endogenous discharge. The more this state of endogenous discharge is approached, the lower is the energy density that suffices to cause a premature discharge. The endogenous oscillation activity extends the working range of a receptor or of a neuron up to the energetic levels of molecular motion[71]. Oscillation centers assimilate exogenously applied frequencies in such a way that a neuron which has been modified in its pacemaker activity synchronizes all the following neurons. In case of magnetically induced coherent electromotive forces those oscillating neuronal systems work as "active receptors" in such a way that the efficiency can get considerably amplified by means of this tendency of synchronization.

Let us take an example, where the electric activity of a neuronal network of 10^6 neurons amounts to the sum of the single potentials. It is known that without synchronization, the generator potential increases only with the squareroot of the cell number. If only 1% of the 10^6 neurons were synchronized by the coherent magnetically-inductive electromotive force and added only 0.01 mV/cell to the postsynaptic potential (EPSP), a total voltage of 100 mV would occur. All the remaining 99% of the cells would then contribute only a voltage of totally 10mV. Consequently, the signal/noise-ratio of the magnetically induced coherent signal takes the value of 100 mV/10 mV = 10. It is worthwhile to note that in such a case, even after switching off the external source, the efficiency of the synchronization remains as the organism continues to work with this new oscillator-frequency. Increasing coherence leads according to Sinz[71] to an improvement of the signal/noise-ratio. This effect described above has been experimentally verified in case of weak magnetic fields applied to populations of neurons[20]. There remains only the question, of how magnetic fields cause a change of the potential of about 0.01 mV.

15.7 Therapeutic Uses of Pulsed Electromagnetic Fields (PEMFs)

PEMFs are used for therapeutical purposes mainly in the frequency range lower than 1 kHz. The magnetic induction of the applied fields leads to current densities up to some 100 nA/cm^2 in the patient's body. The pulse forms are chosen such that they are mostly asymmetric, biphasic, quasi triangular or rectangular. The energy which is absorbed in the body is typically of the order of 10^{-10} W/mm^3. The magnetically induced electric fields compete in the organisms with natural piezo-electric fields in bones and proteins, in the bloodflow as well as in the lymph and in the interstitial fluids, with zetapotentials, contact potentials and Helmholtz double-layer potentials of boundaries between tissue and fluids of different dielectric constants. The biological currents and potentials within the organism including the communication potentials of neurons and muscles, neuronal oscillator networks, all have essential components within the ELF-range. As soon as the articially induced fields become equal or similar to the natural body fields, they are able to substitute or to interfere with the physiological fields. It is obvious that by superposition of the artificial induced and the natural fields, extremely low beat frequencies can occur which, in view of the special low-pass activity of certain receptors, can work as new message or information. In this respect it is worthwhile to note that even genes can be switched on and off by currents which are of the order of the natural biological currents within the cells. This means that displacement currents of the permanent de- and repolarisations of the membranes can interfere locally with genetic activities. It has been known for a long time that as a consequence of membrane depolarization in case of injuries, or communication, anabolic peptides (growth mediators) are produced which stimulate either regeneration (by means of protein synthesis) or the production of specific memory proteins. This mechanism can also provide a simple explanation of healing tendencies in cases where bone fractures or wounds have been treated by using pulsed magnetic fields[42,24,73]. Compared to magnetic sine-waves PEMFs have the advantage that their synergetic efficiency ranges over a broad band, i.e. the influence on the DNA and angiogenesis, while sine-waves may only affect the DNA, however without stimulating capillary-growth and growth of endothelium[90].

According to the literature many effects can occur. Some of these are, stimulation of lysozymes[48], excretion of hormones[29,40,25], regulation of enzymatic activities[46,22], increase of DNA- and collagen synthesis[49], the triggering of the calcium metabolism[10,15], modification of receptors[16], influence on the fluidity of membranes[54], and modulation of messengers like Ca^{++}, adenylcyclase, cAMP, protein kinase, inositol[30,19].

As a consequence, the organism can be influenced in many functions involving cell communication, biosynthesis, endocrine secretion and regulations, reproduction, growth and devolopment, neuronal functions and regulations, as well as behaviour[45].

These have implications for diagnosis and therapy, as, pointed out by Bassett[6] for bone fractures. In contrast to the efficiency of drugs, the effect of external magnetic fields on the body does not depend linearly on the dose[88]. Other dose effect relationships are possible[45]. The efficancy could be proportional to the temporal average of the magnetic field amplitude; to the cumulative field amplitudes; dependent on whether the field is present or absent; and dependent on the number of switching on/off- processes. The effect may arise only if the field has amplitudes and/or frequencies within a definite window, and may dependent on some external cofactors such as light.

The spontaneous effect of pulsed magnetic fields may have therapeutic implication, for instance in the increase of blood flow or improved supply of cells. On the other hand, one has to be wary of 'electrosmog' which may induce, in the long term, undesirable side effects. Sometimes, the organism develops adaptation - or resistance against those influences, probably due to changes in the conformation of membrane proteins. In order to make the therapeutic effects reproduceable, some technical effort for compensating artificial stray fields are necessary. The therapy with PEMFs not only increases the dilatation of the blood vessels, but may lead at the same time to an improved perfusion of the blood vessels by means of a lower rigidity of erythrozytes, an increase of the partial pressure of oxygen causing improved supply of cells and excretion of metabolites, increase of the breath volume and a decrease of the frequency of breathing and heartbeats resulting in a higher efficiency and a better supply of the heart muscle, and change in the pH-value of parts of the body, where macrophages are activated to remove necrotic cells, cell parts, bacteria and virus[81,80].

At the same time, enzymatic coordinates are changed which correlates with an improvement of wound-healing including an antiphlogistic effect. For healing degenerative diseases, moments may become activated. By encouraging H^+ transfer from the blood plasma to the tissues, free fatty acids become bound to albumin, which reduces the oxidation of circulating fatty acids and inhibits the production of detrimental foam cells. The increased concentration of H^+ in the muscles of the blood vessels reduces the constriction of the blood vessels, and, as a consequence of the changed dissociation status of the proteins, leads to an increase of Ca^{++}- extraction. The overshoot of H^+ and Ca^{++} induces, by means of concentration polarization, an increased ionic current towards the interior of the cells which finally change the differentiation and growth of tissue cells (Fig. 15.2).

Medical indications of specific magnetic field applications are, consequently, all regenerations, deteriorations of blood flow and of the cell nutrition.

From empirical experience we now know that the following deseases can be influenced, at least, smoker's leg, Ulcus cruris, Gangrene due to diabetes, posttraumatic oedemes and haematomes, hyper- and hypotony, headaches, pathological lipid metabolism, influenza, tendonitis, morbus bechterew, morbus sudeck, some forms of arthritis, idiopathic hip joint troubler, delayed healing of bone fractures. One has to be careful, however, in cases of patients with cardiac pacemaker and of pregnant women during the first months of pregnancy.

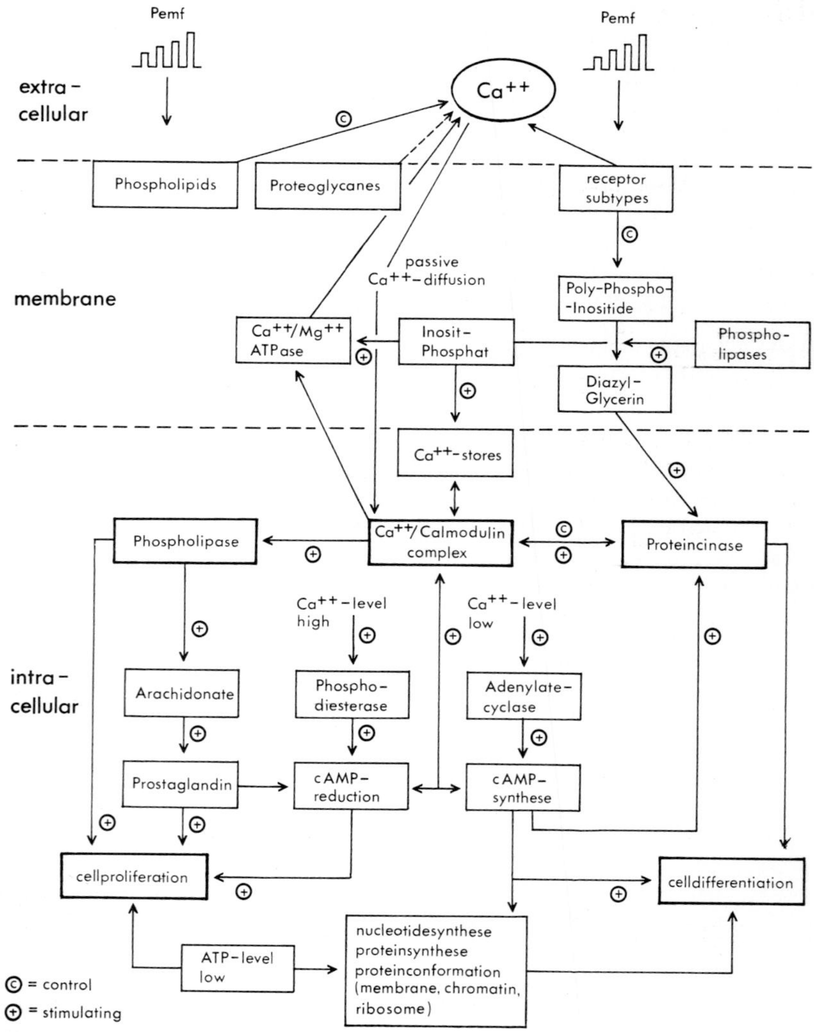

Figure 15.2: *Increased Ca^{++} ionic current towards the interior of the cells caused by pulsating electromagnetic fields (Pemf) and the consequences in different steps for cellproliferation and celldifferentiation.*

Bibliography

[1] J. Arendt, Melatonin and the human circadian system. In: *Melatonin - Clinical Perspectives*, (A. Miles, D.R.S. Philbrick, C. Thompson, eds.), Oxford: Oxford University Press (1988).

[2] K. Arnold, Zusammenhänge zwischen Wetter und Herzinfarkt im Freiburger Raum. *Z. Kreisl. Forsch.* **58**, (1969), 1141.

[3] D. Assmann, *Die Wetterfühligkeit des Menschen.* Ursachen und Pathogenese der biologischen Wetterwirkung, G. Fischer, Jena (1963).

[4] F.E. Barr, J.S. Saloma, A. Buchele, Melanin: The Organization Molecule, *Medical Hypotheses* **11**, (1983), 1-140.

[5] C. Bartsch, H. Bartsch, S.H. Fluchter, A. Attanasio, T. Das Gupta, Evidence for a modulation of melatonin secretion in men with benign and malignant tumors of the prostate: Relationship with pituitary hormones. *J. Pineal Res.* **2**, (1988), 121.

[6] C.A.L. Bassett, Fundamental and Practical Aspects of Therapeutic Uses of Pulsed - Electromagnetic Fields (PEMFs). *Critical Reviews in Biomedical Engineering* **17, 5**, (1989), 451-529.

[7] R.C. Beason, You can get there from here: Responses to stimulated magnetic equator crossing by the bobolink (Dolichonys oryzivorus). *Ethology* **91**, (1992), 75-80.

[8] D.E. Beischer, E.F. Miller, Exposure of man to low-intensity magnetic fields. *NSAM* **823**, NASA Order R-39, Pensacola, Florida (1962).

[9] R. Bittl, K. Schulten, Study of polymer dynamics by magnetic field - dependent biradical reactions. In: *Biophysical Effects of Steady Magnetic Fields* (G. Maret, J. Kiepenheuer, N. Boccara eds.), Springer-Verlag, Berlin (1986).

[10] C.F. Blackman, S.G. Benane, D.E. House, W.T. Joines, Effects of ELF (1-120 Hz) and modulated (50 Hz) RF fields on the efflux of calcium ions from brain tissue in vitro. *Bioelectromagnetics* **6**, (1985), 1.

[11] D.E. Blask, The pineal: An oncostatic gland? In: *Extremely ow Frequency Electromagnetic Fields: The Question of a Cancer* (B.W. Wilson, R.G. Stevens, L.E. Anderson, eds.), Columbus, OH: Batelle Press (1984).

[12] D. Blask, S. Hill, Effects of melatonin on cancer: Studies on MCF-7 human breast cancer cells in culture. *J. Neural Transm.* **21** (Suppl.), (1986), 433-449.

[13] H. Brezowsky, Abhängigkeit des Herzinfarktes von Klima, Wetter und Jahreszeit. *Arch. Kreil. Forsch.* **47**, (1965), 159.

[14] H.R. Brown, O.B. Ilinsky, On the mechanism of changing magnetic field detection by the ampullae of Lorenzini of Elasmobranch. *Nejrofiziologija* **10** (1), (1978), 75-83.

[15] J.J.L Carson, F.S. Prato, D.J. Drost, L.D. Diesbourg, S.J. Dixon, Time-varying magnetic fields increase cytosolic free Ca^{++} in HL-60 cells. *Am. J. of Physiol.-Cell Physiol.* **259/4** 28-4, (1990), C687-C692.

[16] A.F. Chiabreara, M. Grattarola, R. Viviani, Interaction between electromagnetic fields and cells: microelectrophoretic effect on ligands and surface receptors. *Bioelectromagnetics* **5**, (1984), 173.

[17] V.A. Chigirinskii, Possible use of dark adaption to investigate the effect of a magnetic field on the human organism. In: *Materials of 2nd All-Union Conf.* Effect of Magnetic Fields on Biological Objects, Moscow (1969), p.251.

[18] M. Cohen, M. Lippman, B. Chabner, Role of pineal gland in etiology and treatment of breast cancer. *Lancet* **2**, (1978), 814-816.

[19] A. Cossarizza, V. Borghi, F. Bersani, M. Cantini, B. De Rienzo, P. Zucchini, G. Montagnani, D. Mussini, L. Troiano, F. Tropea, E. Grassilli, D. Monti-Biasi, C. Franceschi, Effects of pulsed electromagnetic fields on the proliferation of lymphocytes from aids patients, hivseropositive subjects, and seronegative drug users. *J. Bioelectr.* **8/2**, (1989), 227-237.

[20] V.I. Danilov, V.V. Parshintsev, V.V. Turkin, Influence of a single pulse of a magnetic field on the electrical activity of mollusc neurones. *Biofizika* **29**, 1, (1984), 109-112.

[21] A.P. Dubrov, *The Geomagnetic Field and Life.* Plenum Press, New York & London (1978).

[22] S.K. Dutta, B. Das, B. Ghosh, C.F. Blackmann, Dose Dependence of Acetylcholinesterase Activity in Neuroblastoma Cells Exposed to Modulated Radio-Frequency Electromagnetic Radiation. *Bioelectromagnetics 13*, (1992), 317-322.

[23] R.J. Gavalos, D.O. Walter, J. Hamer, W.R. Adey, Effect of low-level low-frequency electric fields on EEG and behaviour in Macaca menestrina. *Brain Res.* **18**, (1970), 491.

[24] R. Goodman, A.S. Henderson, Patterns of transcription and translation in cells exposed to EM fields: a review. In: *Mechanistic Approaches to Interactions of Electric and Electromagnetic Fields with Living Systems* (M. Blank, E. Findl, eds.), Plenum Press, New York (1987), 217.

[25] Y. Hiraki, N. Endo, M. Takigawa, A. Asada, H. Takahashi, F. Suzaki, Enhance responsiveness to parathyroid hormone and indications of functional differentiation of cultured rabbit costal chondrocytes by pulsed electromagnetic field. *Biochim. Biophys. Acta 931*, (1987), 94.

[26] I.N. Ferrier, J. Arendt, E.C. Johnstone, T.J. Craw, Reduced nocturnal melatonin secretion in chronic schizophrenia, Relationship to body weight. *Clin. Endocrinol. 17*, (1982), 181-186.

[27] E. Jacobi, Untersuchungen zur Pathophysiologie der Thrombozytenadhäsivität. *Habilitationsschrift*, Med. Fakt., Universität Düsseldorf.

[28] E. Jacobi, G. Krüskemper, Der Einfluss simulierter Sferics (wetterbedingte, elektromagnetische Strahlungen) auf die Thrombozytenadhäsivität. *Innere Medizin* **2**, (1975), 73-81.

[29] W.B. Jolley, D.B. Hinshaw, K. Knierim, D.B. Hinshaw, Magnetic field effects on calcium efflux and insulin secretion in isolated rabbit Islets of Langerhans. *Bioelectromagnetics* **4**, (1983), 103-107.

[30] D.B. Jones, The effect of pulsed magnetic fields on cyclic AMP metabolism in organ cultures of chick embryo tibiae. *J. Bioelectr.* **3**, (1984), 427.

[31] M. Kavaliers, K.P. Ossenkopp, Magnetic fields, opoid systems and day-night rhythms of behaviour. In: *Electromagnetic fields and circadian rhythmicity* (Moore-Ede et al. eds.), Birkhauser, Boston, Basel, Berlin (1992).

[32] J.L. Kirschvink, D.S. Jones, B.J. Mac Fadden ed.), *Magnetite biomineralization and Magnetoreception in Organismus: An new Biomagnetism.*, Plenum Press, New York (1985).

[33] J.L. Kirschvink, A. Kobayashi-Kirschvink, B. Woodford, Magnetite biomineralization in the human brain. *Proc. Nat. Acad. Sciences* **256** (1992).

[34] K. Krause, G. Cremer-Bartels, H.J. Kchle, U. Weitkmper, *Der Einfluss schwacher Magnetfeldvariationen auf die menschliche Dämmerungssehschärfe*, (1984).

[35] D.A. Kuterbach, B. Walcott, R.J. Reeder, R.B. Frankel, Iron-Containing Cells in the Honey Bee (Apis mellifera). *Science* **218**, 695-697.

[36] M.J.M. Leask, A physiochemical mechanism for magnetic field detection by migratory birds and homing pigeons. *Nature* **267**, 144-145.

[37] A.M. Leone, R.E. Silman, B.T. Hill, R.D.H. Whelan, S.A. Shellard, Growth inhibitory effects of melatonin and its metabolites against ovarian tumor cell lines in vitro. In: *The Pineal Gland and Cancer* (D. Gupta, A. Attanasio, R.J. Reiter, eds.), London: Brain Research Promotion (1988).

[38] T. Leucht, Interactions of light and gravity reception with magnetic fields in Xenopus laevis. *J. Exp. Biol.* **148**, 325-334.

[39] A.J. Lewy, Biochemistry and regulation of mammalian melatonin production. In: *The Pineal Gland* (R. Relkin, ed), Elsevier Biomedical, Amsterdam (1983).

[40] R.A. Luben, C.D. Cain, M.C.-Y. Chen, D.M. Rosen, W.R. Adey, Effects of electromagnetic stimuli on bone and bone cells in vitro: inhibition of responses to parathyroid hormone by low-energy fow-frequency fields. *Proc. Natl. Acad. Sci. USA* **79**, 4180.

[41] G.J.M. Maestroni, A. Conti, W. Pierpaoli, Role of the pineal gland in immunity: Circadian synthesis and release of melatonin modulates the antibody response and antagonizer immunosuppressive effect of corticosterone. *J. Neuroimmunol.* **13**, (1986), 19-30.

[42] B.R. Mc Leod, S.D. Smith, A. Liboff, K. Cooksey, Marine diatoms and ELF cyclotron resonance. *Abstract Bioelectromagneties Soc.*, Proc. 8th Annual Meeting (1988).

[43] V.N. Mikhailoeskii, N.N. Krasnogorskii, K.S. Voichishin, L.I. Gabar, V.N. Zhegar, Perception of slight variations in strength of magnetic field by people. zit. in A.P. Dubrov: *The Geomagnetic Field and Life*, Plenum Press, New York & London (1973).

[44] M.C. Moore-Ede, S.S. Campell, R.J. Reiter (ed), Electromagnetic Fields and Circadian Rhythmicity. Birkhäuser, Boston-Basel-Berlin (1992).

[45] G.M. Morgan, I. Nair, Alternative Functional Relationships Between ELF Field Exposure and Possible Health Effects: Report on an Expert Workshop. *Bioelectromagnetics* **13**, (1992), 335-350.

[46] J.C. Murray, R.W. Farndale, Modulation of collagen production in cultured fibroblasts by a low-frequency, pulsed magnetic field. *Biochim. Biophys. Acta* **838**, (1985), 98.

[47] T. Narita, H. Kudo, Effect of melatonin on B-16 melanoma growth in athymic mice. *Cancer Res.* **45**, (1985), 4175-4177.

[48] L.A. Norton, Effects of a pulsed electromagnetic field on a mixed chondroblastic tissue culture *Clinical Orthopedics* **167**, (1982), 280-290.

[49] L.A. Norton, Pulsed electromagnetic field effects on chondroblast culture. *Reconstr. Surg. Traumatol* **19**, (1985), 70.

[50] J. Olcese, S. Reuss, Magnetic field effects on pineal gland melatonin synthesis: comparative studies on albino and pigmented rodents. *Brain Res.* **369**, (1986), 365-369.

[51] J. Olcese, S. Reuss, L. Vollrath, Evidence for the involvement of the visual system in mediating magnetic field effects on pineal melatonin synthesis in the rat. *Brain res.* **333**, (1985), 382-384.

[52] R. Philo, A.S. Berkowitz, Inhibition of dunning tumor growth by melatonin. *J. Urol.* **139**, (1988), 1099-1102.

[53] H.A. Pohl, Natural electrical RF oscillation from cells. *J. Bioenerg. Biomembr.* *13*, (1981), 149.

[54] W.R. Rantscht-Froemsdorf, Beeinflussung der nervalen Information durch niederfrequente Schwankungen von Umweltfaktoren. *Z. f. angew. Bäder- und Klimaheilk.* **9**, (1962), 463-477.

[55] M.S. Raybourn, The effects of direct-current magnetic fields on turtle retinas in vitro. *Science* **220**, (1983), 715-717.

[56] R. Reiter, *Meteorobiologie und Elektrizität der Atmosphähre.* Akademische Verlagsgesellschaft, Geest & Portig K.-G., Leipzig (1960).

[57] R.J. Reiter, The pineal and its hormones in the control of reproduction in mammals. *Endocrine Rev.* **1**, (1980), 109-131.

[58] R.J. Reiter, Pineal gland, cellular proliferation and neoplastic growth: An historical account. In: *The Pineal Gland and Cancer* (D. Gupta, A. Attanasio, R.J. Reiter, eds.), Brain research Promotion, Tübingen, (1988), 41-64.

[59] S. Reuss, J. Olcese, Magnetic field effects on the rat pineal gland: Role of retinal activation by light. *Neurosci Lett.* **64**, (1986), 97-101.

[60] G. Ruhenstroth-Bauer, E. Ruther, Th. Reinertshofer, Dependence of a sleeping parameter from N-S or E-W-sleeping direction. *Z. Naturforschung* **42c**, (1987), 1140-1142.

[61] G.V. Ryzhikov, V.A. Kuz'menko, A.B. Bulnev, Effect of disturbances of the geomagnetic field on the circadian rhythm of physiological functions. *Human Physiology* **8,2**, (1983), 87-93.

[62] P.V. Sanker Narayan, M. Sarada Subrahmanyam, K. Satyanarayana, Rajeswari & T.M. Srinivasan, Effects of pulsating magnetic fields on the physiology of test animals and man. *Current Science* **53, 18**, (1984), 959-965.

[63] D.E. Schmitt, H.E. Esch, Magnetic orientation of honeybees in the laboratory. *Naturwissenschaften* **80**, (1993), 41-43.

[64] TH. Schneider, P. Semm, Influence of artificial magnetic fields (earth field intensity) in birds and mammals: Experience on behaviour and neurobiological aspects. Fachausschuss "Biomedizinische Informationstechnik", Universität Erlangen, (1992).

[65] L. Schua, Influence of meterologic elements on the behaviour of honeybees. *Z. vergl. Physiol.* **34**, (1952), 258-2.

[66] K. Schulten, Spin Polarization and Magnetic Effects in Radical Reactions. *Advances in Solid-State Physics* **22**, (1982), 61-83.

[67] K. Schulten, A. Weller, Exploring fast electron transfer process by magnetic fields. *Biophysical Journal* **24,1**, (1978), 295-305.

[68] K. Schulten, A. Windemuth, Model for a physiological magnetic compass. In: *Biophysical Effects of Steady Magnetic Fields* (G. Maret, J. Kiepenheuer, N. Boccara, eds.), Springer-Verlag, Berlin (1986).

[69] P. Semm, D. Nohr, C. Demaine, W. Wiltschko, Neural basis of the magnetic compass: Interactions of visual, magnetic and vestibular inputs in the pigeons's brain. *J. Comp. Physiol.* **A 155**, (1984), 283-288.

[70] R. Sinz, *Zeitstrukturen und organismische Regulation.*, Akademie Verlag, Berlin (1978).

[71] B.M. Vladimirskii, A.M. Volynskii, Effect of short-period fluctuations (SPF) of geomagnetic field of pc1 type on cardiovascular and nervous system of animals. zit. in A.P. Dubrov, *The Geomagnetic Field and Life*, Plenum Press, New York & London (1971).

[72] L. Vodovnik, R. Karba, Treatment of chronic wounds by means of electric and electromagnetic fields. Part 1 Literature review. *Med. & Biol. Eng. & Comput.* **30**, (1992), 257-266.

[73] E.YA. Voitinskii, B.S. Gendel's, B.I. Gol'tsman, N.I. Muzalevskaya, M.E. Livshits, Effect of a weak magnetic field on electrical activity of brain. zit in A.P. Dubrov, *The Geomagnetic Field and Life*, Plenum Press, New York & London (1974).

[74] A.M. Volynskii, Change in cardiac and nervous activity in animals of different age due to an electromagnetic field of low frequency and low strength, *Communication I.- trudy Krymskogo Med. Inst.* **53**, (1973), 7-13.

[75] M.M. Walker, J.L. Kirschvink, G. Ahmed, A.E. Dizon, Evidence that fin whales respond to the geomagnetic field during migration. *J. exp. Biol.* **171**, (1992), 67-78.

[76] U. Warnke, Physikalisch-physiologische Grundlagen zur luftelektrisch bedingten "Wetterfühligkeit" der Honigbiene. Diss. Math. Nat. Fak. Univ. Saarbrücken.

[77] U. Warnke, Avian Flight Formation with the Aid of Electromagnetic Forces: A new theory for the formation alignement of migrating birds. *J. Bioelectricity* **3,3**, (1984), 493-508.

[78] U. Warnke, Information transmission by means of electrical biofields. In: *Electromagnetic Bio-Information* (F.A. Popp, eds.), 2nd ed., Urban & Schwarzenberg, München, (1989), 74-101.

[79] U. Warnke, Survey of some working mechanisms of pulsating electromagnetic fields (PEMF). *Bioelectromagnetics and Bioenergetics* **27**, (1992), 313-320.

[80] U. Warnke, G. Altmann, Die Infrarotstrahlung des Menschen als physiologischer Wirkungsindikator des niederfrequent gepulsten schwachen Magnetfeldes. *Zeitschrift f. Physikal. Medizin*, **8,3**, (1979), 166-174.

[81] H. Weiss, *Umwelt und Magnetismus.*, Deutscher Verlag der Wissenschaften, Berlin (1991).

[82] R. Wehner, Hunt for the magnetoreceptor. *Nature* **359**, (1992), 105-106.

[83] H.A. Welker, P. Semm, R.P. Willig, J.C. Commentz, W. Wiltschko, L. Vollrath, Effects of an artificial magnetic field on Serotonin N-acetyltransferase activity and Melatonin content of the rat pineal gland. *Exp. Brain Res.* **50**, (1983), 426-432.

[84] R. Wever, Über die Beeinflussung der circadianen Periodik des Menschen durch schwache elektromagnetische Felder. *Z. Vergl. Physiol.* **56**, (1967), 111-128.

[85] R. Wever, Einfluss schwacher elektromagnetischer Felder auf die circadiane Periodik des Menschen. *Naturwissenschaften* **55**, (1968), 29-32.

[86] R.A. Wever, Circadian rhythmicity of man under the influence of weak electromagnetic fields. In: *Electromagnetic fields and circadian rhythmicity* (M.C. Moore-Ede, et al eds.), Birkhauser, Boston-Basel-Berlin (1992).

[87] B.W. Wilson, R.G. Stevens, L.E. Anderson, Effects of electromagnetic field exposure on neuroendocrine function. In: *Electromagnetic fields and circadian rhythmicity* (M.C. Moore-Ede, S.S. Campbell, R.F. Reiter, eds.), Birkhauser, Boston-Basel-Berlin (1992).

[88] W. Wiltschko, R. Wiltschko, Migratory Orientation: Magnetic compass orientation of garden warblers (Sylvia borin) after a simulated crossing of the magnetic equator. *Ethology* **91**, (1992), 70-74.

[89] G.P.A. Yen-Patton, W.F. Patton, D.M. Beer, B.S. Jacobson, Endothelial cell response to pulsed electromagnetic fields: stimulation of growth rate and angiogenesis in vitro. *J. Cell. Physiol.* **134**, (1988), 37.

Chapter 16

Fröhlich's Theory of Coherent Excitation – A Retrospective

T.M. Wu

16.1 Introduction

Many biological systems have been extensively studied with regard to their structure, function and detailed biochemical and chemical content. However, very little work has been done on the effects of microwave electromagnetic radiation on biological systems. Such work has recently taken on new importance due to the development of novel methods of measurements, which generated widespread reports on suspected effects of microwaves on biological systems including membranes, proteins, nucleic acids and cells. At the same time, theoretical interpretations have been presented in terms of collective excitations in biological systems.

Most large non-biological, physical systems (such as crystals, atomic nuclei, large molecules, spin arrays) which are built up from smaller, more fundamental constituents, possess vibrational modes characterized by a coherent motion of many constituent parts of the large system. It is likely that biological systems such as cell membranes, large biological macromolecules, or intact cells, also possess vibrational modes which will couple weakly or strongly to electromagnetic radiation. To illustrate this, let us consider a cell membrane which, in the simplest model, consists of a bilayer of phospholipid molecules interspersed with protein. These macromolecular assemblies contain dipolar molecules arranged in such a way as to give rise to an ordered array of dipoles. Each dipole is embedded in a complex structure, and it is possible that the interaction between the dipole and the underlying superstructure manifests itself in a vibrational excitation. The charge groups of the dipoles are displaced relative to each other with some frequency f. The detailed nature of this excitation is very complex, involving deformations of the underlying structure.

In addition, each dipole also interacts with all other dipoles via electromagnetic forces. Other types of interactions mediated by the structure are also possible. The net result of these mutual interactions is to spread the frequency f into a narrow band and to provide for energy sharing between the individual dipoles. If a particular dipole is perturbed, the perturbation propagates to other dipoles until the whole array of dipoles is excited to some collective quantum state, which we will call a 'dipole wave' or 'electromagnetic oscillation'. The excitation energy of this electromagnetic oscillation is expected to lie in a narrow band close to hf where h is Planck's constant.

Fröhlich[1–3] has estimated the frequency of the oscillation to be in the order of $10^{11} Hz$. His estimation is based on the relation $f = v/\lambda$, where v is the velocity of sound in organic material and λ is the wavelength. If we take $v \sim 10^5 cm/sec$, which is the approximate velocity of sound in water and many organic liquids, and $\lambda \sim 10^{-6} cm$, which is a typical dimension of a large biological molecule, then the frequency is equal to $10^{11} Hz$. In general, the expected frequencies are $10^{10} - 10^{11} Hz$ for membranes, $10^{12} - 10^{14} Hz$ for proteins or more general for certain bond-stretching groups, and $10^9 Hz$ for DNA or RNA molecules. There is a lot of experimental evidence in support of the existence of these millimeter waves in biological systems. We will only mention a few cases.

Kiselev and Zalyubovskaya[4] have examined the influence of millimeter band electromagnetic waves on isolated human and animal cells. Their experiments, as all of the others described in this section, were performed using low intensity microwaves so that thermal effects of irradiation could be excluded. Individual cells were arranged in a monolayer readily accessible to microwave exposure. This also facilitated the subsequent examination of the effects of microwave exposure. Their results showed a decrease in the viability of cells after irradiation at certain electromagnetic wavelengths (see Fig. 16.1). Within the range 5.9 to 7.5 mm, the wavelength 6.50 mm gave a conspicuously higher effect in all three cell lines. The dependence of the biological effect on the frequency of the radiation is thus of a resonant nature as predicted by Fröhlich.

The studies also indicated that millimeter wave irradiation of isolated cells resulted in damage to the cell membrane, the degeneration of protoplasm, an increase in the size of the cells and cell nuclei, and an increase in the total nucleic acid and albumin contents. All of these effects are specific to the resonant wavelength of 6.5 mm. Similarly, microwave irradiation of several viruses (such as adenoviruses, measles virus, and vesicular stomatitis virus) caused a reduction in the number of virus particles by a factor of 2 to 3. The lowered infectious activity of irradiated adenoviruses and the measles virus manifested itself in a delay of the cytopathogenic effect in a tissue culture.

Smolyanskaya and Vilenskaya[5] studied the effect of millimeter waves on the *col*-factor of *E. coli*. *Col*-factor is an extra-chromosomal genetic element whose activity is normally repressed in *E. coli*. The suppression of the *col*-factor results in the

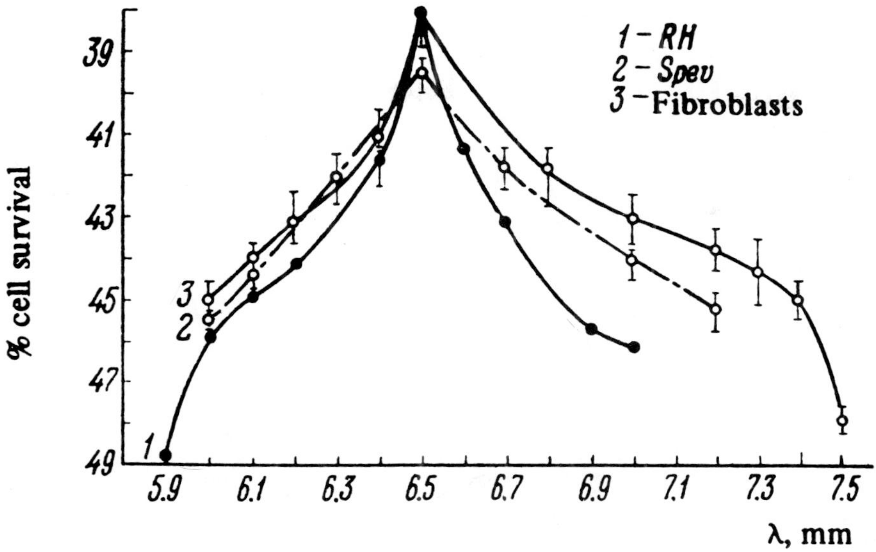

Figure 16.1: *Influence of millimeter wave irradiation on survival of tissue cultures.*

synthesis of a proteic substance called *colicin*, which causes the cells to die. The activity of colicin synthesis was determined by using the induction coefficient of colicin synthesis, $K_i = \frac{L_e/K_e}{L_c/K_c}$, where L_e and L_c are the number of cells forming colicin in the experiment and in the control respectively; K_e and K_c are the total number of colicinogenic cells in the experiment and in the control respectively.

As illustrated in Figure 16.2, the induction coefficient, K_i, of colicin synthesis has a strong correlation with certain electromagnetic wavelengths, representing a further example of resonance phenomena in biological systems. In Figure 16.3, the induction coefficient is plotted against the power flux density of the radiation. There is no noticeable variation in colicin synthesis compared to the control as the power fluxdensity increases from zero to 0.01 mW/cm^2 (not shown in Fig. 16.3). However, when density was raised to above 0.01 mW/cm^2 , the induction coefficient increases abruptly from 1.0 to more than 3.0, and remains at the same value as the power density is further increased. The power flux density at which the induction coefficient of colicin synthesis rapidly increases, ie, 0.01 mw/cm^2, can be regarded as the energy density threshold for the biological effect.

Sevast'yanova and Vilenskaya[6] also examined the effects of millimeter waves on the bone marrow of mice. They counted mouse bone marrow cells that remained undamaged by x-rays (700 rad) after prior irradiation with the millimeter waves at 10 mW/cm^2. The microwave field was turned on 60 minutes before the x-rays.

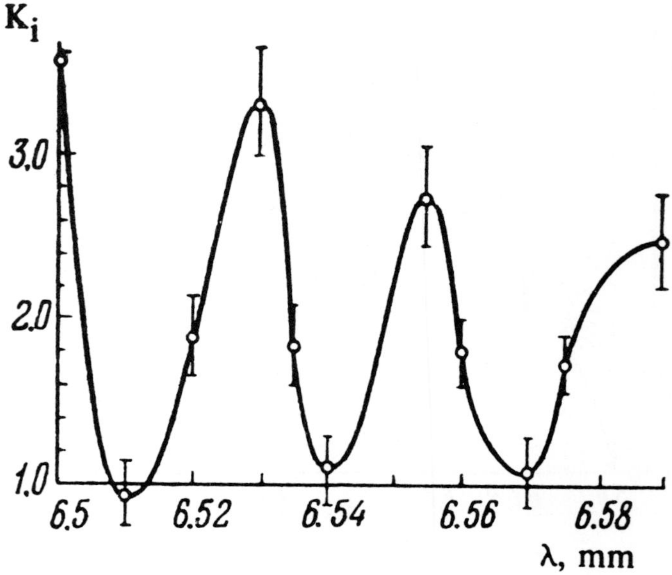

Figure 16.2: ***Induction coefficient K_i of colicin synthesis as a function of wavelength.***

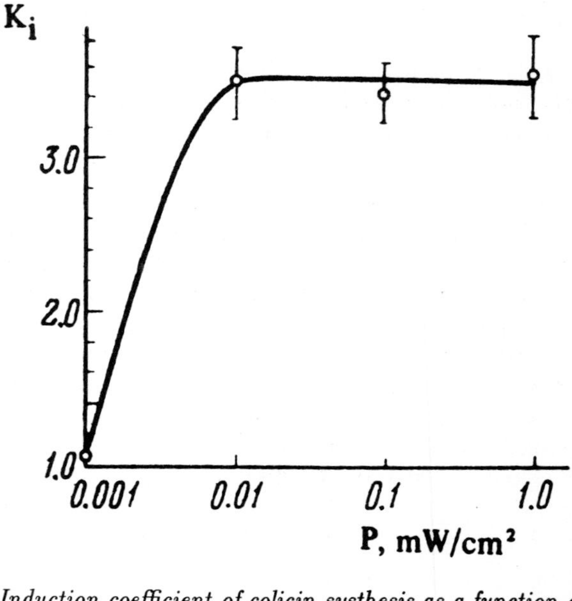

Figure 16.3: ***Induction coefficient of colicin systhesis as a function of power density.***

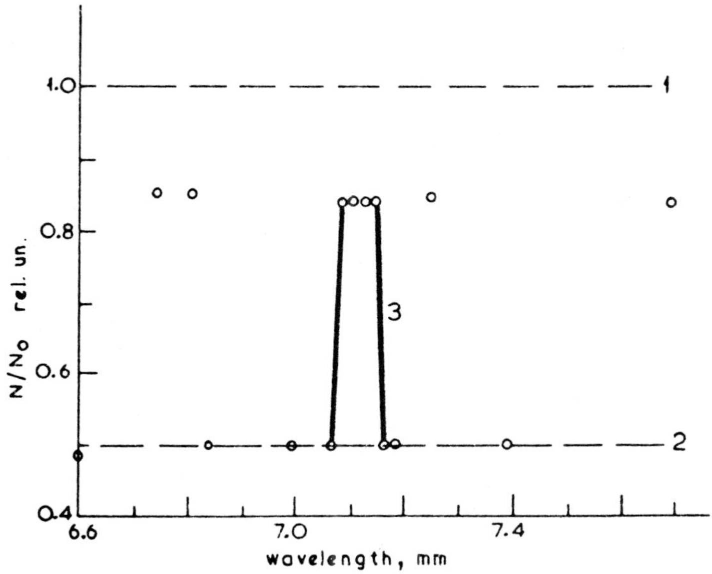

Figure 16.4: *Variation of the relative number of bone-marrow cells, N/N_o, as a function of the wavelength of irradiation. N, number of undamaged cells; N_o, number of cells without radiation. 1, control; 2, X-ray irradiated; 3. X-ray and microwave irradiated.*

Despite the fact that microwaves are absorbed in the surface skin layer of the animals up to a depth of about 3 x 10^{-2} cm, they observed a decrease in the number of bone marrow cells that were damaged by the x-rays when the animals were pre-exposed to microwaves (see Fig. 16.4).

The protective effect of the pre-exposure of the animals to millimeter waves is strongly dependent on the wavelength. The normalized undamaged cell count rises from 0.5 to 0.85 at λ values of 6.7 - 6.82 mm, 7.09- 7.16 mm, and 7.26 - 7.7 mm, whereas no protective effect appeared at the same power density at other wavelengths. This dependence of protective effects on millimeter wavelength again suggests a resonant mechanism for the action of millimeter fields on biological systems. In Figure 16.5, changes in the number of bone marrow cells of irradiated animals is plotted as a function of microwave power density. The plot shows that pre-irradiation of the animals has no influence on undamaged cell count up to a power density of 9 mw/cm^2. Thus, there is a threshold power density below which the millimeter field has no effect. As the power density is increased beyond 9 mW/cm^2, the normalized undamaged cell increases rapidly to about 0.85 and stays almost constant thereafter.

Grundler *et al*[7,8] have observed resonant behavior in the growth of yeast cells exposed

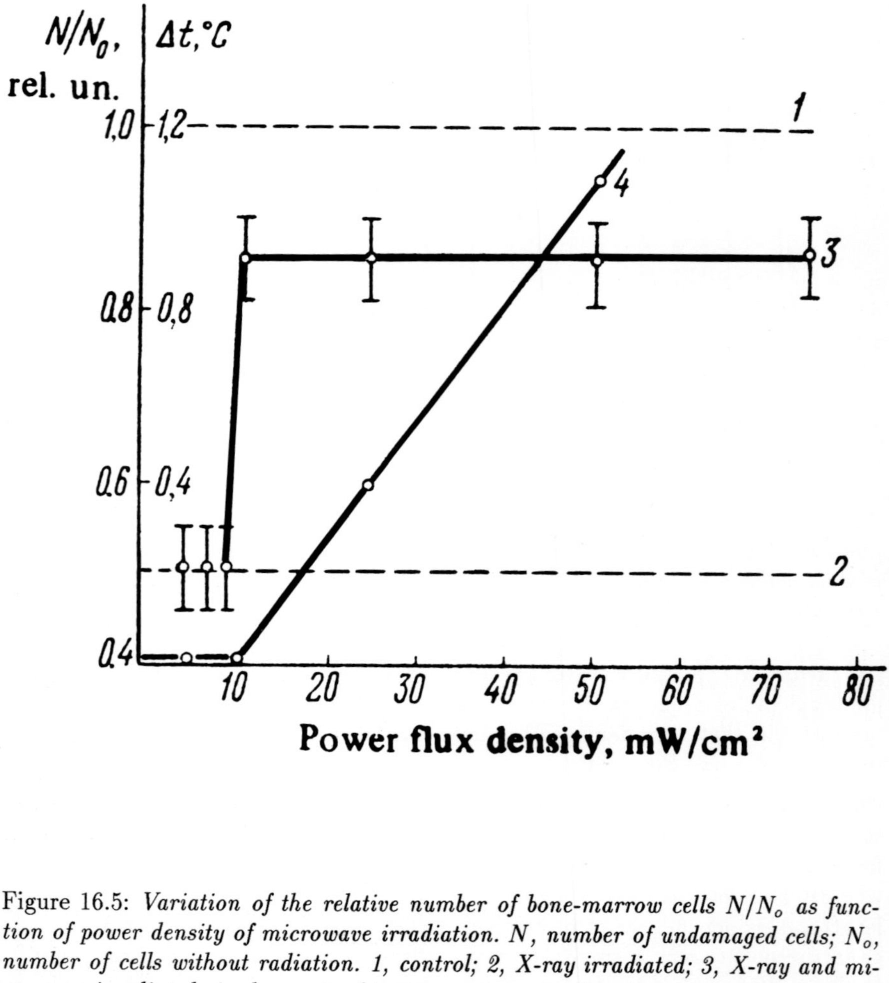

Figure 16.5: *Variation of the relative number of bone-marrow cells N/N_o as function of power density of microwave irradiation. N, number of undamaged cells; N_o, number of cells without radiation. 1, control; 2, X-ray irradiated; 3, X-ray and microwave irradiated; 4, change in skin temperature as a function of microwave power density.*

to millimeter electromagnetic radiation. They monitored the intensity of a light beam which passed through a sample of yeast cells placed in a glass cuvette; as the yeast cells multiplied, the transmittance decreased. The growth rate of the sample irradiated with microwaves was divided by the rate of a sample not irradiated. The ratio obtained showed resonance in the region 41.63 to 41.96 GHz.

16.2 Bose-Einstein Condensation in Biological Systems

In order to demonstrate the possibility of coherent behaviour in biological systems, Fröhlich[9–13] wrote down rate equations and showed that if energy is supplied above a critical rate to the branch or branches of electromagnetic vibration modes, Bose-Einstein condensation into the lowest energy state occurs. The general forms of the rate equations were dictated by requiring a Bose-Einstein distribution for thermal equilibrium when there is no energy supplied. Using microscopic theory and perturbation calculations, Wu and Austin[14–18] were able to obtain the rate equations of Fröhlich from the Hamiltonian of the biological system under study.

In this section, we consider a simple model suggested by Fröhlich. It is presumed that the biological system consists of three parts: (i) the main oscillating units of giant dipoles occurring approximately along the length of the macromolecule, (ii) the rest of the biosystem constituting a heat bath, and (iii) an external energy source which couples to the oscillating units. The interaction between the dipoles will produce a narrow band of energy, $\omega_i(\omega_o \leq \omega_i \leq \omega_{max})$, which corresponds to the normal modes of the electromagnetic vibrations. The heat bath is, of course, a very complex system. Interaction with the heat bath will lead to emission and absorption of quanta by these oscillating electromagnetic modes, and we consider processes involving one and two quanta only. The interaction involves several factors: dipoles of water and other molecules, mobile ions, certain electronic degrees of freedom, and, to some extent, elastic displacements.

Instead of rate equations, we will approach this theoretical problem with microscopic techniques[19] which are used extensively in quantum field theory. To each mode, ω_i, we assign a creation operator, a_i^+, and a destruction operator, a_i. The normal modes of oscillation will interact with the remainder of the biological system (the heat bath) which is represented by a set of independent excitation, Ω_k, associated with creation and destruction operators, b_k^+ and b_k respectively. Furthermore, the external energy supply, associated with creation operator, p_f^+ and destruction operator p_f, and excitation energy, θ_f, feeds into the electromagnetic oscillatory units and acts as impetus for the initiation of the biological effects. The Hamiltonian of the biosystem can then be written as:

$$\begin{aligned} H &= \sum_i \omega_i \, a_i^+ \, a_i + \sum_k \, \Omega_k \, b_k^+ \, b_k + \sum_f \, \theta_f \, p_f^+ \, p_f \\ &+ \frac{1}{2} \sum_{ijk} (\varphi b_j \, b_k^+ \, a_i + \varphi^* \, b_j^+ \, b_k \, a_i^+) + \frac{1}{2} \sum_{ijk} (\chi \, a_j^+ \, a_i \, b_k^+ \, + \chi^* \, a_j \, a_i^+ \, b_k) \\ &+ \sum_{if} (\varsigma \, p_f \, a_i^+ \, + \varsigma^* \, p_f^+ \, a_i) \end{aligned} \tag{16.1}$$

where φ, χ and ς are the coupling constants for the one quantum process, the two quantum process, and the energy source with the vibration modes respectively. Strictly speaking, we should consider higher order and possibly anharmonic terms; however, it can be shown[15,16] that accounting for those terms have only minor effects and, within certain limits, they may be ignored altogether. The electromagnetic oscillations are Bosons. The excitations within the heat bath and the energy source can be either Bosons, or fermions, e.g. phonons or electrons. The operators a_i^+, a_i, b_k^+, b_k, p_f^+ and p_f satisfy the commutation relations:

$$\begin{aligned} a_i \, a_j^+ - a_j^+ \, a_i &= \delta_{ij} \\ a_i^+ \, a_j^+ - a_j^+ \, a_i^+ &= a_i \, a_j - a_j \, a_i = 0 \\ b_k \, b_e^+ \pm b_e^+ \, b_k &= \delta_{ke} \\ b_k \, b_e \pm b_e \, b_k &= b_k^+ \, b_e^+ \pm b_e^+ \, b_k^+ = 0 \\ p_f \, p_g^+ \pm p_g^+ \, p_f &= \delta_{fg} \end{aligned} \tag{16.2}$$

(+ sign for fermions and - sign for bosons).

The rate of change of the number of quanta in the ith mode is given by

$$\dot{n}_i = \frac{1}{\hbar}[n_i, H] = \frac{1}{\hbar}(n_i H - H n_i) \tag{16.3}$$

with $n_i = a_i^+ a_i$; the number of quanta in the ith mode and $\hbar$ is the Planck's constant divided by 2π. Using the commutation relations in Equation (16.2) and the finite Green functions[18], the expectation value of the rate of change of the number of quanta to infinite order of interactions is

$$\begin{aligned} < \dot{n}_i > &= s_i - \Phi(T, \omega_i) \left[< n_i > e^{\beta\omega_i} - (1 + < n_i >) \right] \\ &- \sum_j \Lambda(T, \omega_i, \omega_j) \left[< n_i > (1 + < n_j >) e^{\beta\omega_i} - < n_j > (1 + < n_i >) e^{\beta\omega_j} \right] \end{aligned} \tag{16.4}$$

where the angular brackets refer to the thermal averaging in the grand ensemble appropriate to the entire system; $s_i = 4\pi \mid \varsigma \mid^2 < p_i^+ p_i >$ is the energy density supplied to the ith mode from the external source, and $\Phi(T, \omega_i)$ and $\Lambda(T, \omega_i, \omega_j)$ are given by

$$\Phi(T, \omega_i) = 4\pi \mid \varphi \mid^2 e^{-\beta\omega_i}[1 \pm N(\omega_i)] \tag{16.5}$$

$$\Lambda(T, \omega_i, \omega_j) = 2\pi \mid \chi \mid^2 \; e^{\beta(\omega_j - \omega_i)} \begin{cases} N(\omega_j - \omega_i) & \text{for } \omega_j > \omega_i \\ [1 \pm N(\omega_i - \omega_j)] & \omega_i > \omega_j \end{cases} \tag{16.6}$$

where $N(\omega)$ is the number of exitations with energy ω in the heat bath, and the plus or minus sign corresponds to bosons or fermions respectively. Note that the second and third terms in Eq. (16.4) possess exactly the same forms as Fröhlich's Ansätze for loss rate of the ith quanta in one and two quanta processes respectively.

For the stationary state, one requires that $< \dot{n}_1 >= 0$. As a result, the average of the ith quanta is

$$< n_i >= \left[1 + \frac{s_i}{\Phi(T, \omega_i) + \sum\limits_j \Lambda(T, \omega_i, \omega_j) < n_j > e^{\beta\omega_j}} \right] (e^{\beta(\omega_i - \mu_i)} - 1)^{-1} \tag{16.7}$$

where

$$e^{\beta\mu_i} = 1 + \frac{\sum\limits_j \Lambda(T, \omega_i, \omega_j)(e^{\beta\omega_j} - 1)}{\Phi(T, \omega_i) + \sum\limits_j \Lambda(T, \omega_i, \omega_j)(1+ < n_j >)} \geq 1 \tag{16.8}$$

and μ_i is the chemical potential of the ith mode.

The inequality of Eq. (16.8) together with the requirement that $< n_i > \geq 0$ dictates that $\omega_i \geq \mu_i \geq 0$. Equations (16.7) and (16.8) are exactly the same form as those derived by Fröhlich. (For details see Appendix I.) If there is no energy supplied, that is $s = 0$, then $\mu = 0$, and Eq. (16.7) becomes the thermal equilibrium distribution as required. From Eqs. (16.7) and (16.8), one notices that as s increases, μ will increase. When s exceeds a critical value s_o, μ approaches ω_o, where ω_0 is the lowest energy in the excitation band. Therefore, a large number of quanta are condensed into the lowest energy state. This is exactly the Bose-Einstein condensation in a Bose gas system when the temperature is lower than a certain critical value. In our case, the corresponding phase transition is not induced by lowering the temperature. Rather, it occurs by keeping the temperature constant and increasing the energy supply beyond the critical value s_o. From Eq. (16.8), it should be noted that when $\Lambda(T, \omega_i\omega_j) = 0$, $\mu = 0$. Thus, it is the two-quanta processes that are responsible for Bose-Einstein condensation; the larger the $\Lambda(T, \omega_i\omega_j)$, the greater the effect.

16.3 One- and Two-Dimensional Cases

In this section we will use two simple cases to demonstrate the enhancement of excitations in biological systems by external stimulation.

16.3.1 One-Dimensional Case

Let us consider a linear electric dipolar chain; this could correspond to the long chain of a protein, or a DNA chain, etc. Using the Debye model for the phonon in the heat bath, we are able to solve the coupling equations, Eqs. (16.7) and (16.8) numerically and obtain the values of the chemical potential μ and $< n_i >$ as a function of the rate of energy supply s_i. For simplicity as before, we assume $s_i = s$, for all modes, and the coupling constants $\Phi(T, \omega_i)$ and $\Lambda(T, \omega_i, \omega_j)$ are assumed to be mode independent. Thus, for a biological system at a stable temperature, these coefficients Φ and Λ can be treated as constants. In Figure 16.6, we plot μ/ω_o as function of the energy supply rate s using $\omega_o/kT = 0.1$ for a constant χ and several values of ϕ for temperature $T = 300^o K$.

One sees that μ/ω_o approaches 1 when s increases to infinity. This clearly shows that for the one-dimensional case there is no Bose-Einstein condensation as expected. Figure 16.7 shows how the number of excited phonons in the biosystem increase with s.

Though there is no Bose-Einstein condensation, as s increases, the total numbers of the excited phonons in the biosystem is greatly increased. The largest enhancement of the phonon excitation in the biosystem occurs in the very low energy modes which is just above the lowest energy ω_o. Therefore, even if there is no phase transition, the biological system will be greatly affected due to the enhancement of induced phonons when the energy supply rate is sufficiently large.

16.3.2 Two-Dimensional Case

The two-dimensional case could correspond to cell membranes. Using the same assumptions as in the one-dimensional case, we plot μ/ω_o as a function of s in Figure 16.8.

One notices that when s exceeds a certain value s_o, μ/ω_o becomes 1. Thus, the phase transition surely occurs and the phonon will begin to condense to the lowest energy

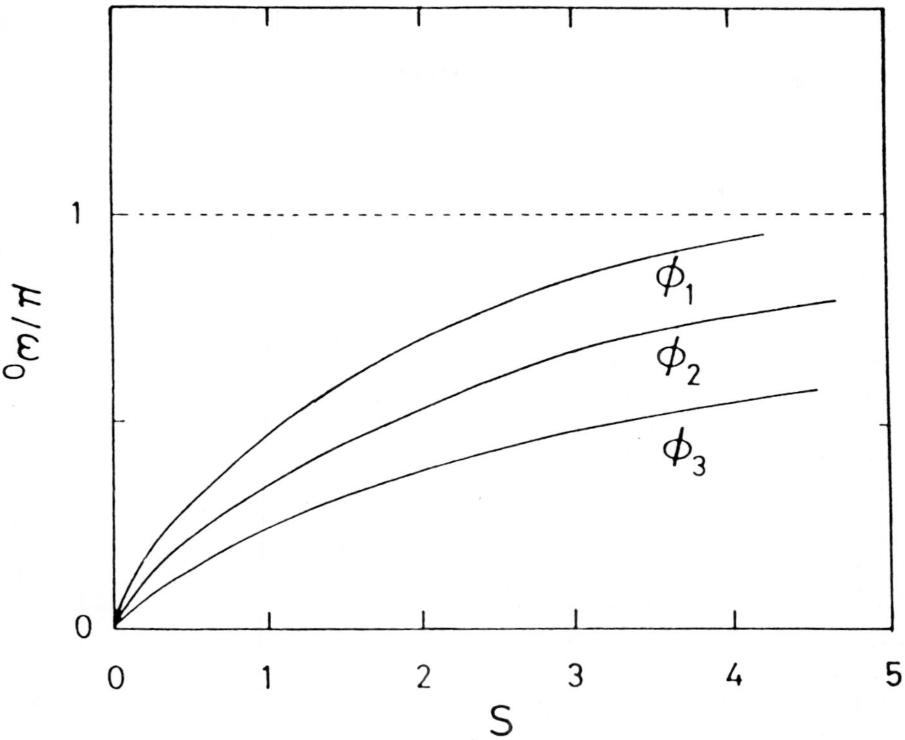

Figure 16.6: *The normalized chemical potential vs. energy supply rate S for* $\chi = 0.1, \phi_1 = 0.05, \phi_2 = 0.1$ *and* $\phi_3 = 0.2$ *respectively. The energy is in* ω_o *units.*

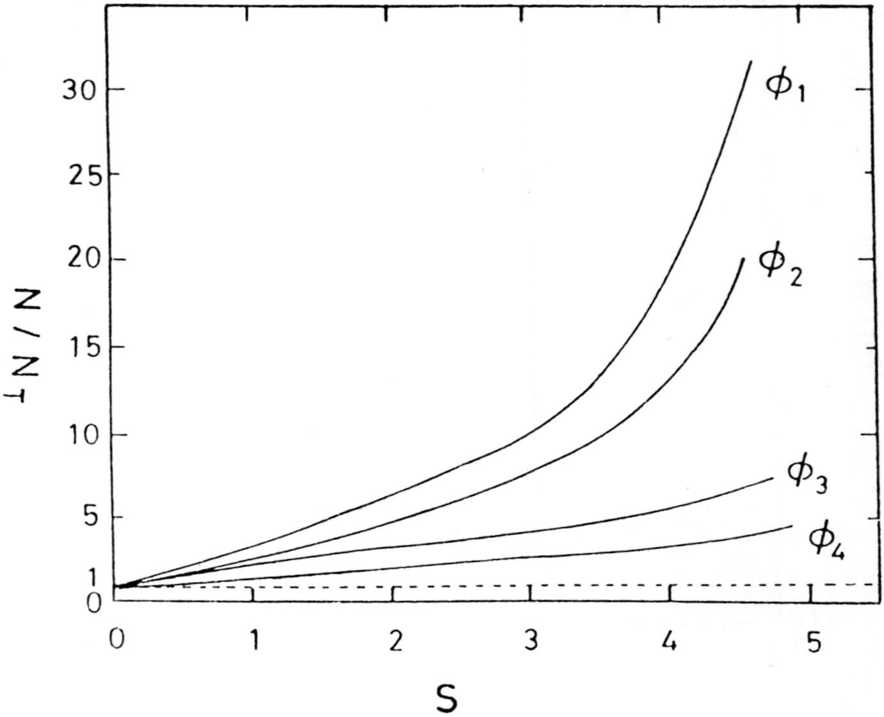

Figure 16.7: *The ratio of the number of enhanced phonons in the biosystem to that of the thermal equilibrium phonons vs. energy supply rate* S *for* $\chi = 0.1, \phi_1 = 0.05, \phi_2 = 0.07$ $\phi_3 = 0.1$ *and* $\phi_4 = 0.2$.

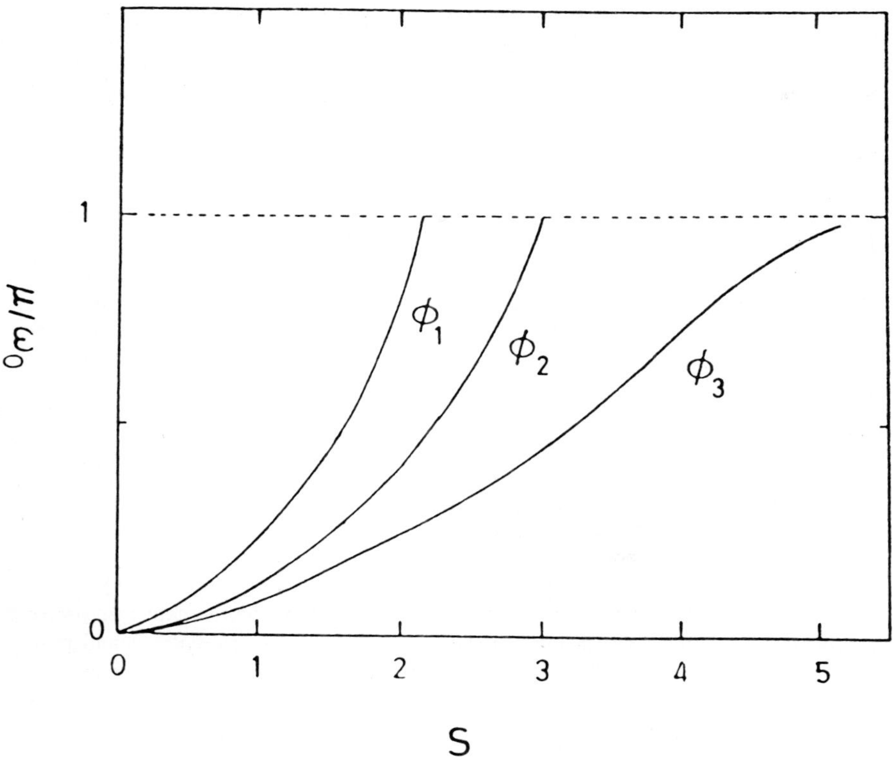

Figure 16.8: *The normalized chemical potential vs. energy supply rate* S *for* $\chi = 0.1, \phi_1 = 0.05, \phi_2 = 0.1$ *and* $\phi_3 = 0.2$.

mode as s increases beyond the value of s_o. The critical rate of energy supply, s_o, is strongly dependent on the coupling strengths ϕ and χ. Note that as the relative magnitude between ϕ and χ increases, the critical energy supply rate s_o increases. This shift to higher critical energy supply can easily be explained as follows. Recall that χ describes the degree to which the interaction processes channel absorbed energy between the different phonon modes in the biosystem and ϕ describes the tendency of the mode's energy to be channelled to the heat bath. As long as the modes have an external energy source, if ϕ gets larger for a fixed χ, then more energy will be transmitted to the heat bath and less energy will be retained in the system. Therefore, the Bose-Einstein condensation will occur at high energy supply rate s_o. If s is smaller than s_o, there is no Bose-Einstein condensation, but the number of excited phonons will increase as s increases, just as in the one dimensional case. When s is larger than s_o, one expects not only the number of excited phonon to increase, but also the system to undergo a phase transition, i.e. a large number of the excited phonons in the system will be accumulated in the lowest energy state, ω_o.

In the examples above, we kept the coupling constant χ fixed, and ϕ variable. Actually we can change the values of χ; the results are similar. The only difference is that when χ increases, the number of phonons of lower energy will increase faster and the phase transition in the two-dimensional case will occur at lower s_o.

The experiments for Stokes and anti-Stokes Ramman spectra has been done by Webb *et al*[20–22] on synchronized active cells of *E coli* B.The incident microwave of frequency f_o will force the dipoles in the system to undergo oscillations and re-emit radiation. The spectrum of the oscillation - the Raman spectrum - will be composed of three lines. The central line has the frequency of the incident microwave, f_o, and is due to Rayleigh scattering. The two shifted lines are the result of Stokes and anti-Stokes scattering; they will have frequencies of $f_o - f$ and $f_o + f$ respectively. The frequency shift is equal to the vibrational frequency f, of the system. Figures 16.9-16.11 show the spectra taken at 40, 50, and 60 min. respectively after incubation.

Note that the line near 120 cm^{-1} range between 118 and 125 cm^{-1}. The experimental value of the ratio of the intensities anti-Stokes and Stokes is between 0.9 to 1. The theoretical value of this ratio is given by $n/(1+n)^{22}$, where n is the excitation number. At these frequencies and at room temperature, the thermal equilibrium distribution n^T is smaller than one; this gives values between 0.55 and 0.57 for the ratio of anti-Stoke and Stoke intensities. However, these values are much too small compared to the values obtained experimentally. In order for the ratio of these intensities to approach 1 as shown in the experiments, it is required that the number of excitation, n, should be much larger than 1. Thus, the Stokes and anti-Stokes lines in the active cell spectra may arise from a condensation or enhancement of the excited phonons in the biosystem induced by the energy supply from metabolic processes.

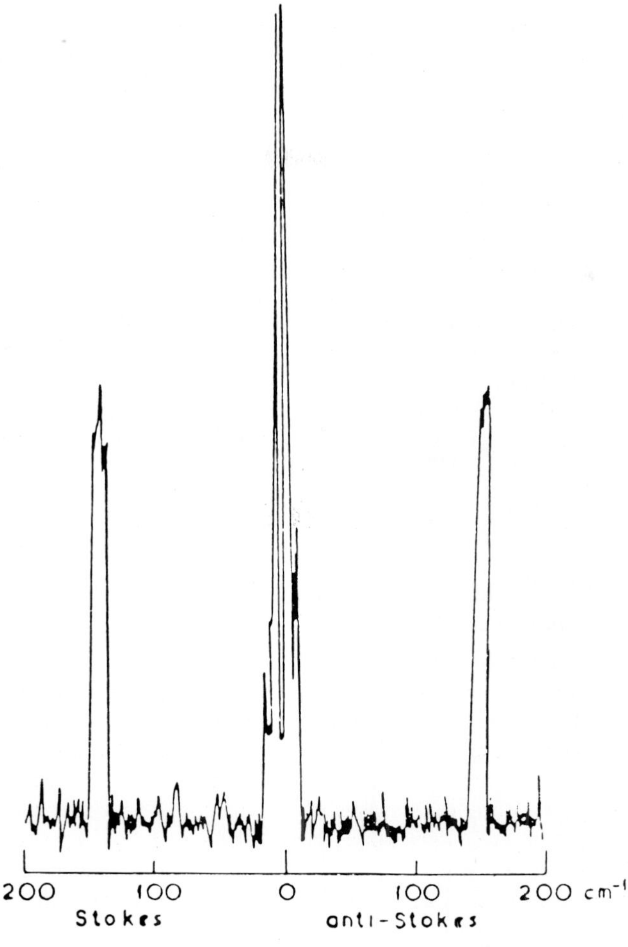

Figure 16.9: *Stokes and anti-Stokes Raman spectra of synchronized active cells of E. coli B bacteria, taken at 40 min. after incubation.*

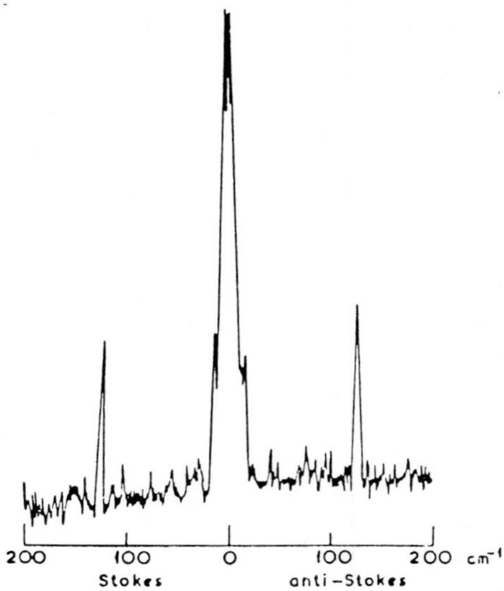

Figure 16.10: *Stokes and anti-Stokes Raman spectra of active cells of E. coli bacteria, taken at 50 min. after incubation.*

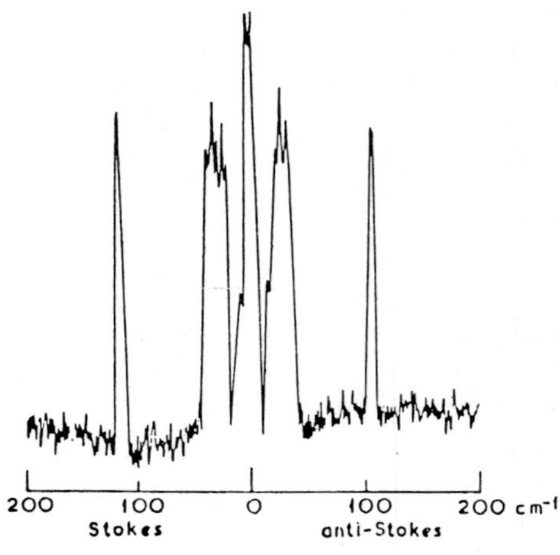

Figure 16.11: *Stokes and anti-Stokes Raman spectra of active cells of E. coli B bacteria, taken at 60 min. after incubation.*

16.4 Time Threshold for Biological Effects

Frölich's model has been particularly useful in understanding recent experiments revealing the effects of low level microwaves on various biological systems. One important general characteristic reported in these experiments is the existence of a time threshold for the initiation of the differing biological effects. Recall Sevastyanova and Vilenskaya's experiment[6] involving the simultaneous irradiation of mice bone marrow cells. The time threshold for a protective effect was 30 minutes of microwave exposure. On the induction coefficient of colicin synthesis[5], they also found that a minimum irradiation time is needed to produce measurable biological effects, again of the order of 30 minutes. The threshold will depend on the particular biological activity being monitored as well as the temperature. An example of the latter is clearly demonstrated in Smolyanskaya and Vilenskaya's experiment on *E coli.* Irradiation for 30 minutes at 20° produced no change in the rate of colicin synthesis, whereas at 37°, colicin synthesis increased. Therefore, it is expected that the analysis of the time threshold will not be a trivial matter. Both the detailed microscopic structure of the biological system and its thermodynamic behaviour must be considered in order to provide a full understanding of the specific time thresholds encountered for different biological systems.

Using Fröhlich's model, we intend to obtain an approximation for a segment of this time threshold.[23] In applying the Fröhlich hypothesis, one sees that the biological effects cannot begin until condensed phase occurs. It has not been determined how long this condensation has to be maintained in order to produce biological effects. Generally, the time needed to initiate biological effects will consist of two parts, τ_1 and τ_2. τ_1 represents the time from the beginning of irradiation to the onset of condensation; τ_2 represents the time elapsed from the onset of the condensed phase to the production of the actual biological effects. Thus, the total time, τ, elapsed before biological effects occur from the initiation of irradiation, is given by $\tau = \tau_1 + \tau_2$ In the following analysis, we will evaluate τ_1 using the Hamiltonian given in Equation (16.1). τ_1 represents the minimum time required for the biological system to exhibit effects after irradiation and is the lifetime of the collective excitations in the biological system. An explicit form of τ_2 requires an in-depth biochemical analysis of the particular system under scrutiny and is beyond the scope of the present study. However, τ_2 represents an important part and should be studied in future.

As mentioned in the previous section, the heat bath is a very complex system. The excitation modes in the heat bath can be either Bosons or Fermions. For simplicity, we will only consider the case that the excitations in the heat bath are also Bosons and only to the first order in interaction. This method can also easily be extended to other kinds of excitations. The lifetime of τ_1 is given by the relation
$\tau_1 = h/ \mid Im\Sigma \mid \; where Im\Sigma$ is the imaginary part of the self-energy of vibration mode in the energy band.

It should be noted that the lifetime (and thus the self-energy) of interest is due to terms with coupling constants χ and χ^* because these terms are responsible for creating the condensed phase. The rest of the terms in the Hamiltonian do not contribute to the condensation in the lowest order of coupling constant and therefore are not relevant to the present calculation.

Using the finite temperature Green's function, the self- energy, $\Sigma(p_i\omega_i)$ for the biological system phonon associated with operator a_i and a_i^+ yields

$$\sum(p_i,\omega_i) = |\chi|^2 \sum_j \left[\frac{1+n(\Omega_{p_j})+n(\omega_{p_i-p_j})}{\omega_{p_i}-\Omega_{p_j}-\omega_{p_i-p_j}+i\delta} + \frac{n(\Omega_{p_j})-n(\omega_{p_i-p_j})}{\omega_{p_i}+\Omega_{p_j}-\omega_{p_i-p_j}+i\delta}\right] \quad (16.9)$$

where $n(\omega) = (e^{\beta\omega}-1)^{-1}$, Ω_p is the energy of a phonon with momentum p in the heat bath and ω_p is the energy of a phonon with momentum p in the biological system. At room temperature, these energy bands are much smaller than kT, therefore $n(\omega_{p_i})$ and $n(\Omega_{p_j})$ are much larger than 1. Also, $\Omega_{p'} = \pm(\omega_{p_i} - \omega_{p_i-p_j})$ is smaller than the width of the narrow energy band of biological system under consideration, which in turn is smaller than a typical energy in the set of vibration modes (i.e., $\Omega_{p_j} < \omega_p$). Thus, $n(\Omega_{p_j}) > n(\omega_{p_i-p_j})$. Therefore, the self-energy can be approximately written as

$$\sum(p,\omega_p) = \frac{|\chi|^2}{4\pi^2\hbar\beta} \int_{-1}^{1} dx \int_0^\infty \frac{p'^2 dp'}{\Omega_{p'}} \left(\frac{1}{\omega_p - \Omega_{p'} - \omega_{p-p'} + i\delta} + \frac{1}{\omega_p + \Omega_p - \omega_{p-p'} + i\delta}\right) \quad (16.10)$$

where the summation over heat bath momenta has been converted to an integration. Using the Debye model for the excitations in the heat bath, $\Omega_{p'} = v'p'$, where v' the velocity of sound in the heat bath.

The form of the dispersion relation for the biological system is unknown at present and depends in an intricate manner upon the biological structure. However, to obtain an approximate result, it is assumed that $\omega_p - \omega_{p-p'} = \pm\gamma\Delta$ where Δ is the width of the energy band which is very narrow and γ is a positive quantity less than 1. The value of γ depends on the form of the dispersion relation for the biological system. The real part of self-energy in Eq. (16.10) is impossible to evaluate without knowing the form ω_p. Fortunately we are only interested in the imaginary part which can be obtained easily from Eq. (16.10) as

$$Im\Sigma = -|\chi|^2 \frac{\gamma\Delta}{\beta\pi\hbar^3 v'^3} \quad (16.11)$$

Therefore, the lifetime τ_1 becomes

$$\tau_1 = \pi\hbar^2 v'^3 / \mid \chi \mid^2 \gamma k T \Delta \tag{16.12}$$

which is inversely proportional to the temperature T and to the square of the coupling constant. Thus, as the temperature is raised, the condensed phase should occur sooner, in turn causing the biological effects to begin sooner. This agrees, at least qualitatively, with Smolyanskaya and Vilenskaya's experiment where changing the temperature from 20°C to 37°C brought the biological effects into play, while holding the irradiation time fixed at 30 minutes.

The τ_1 given in Eq. (16.12) represents the irradiation time needed to produce the condensation and is the minimum time needed to produce biological effects. In order to apply this equation to specific circumstances, model studies and more experiments need to be performed to obtain expressions for the coupling constants and the dispersion relations in the energy bands. Given this information, a more complete understanding of the Fröhlich model and all its implication can be obtained.

16.5 Long Range Interaction

Further understanding of cooperative behavior of biological systems requires the development of models which represent the specific biological substance being analyzed. This is a difficult but very important problem which should be investigated carefully in order to understand the implications of the theory and make meaningful experimental comparisons. As an example, consider two biological systems at a distance R, larger than their linear dimensions, capable of giant dipole vibrations with frequencies f_1 and f_2 respectively (assuming $f_1 > f_2$). Through the coulombic interaction of these giant dipoles, the combined system then has two normal modes f_+ and f_- which can be expressed as[3],

$$f_\pm^2 = \frac{1}{2}(f_1^2 + f_2^2)\left[1 \pm \left(A^2 + \frac{B^2}{\epsilon_\pm}\right)^{1/2}\right] \tag{16.13}$$

and

$$A = \frac{f_1^2 - f_2^2}{f_1^2 + f_2^2} \geq 0 \tag{16.14}$$

$$B = \left(\frac{be^4 z_1 z_2}{M_1 M_2 R^6}\right)^{1/2} \tag{16.15}$$

where $\epsilon_\pm$ is the dielectric constants; z_1, z_2 are the numbers of the bound ions of charge e; M_1, M_2 are the molecular masses; and b is a numerical constant of order unity but depends on various angles. The interaction energy, U, is defined as the difference in the free energy F at a distance R from its value at R going to infinite. The interaction energy is then given as

$$\begin{aligned} U_{\pm} &= F_{\pm}(R) - F_{\pm}(\infty) \\ &= n_{\pm}h[f_{\pm}(R) - f_{\pm}(\infty)] \\ &= \frac{n_{\pm}h}{\sqrt{2}}\left(f_1^2 + f_2^2\right)^{1/2}\left[\left(1 \pm \sqrt{A^2 + B^2/\epsilon_{\pm}^2}\right)^{1/2} - (1 \pm A)^{1/2}\right] \end{aligned} \tag{16.16}$$

where $n_{\pm}$ is the number of quanta of the coherently excitated state. If $f_1 \neq f_2$, and at sufficient large distance, then $B^2 < A^2$. Using Equations (16.14) and (16.15), we find that

$$U_{\pm} \doteq \frac{n_{\pm}h}{\sqrt{2}}(f_1^2 + f_2^2)^{1/2}\frac{B^2}{\epsilon_{\pm}^2 A^2(1 \pm A)^{1/2}}\ \alpha\ \frac{1}{R^6} \tag{16.17}$$

Thus $U_{\pm}$ is proportional to R^{-6}, i.e., the interaction is short range regardless of which coherent state.

However, if $f_1 \doteq f_2$, then

$$U_{\pm} \doteq \pm\frac{nh}{\sqrt{2}}(f_1^2 + f_2^2)\frac{B}{\epsilon_{\pm}}\ \alpha\ \pm\frac{1}{R^3} \tag{16.18}$$

In this case of near resonance, the interaction has very long range. The interaction is attractive for the lower frequency f_- and repulsive for the higher frequence f_+ respectively.

The long-range interaction from the above discussion may have considerable biological significance. Many biological processes depend on a certain molecule 'recognizing' another molecule. An enzyme and its substrate(s) must interact in a very specific manner both spatially and temporally. A similar situation exists for the antibodies of immune systems; antibodies must distinguish whether a molecule is foreign or belongs to the host. The fundamental biological question is "How does an enzyme or antibody recognize its own very specific target against the enormously high background found in living systems?" Each of these biological molecules can be considered as a unique entity, each with its own characteristic frequency or frequencies. How these molecules interact with each other may be governed by the state of the biological system. One notices that if the system condenses into f_- state, then the interaction is attractive; conversely, if the system condenses into f_+ state, then the interaction is repulsive. This provides a possible mechanism by which an enzyme or antibody can distinguish its target from other molecules. It is possible that the biological system can control which way it will condense into by supplying sufficient chemical energy to one normal mode of the vibration or the other. It can also be perturbed by external microwave radiation which acts by pumping energy into one or the other normal mode of the oscillation.

16.6 Appendix

The Bose-Einstein condensation in biosystems can be demonstrated explicitly as follows: summing over j in Equation (16.4) and set $<\dot{n}_i>=0$ for stationary state, one gets

$$S=\sum_i s_i=\sum_i \Phi(T,\omega_i)\left[<n_i>\ e^{\beta\omega_i}-(1+<n_i>)\right] \tag{16.19}$$

On introducing the excess m_i over the thermal equilibrium distribution $<n_i>^T$

$$<n_i>=<n_i>^T+m_i, \qquad <n_i>^T=(e^{\beta\omega_i}-1)^{-1} \tag{16.20}$$

we obtain

$$S=\sum_i \Lambda(T,\omega_i)m_i(e^{\beta\omega_i}-1)<N\Phi_{max}(e^{\beta\omega_i}-1) \tag{16.21}$$

where

$$N=\sum_i m_i \tag{16.22}$$

and Φ_{max} is the largest of the $\Phi(T,\omega_i)$. Thus we have a lower limit of the total number N of excess quanta, which increases proportionally to the total rate of supply of energy. Using Eq. (16.8) and Eq. (16.20), Eq. (16.7) can be written as

$$<n_i>=\left[1+s_i\frac{1-e^{-\beta\mu_i}}{\sum_j \Lambda(T,\omega_i,\omega_j)m_j(e^{\beta\omega_j}-1)}\right](e^{\beta(\omega_i-\mu_i)}-1)^{-1} \tag{16.23}$$

We now consider the simple case such that all $\Phi(T,\omega_i)$ are equal to Φ, and all $\Lambda(T,\omega_i,\omega_i)$ are equal to Λ. Equation (16.19) with Equation (16.20) then becomes

$$S=\Phi\sum_i m_i(e^{\beta\omega_i}-1) \tag{16.24}$$

From Eq. (16.8), it follows that all μ_i are equal to μ such that,

$$\omega_o \geq \mu \geq 0 \tag{16.25}$$

Eq. (16.23) for $< n_i >$ then becomes

$$< n_i >= \left[1 + s_i \frac{1 - e^{-\beta\mu}}{\Lambda \sum_j m_j (e^{\beta\omega_j} - 1)} \right] (e^{\beta(\omega_i - \mu)} - 1)^{-1} \tag{16.26}$$

or making use of Eq. (16.24), $< n_i >$ becomes

$$< n_i >= \left[1 + \frac{s_i \Phi}{S\Lambda} (1 - e^{-\beta\mu}) \right] (e^{\beta(\omega_i - \mu} - 1)^{-1} \tag{16.27}$$

Using Eqs. (16.20), (16.21), and $N^T = \sum_i < n_i >^T$, we find

$$N + N^T = = \sum_i \left[1 + \frac{s_i \Phi}{S\Lambda} (1 - e^{-\beta\mu}) \right] (e^{\beta(\omega_i - \mu)} - 1)^{-1} \tag{16.28}$$

Substituting s_i by S and μ by ω_o, and because $s_i \leq S,\ \mu < \omega_o$, we obtain

$$N + N^T \leq \left[1 + \frac{\Phi}{\Lambda} (1 - e^{-\beta\omega_o}) \right] \sum_i (e^{\beta(\omega_i - \omega_o)} - 1)^{-1} \tag{16.29}$$

In the particular case where the energy supply s_i to all modes is the same, $S = zs_i$, where z is the number of states in the energy band, then Eq. (16.28) becomes

$$N + N^T = \left[1 + \frac{\Phi}{z\Lambda} (1 - e^{-\beta\mu}) \right] \sum_i \frac{1}{exp[\beta(\omega_i - \mu)] - 1} \tag{16.30}$$

The dependence of both Eqs. (16.29) and (16.30) on S is implicit through μ only. Furthermore, N increases with increasing energy supply S, as seen from Eq. (16.21). One also notices that, by Eqs. (16.7) and (16.8), when S has surpassed a critical value S_o, μ approaches ω_o. Therefore, in both Eqs. (16.29) and (16.30), N becomes very large, as a large number of quanta are condensed into the lowest energy state. This is exactly the Bose-Einstein condensation in a Bose gas when the temperature is lower than a certain critical temperature. In our case the corresponding phase transition is not by lowering temperature but is enforced by increasing the energy supply beyond the critical value S_0 and keeping the temperature constant.

Bibliography

[1] H. Fröhlich,*Riv. Nuovo Cimento* **7** (1977) 399.

[2] H. Fröhlich,*Biosystems* **8** (1977) 193.

[3] H. Fröhlich,*Adv. Electronics and Electron Physics* **53** (1980) 85.

[4] R.I. Kiselev and N.P. Zalyubovskaya, *Soviet Phys. USPEKHI* **16** (1974) 576.

[5] A.Z. Solyanskaya and R.L. Vilenskaya. *Soviet Phys. USPEKHI* **16** (1974) 571.

[6] L.A. Sevast'yanova and R.L. Vilenskaya. *Soviet Phys. USPEKHI* **16** (1974) 570.

[7] W. Grundler and F. Keilmann, *Z. Naturforch.* **C33** (1978) 15.

[8] W. Grundler, F. Keilmann and H. Fröhlich, *Phys. Letters* **A62, 463** (1977) 463.

[9] H. Fröhlich, *Phys. Letters* **26A** (1968) 402.

[10] H. Fröhlich, *Int. J. Quant. Chem.* **2** (1968) 641.

[11] H. Fröhlich, *Nature* **228** (1970) 1093.

[12] H. Fröhlich, *Phys. Letters* **39A** (1972) 153.

[13] H. Fröhlich, *Phys. Letters* **51A** (1975) 21.

[14] T.M. Wu and S. Austin, *Phys. Letters* **64A** (1977) 151.

[15] T.M. Wu and S. Austin, *Phys. Letters* **65A** (1978) 74.

[16] T.M. Wu and S. Austin, *J. theor. Biol.* **71** (1978) 209.

[17] T.M. Wu, in *The Living State II* R.K. Mishra, ed., (1985) 318.

[18] T.M. Wu *Molecular and Biological Physics of Living Systems* **19** (1990).

[19] See, for example, A.A. Abridosov, L.P. Gor'kov and I.Y. Kzyaloshinskii, *Quantum Field Theoretical Method in Statistical Physics* (Pergamon Press, Oxford)

[20] S.J. Webb and M.E. Stoneham, *Phys. letters* **60A** (1977) 267.

[21] S.J. Webb, *Phys. letters* **73A** (1979) 145.

[22] S.J. Webb, R. Lee and M.E. Stoneham, *Int. J. Quant. Chem. Quantum Biology Symposium* **4** (1977) 277.

[23] T.M. Wu and S. Austin, *Phys. Letters* **73A** (1979) 74.

Chapter 17

Energy and Electron Transport in Biological Systems

A.S. Davydov

17.1 Energy Transfer along Alpha-helical Proteins

17.1.1 Solitons and Excitons in Protein Molecules

The energy released in the hydrolysis of an adenosine triphosphate (ATP) molecule is the universal energy unit used in a great number of energy transformations in living organisms. All these transformations involve protein molecules, which are generally of a very large size. And the site at which ATP hydrolysis occurs is often separated by a considerable distance from the site at which the energy is used. The question arises concerning the mechanism responsible for the transduction of the energy released in ATP hydrolysis along large protein molecules.

Theoretical investigation conducted since 1976[1–5] at the Institute for Theoretical Physics in Kiev showed that energy from ATP hydrolysis can be transferred without loss along alpha-helical protein molecules as special collective excitations - *molecular solitons.* All protein molecules are formed by a linear polymerization of 20 different aminoacids and represent very long polypeptide chains with periodic repetition of peptide groups (PGs) which contain the H,N,C,O atoms. These polypeptide chains can exist in different configurations, among which the alpha-helices are of especial interest.

In alpha-helical proteins, the polypeptide chains are coiled into long helices, stabilized by hydrogen bonds between the PGs, in addition to the peptide bonds between neighbouring carbon atoms within the primary polypeptide chain. In each helical protein molecule, the peptide groups are situated along three chains of hydrogen bonds equidistant from one another, and forming periodic structures. The energy of the hydrogen bond is approximately an order of magnitude less than that of ordinary chemical bonds. Therefore, the displacements of peptide groups from their equi librium positions along the chains of hydrogen bonds occur much more readily than their displacements along the main polypeptide chain.

The quasi-periodic structure of an alpha-helical protein molecule provides the basis for the collectivization of vibrational excitations of individual PGs. The basic vibrational excitation of the PG corresponds to vibrations of the C=O bond. It has an energy of $0.21ev$, and a dipole moment of about 0.3 Debye directed along the chain of hydrogen bonds. Owing to the dipole resonance interactions, the vibrations in C=O of individual PGs are collectivized. As shown by Davydov *et al.*[1-4], it is very important to take into account the displacements from equilibrium positions of the PGs along hydrogen bonds in theoretical considerations of such collective excitations. In the continuum approximation, the collective excitations are described by a set of differential equations. If we consider the excitation of each chain of PG's in the protein molecule independently, then the collective excitation corresponding to the excitations of C=O in the peptide groups spreading along the chain with a constant velocity V, is described by a non-linear Schrödinger equation with a non-linearity coefficient,

$$G = 4\chi^2/\kappa(1 - s^2), \qquad s = V/V_0 \tag{17.1}$$

depending on the value of the longitudinal elasticity, κ, of the peptide chain, on the parameter χ characterizing the connection of PG vibrations with the displacements from their equilibrium positions, and on the ratio, s, between the excitation velocity V and the longitudinal sound velocity V_0.

Two types of collective excitations, namely, *excitons* and *solitons* may appear in each chain of PGs[1-4]. The excitons are collective excitations spreading along the chain with a group velocity exceeding the velocity of longitudinal sound waves ($s > 1$). They are described by wave packets smearing with time. Owing to the rapid motion of an exciton, the local deformation cannot keep up. Therefore, the exciton carries only the energy of an intrapeptide vibration which can be expressed by the formula,

$$E_{ex}(V) = E_{ex}(0) + \frac{1}{2}m_{ex}V^2 \tag{17.2}$$

Here $E_{ex}(0)$ is the internal energy of an exciton which is slightly smaller than the energy of an isolated vibration C=O; the second term in (17.2) determines the exciton kinetic energy proportional to the squared velocity. The proportionality coefficient m_{ex} is called an *effective exciton mass.* It is inversely proportional to the energy of resonance interaction, J, between the neighbouring PGs in the chain. While travelling, the excitons generate phonons, therefore, they are rapidly decelerated.

The solitons are collective excitations formed as a result of the combination of intrapeptide C=O vibrations with a local deformation of the chain. They always move with velocity V less than that of longitudinal sound. The energy of a soliton travelling along the chain with velocity V is expressed by the formula,

$$E_{sol}(V) = E_{sol}(0) + \frac{1}{2}m_{sol}V^2 \, , \quad (s^2 \ll 1) \tag{17.3}$$

In this case it is very important that the internal soliton energy,

$$E_{sol}(0) = E_{ex}(0) - \chi^4/3\kappa^2 J \tag{17.4}$$

is less than the internal exciton energy. In other words, when the vibrations are 'connected' with a local deformation there appears an excitation with an energy less than the sum of the energies of excitations of its constituents. As the internal soliton energy is less than the internal exciton energy, soliton formation is energetically more advantageous than the formation of an exciton. The formula (17.3) is valid only at small soliton velocities ($s^2 < 1$). The effective soliton mass considerably exceeds the effective exciton mass,

$$m_{sol} = m_{ex}(1 + 8M\chi^4/3\kappa^3\hbar^2) \tag{17.5}$$

where M is the mass of the peptide group. The large soliton mass is due to the fact that its motion is accompanied by the motion of a local deformation of the chain. The soliton of a large mass can transfer a great amount of kinetic energy even at small velocities of its motion.

If the position of PG in the chain is characterized by the coordinate $z = na$ where n is an integer and a is the distance between neighbouring PGs, then in the soliton state the probability of distribution of the vibrational excitation in PGs will be expressed by a bell-like function

$$W(z,t) = \frac{aQ}{2\cosh^2[Q(z - z_0 - Vt)]}, \tag{17.6}$$

where $2\pi a/Q$ is the linear size of a soliton.

In real alpha-helix proteins molecules, it is neccesary to take into account three parallel chains of hydrogen-bonded PGs, which can lead to the formation of three types of solitons. The first - the symmetric soliton - corresponds to a synchronous transfer of a local excitation of the type described by Eq. (17.6) along all three chains of PGs. In this case, the helix diameter in the excitation region increases and the distances between neighbouring PGs decrease. The two other types of solitons are nonsymmetric. They correspond to the transfer of local excitations along the three chains of PGs with some phase shifts. In that case, the excitation region undergoes a slight increase in diameter. One type of nonsymmetric solitons corresponds to the transfer of local excitation only along two chains. Solitons of this type have the lowest internal energy, and are therefore the most stable.

As the formation of solitons in protein molecules is accompanied by a local deformation, i.e. by a displacement of PGs from their equilibrium positions, they cannot be excited by light, on account of the Frank-Condon principle, which states that light absorption by molecular systems does not cause the change in coordinates of heavy particles at the moment of the quantum transition. By the same principle, the probability of the light emission by solitons is small[5] (see section 17.2.3). The solitons may be excited by local external effects, for example, by chemical reactions. The probability of an excitation will be the greatest when the edge of a molecule is acted upon. Such an excitation may be realized by the hydrolysis of an ATP molecule attached to the end of the protein molecule.

Although the excitons are less stable, they are easily excited by light of a corresponding frequency, because the transition into this excited state is not accompanied by the displacement from equilibrium positions of PGs. As shown by Nevskaya and Chirgadze[6], the study of infrared spectra when excitons are generated in protein molecules, is an effective means for analyzing protein structure.

The above analytical expressions are obtained in the continuum approximation when discrete chains of PGs are replaced by continuous chains of an infinite length. In 1981, Hyman, McLaughlin and Scott[7] obtained numerical solutons for the equations proposed by Davydov et al.[1–5] for discrete chains using the computer at Los Alamos Scientific Laboratory. They have considered chains containing 200 PGs. It was assumed that the edge PGs were initially excited and the propagation of the excitation along the chain was studied. The numerical calculations showed that if the coupling parameter of vibrational excitations with displacements of PGs exceeds a certain critical value, then an excitation of the soliton type is propagated along the chain. The authors[6] concluded that the numerical calculations support the results obtained analytically[1–4].

17.1.2 The Lifetime of Molecular Solitons

The problem of calculating the lifetime of a soliton has arisen in recent years, specifically: whether the lifetime of the Davydov soliton at nonzero temperatures is long enough for it to be useful in biology. Gottingham and Schweitzer [8] (and others), using perturbation analysis at 300 K, obtained an estimate for the lifetime of a Davydov soliton of $10^{-14} - 10^{-12}$s, which is obviously too short to be useful.

The same problem was discussed in detail by Davydov[9], and in 1979, J.C.Eilbeck made a computer film -*Davydov solitons*[10] - which demonstrated the stability of solitons in their interaction with the packet of acoustic waves which was excited simultaneously with the soliton. The velocity of a packet was higher than the velocity of a soliton. So it was reflected several times by the ends of the chain, and in so doing, pass through the soliton and leaving it unchanged.

17.1.3 A New Hypothesis of the Mechanism of Muscle Contraction

One of the most interesting problems in bioenergetics is the molecular basis of converting the chemical energy released in ATP hydrolysis into the mechanical energy of motion. Most of the studies have concentrated on the striated muscles of vertebrates and some invertebrates. I have used the concept of soliton propagation in helical protein molecules[9–11] to elucidate the mechanism of contraction of striated muscles at the molecular level. A muscle fiber is a bundle consisting of several thousands of densely packed parallel myofibrils. Each myofibril consists of repeating segments, or sarcomeres, separated by transverse membranes. From both sides of the transverse membrane, thin filaments of actin project into the sarcomere, interdigitating with the thick myosin filaments situated in the middle of the sarcomere. Thin and thick filaments are bathed in the sarcoplasm which contains ATP molecules and Mg^{2+} ions, among other things.

The shortening of the sarcomere and, consequently, the contraction of a muscle fibre, occurs when a nerve impulse causes the release of calcium ions from the sarcoplasmic reticulum into the sarcoplasm. There, calcium ions stimulate ATP hydrolysis where thick and thin protein filaments come into contact. Investigations with the electron-microscope established that the shortening of the sarcomere is caused by an increased overlap between thin and thick filaments without any change in the length of the filaments themselves. The problem arises in the mechanism responsible for the sliding of thin filaments relative to thick ones at the molecular level. What forces are involved? How is the energy from ATP hydrolysis transformed into sliding energy? The currently most widely held view is that the sliding of thin filaments relative to the thick is due to an active motion of the heads of myosin molecules. It is supposed that on ATP hydrolysis, the head of the myosin molecule lengthens to form a link ('bridge') with a globular actin molecule of the thin fibre, the head turns, displacing the thin fibre towards the sarcomere centre, and then detaches itself from the actin molecule to regain its previous size and position for the next cycle.

The idea of cross-bridges somehow forming between thick and thin filaments - which somehow pull the thin fibres toward the sarcomere center before breaking off to start the cycle again - says nothing conerning the molecular nature of the phenomenon. The question remains: how is the energy from ATP hydrolysis used in the lengthening of the myosin head to form the cross-bridge, to cause the displacement, and then to break the bridge? What molecular mechanisms are responsible for this sequence of events? And, finally, why is it that only the head of the huge myosin molecule (molecular weight 500.000) should take an active part in the sliding mechanism, when, in addition to the 'head', the myosin molecule has a long 'tail' of about 1700 amino-acids consisting of an alpha-helical domain?

Using theoretical investigations of solitons in helical proteins, I have proposed a new hypothesis on the mechanism of shortening of the sarcomere. According to this hypothesis, calcium ions, diffusing to the first series of myosin heads at the ends of thick filaments, initiate the hydrolysis of ATP molecules attached to them. The energy released generates, in the long helical parts of the myosin molecules, nonsymmetric solitons which move from the heads of the molecules to their tails in the sarcomere centre. The motion of a nonsymmetric soliton is accompanied by a local bending and swelling of the molecule. The heads of myosin molecules attached to the thin filaments cause a slight displacement of the latter in the direction of motion of the solitons, that is, towards the centre of the sarcomere.

According to this model, the myosin heads attach to thin filaments and detach themselves from thin filaments (as in the model of 'cross-bridge' formation and breaking). However, this movement is due, not to the lengthening, turning and contraction of the head itself, but to the movement of a soliton inside the thick fibre accompanied by the motion of bend in the helical domain of the myosin molecule. In other words, the kinetic energy of solitons is converted into the contraction energy, or the energy of tension, if a load is applied to the muscle. In this model, all parts of the myosin molecule, and not only its head, are active contractile elements.

17.2 Electron Transfer along Protein Molecules

17.2.1 Electron Transfer by Solitons

Electron transfer from donor to acceptor molecules along protein molecules play an important role in the bioenergetics of living organisms. It has been established experimentally that electrons can move without energy loss along protein molecules over distances of tens of interatomic distances. Protein molecules are dielectrics. The conduction band of the electrons of such a molecule is situated relatively high at 4 - $6eV$. The transfer process involves 'excess' electrons that come to a protein molecule from a donor molecule. Electron transfer appears to be realized by the alpha-helical parts of the molecules. The alpha-helical parts of protein molecules involve periodic peptide groups that consist of four atoms, H, N, C and O. As their electric charges are situated non-symmetrically, the peptide groups have constant electric dipole moments (about 3-4 Debye).

It was shown by Turner *et al.*[14] that for external electrons, these dipole moments play the role of potential wells, capturing excess electrons at the ground energy levels with a binding energy of about 1 eV. The overlapping of the wave functions of the ground states of electrons in neighbouring potential wells results in the formation of a conduction band for an excess electron. The motion of an electron in this conduction band is characterized by the effective mass,

$$m = \hbar^2/2a^2J. \tag{17.7}$$

where $\hbar$ is Planck's constant $h/2\pi$, a is the interpeptide group distance, and J, the energy of resonance interaction. In alpha-helical protein molecules, the distance between peptide groups is $a = 5.4\AA$, $J \approx 10^{-3}$eV. The effective mass of a quasi-particle in the band is 120 free electron masses

The excess electrons that come to a protein molecule from a donor give rise to a local deformation of interpeptide distances resulting from the additional interaction of electrons with the molecule. The motion of an electron under the influence of constant electric dipole moments and the field of the displacements from equilibrium positions of the peptide groups is described by a set of coupled nonlinear differential equations for the function $\psi(x,t)$, which characterizes the probability of electron distribution, and the function $\rho(x,t)$, which characterizes the relative change in interpeptide distances. The nonlinear local interaction of the field $\psi(x,t)$ with the field $\rho(x,t)$ generates a bound collective field that moves as a single whole along a protein molecule. This bound field is called a *two-component soliton.*

When the interaction with a local deformation is disregarded, free quasi-particles could propagate in the conduction band with a velocity that does not exceed the maximum group velocity:

$$v_g = 2aJ/\hbar \approx 1.8 \times 10^8 \text{cm s}^{-1}. \tag{17.8}$$

The velocity of a free local deformation is the same as that of longitudinal sound:

$$V_0 \approx 3.5 \times 10^5 \text{cm s}^{-1}. \tag{17.9}$$

The velocity, V, of the moment of a two-component soliton cannot exceed the lower of these two velocities. In an alpha-helical molecule, two-component solitons can move with a velocity that satisfies the inequality,

$$V \leq V_0. \tag{17.10}$$

At velocities above V_0, the local deformation cannot follow the movement of a quasi-particle.

In the continuum approximation[14–16], the motion of a two-component soliton with constant velocity V in a coordinate system,

$$\xi = (x - x_0 - Vt)/a \tag{17.11}$$

is described by the functions,

$$\psi(x,t) = a^{-1/2}\Phi(\xi) exp\frac{i}{\hbar}[mxV - (\zeta + \frac{1}{2}mV^2)t] \tag{17.12}$$

and

$$\rho(\xi) = a^2\sigma\Phi^2(\xi)/MV_0^2 \ ; \qquad \int \Phi^2(\xi)\ d\xi \ = \ 1. \tag{17.13}$$

The real function $\Phi(\xi)$ satisfies the Schrödinger equation:

$$\left[\frac{d^2}{d\xi^2} + \varepsilon + 2g\Phi^2(\xi)\right]\gamma(\xi) \ = \ 0 \tag{17.14}$$

with the parameters,

$$g = a^2\sigma^2/2JMV_0^2; \quad \varepsilon = (\zeta - \zeta_0)/J; \quad MV^2 = a^2\kappa. \tag{17.15}$$

The function $\rho(\xi)$ characterizes the local decrease in interpeptide distances: $a \rightarrow a - \rho(\xi)$. In all the preceding expressions, we have used the following notation: ζ^0 is the energy of the conduction band bottom of a free quasi-particle; ζ is its energy in the field of a local deformation; σ is the parameter of electron-phonon interaction and M is the mass of a peptide group.

The spatially localized solution of Eq. 17.10 have the form

$$\Phi_s(\xi) = \frac{1}{2}\text{sech}(g\xi/2); \qquad \int \Phi^2(\xi)\ d\xi \ = \ 1\ . \tag{17.16}$$

These solutions correspond to the soliton energy (measured from the energy at the bottom of the conduction band ζ^0) that includes the energy of a local deformation:

$$\Delta E_s(V) = \Delta E_s(0) + \frac{1}{2}M_sV^2; \quad V^2 \ll V_0^2\ , \tag{17.17}$$

where $\Delta E_s(0)$ is the energy of a soliton at rest:

$$\Delta E_s(0) \ = \ - \ g^2J/12\ . \tag{17.18}$$

The effective mass M_s of a soliton exceeds the mass, m, of a free quasi-particle, because its movement is accompanied by the movement of a local deformation:

$$M_s = m(1 + g^2J/3mV_0^2)\ . \tag{17.19}$$

According to Eqs. (17.13) and (17.14), the field of a local deformation in a soliton state is defined by

$$\rho(\xi) = \frac{a^2\sigma g}{4MV_0^2}\text{sech}^2(g\xi/2)\ . \tag{17.20}$$

The effective potential well in which a quasi-particle moves is defined, in a system of coordinates ξ, by

$$U_s(\xi) = -Jg^2\text{sech}^2(g\xi/2)\ . \tag{17.21}$$

The soliton is very stable on account of the following three factors: (1) its energy is below the bottom of the conduction band of a free quasi-particle by the value,

$$\mid \Delta E_s(0) \mid = \hbar^2 g^2/24ma^2 = g^2 J/12, \tag{17.22}$$

(2) the soliton always moves with a velocity less than that of sound and therefore it emits no phonons, i.e. its kinetic energy does not transform into thermal energy; and (3) the soliton has a topological stability. After the soliton has passed, the equilibrium positions of all peptide groups are displaced, but in front of the soliton they remain unchanged. To destroy the soliton, it is necessary to return all peptide groups into their initial positions. Hence the interaction of electrons with the field of peptide group displacements in alpha-helical protein molecules stabilizes their motion.

17.2.2 Electron Pairing that Generates a Bisoliton

In soft molecular chains capable of producing solitons when the electron interacts with local deformation, two electrons with opposite spins can form a paired state, giving rise to *a bisoliton*. The simplest theory of electron pairing in quasi-one-dimensional molecular systems was developed by Brizhik and Davydov[18,19]. Two excess electrons with opposite spins form a common potential well through a local displacement from equilibrium positions in a chain of molecules (peptide groups). The potential well generated under a local deformation of a chain by one electron attracts a second electron which, in turn, causes a greater deformation, deeping the well.

When the Coulombic repulsion of electrons is disregarded, the energy of bisoliton at very low velocity is defined by,

$$\Delta E_{ss}(V) = \Delta E_{ss}(0) + \frac{1}{2}M_{ss}V^2; \quad V^2 < V_0^2 \tag{17.23}$$

where $\Delta E_{ss}(0)$ is the energy of a bisoliton at rest:

$$\Delta E_{ss}(0) = -2g^2 J/3 = -\sigma^4/2\kappa^2 J \tag{17.24}$$

and,

$$M_{ss} = 2m(1 + 4Jg^2/3mV_0^2) \tag{17.25}$$

is the effective mass of a bisoliton. The latter is greater than the effective mass of two isolated solitons, because the motion of a bisoliton is connected with the movement of a deeper local deformation.

According to Eq. (17.24), the energy level of a bisoliton at rest is below the conduction band level of a quasi-particle by a distance eight time larger than the energy of a soliton. Therefore, when a bisoliton is formed by two rest solitons, the energy released is,

$$2\Delta E(0) - \Delta E_{ss}(0) \;=\; \frac{1}{2}g^2 J \;, \tag{17.26}$$

which is six times greater than the energy released when one soliton is formed.

If solitons move with velocity V, bisoliton generation is accompanied by the release of energy:

$$2\Delta E_s(V) - \Delta E_{ss}(V) = \frac{g^2 J(1 - 5s^2)}{2(1 - s^2)}; \quad s \;\equiv\; V/V_0 \;. \tag{17.27}$$

Consequently, pairing is violated at high velocities when $s^2 > 1/5$.

Both quasi-particles in a bisoliton state move in a coordinate system inside the common effective potential well,

$$U_{ss}(\xi) \;=\; -2g^2 J\text{sech}^2(g\xi) \tag{17.28}$$

the radius of which is half that of a potential well (Eq. 17.21) of one soliton. The motion of quasi-particles inside the potential well in Eq. (17.28) is described by the wave function,

$$\Phi_{ss}(\xi_1, \xi_2) \;=\; \frac{1}{2}g\text{sech}(g\xi_1)\text{sech}(g\xi_2) \tag{17.29}$$

Quasi-particles are paired in a singlet spin state. Consequently, bisolitons are particles with zero spin. They carry a double electric charge.

The above results have not taken into account the Coulombic repulsion between electrons. In neutral systems where electric charges are screened, such a repulsion, as shown previously, impedes the generation of bisolitons only at very low values of the dimensionless coupling parameter g. The transfer of electron pairs (experiencing no resistance) in a singlet spin state as a bisolitons appears to be realized even at 310 K in alpha-helical protein molecules of living organisms when ATP molecules are synthesized in conjugating membranes of mitochondria and chloroplasts[19] and under some redox reactions that play a basic role in the functioning of living organisms. This 'high-temperature superconductivity phenomenon' is thus realized in living organisms.

The mechanism of soliton and bisoliton generation in electron transfer is investigated in Appendix I. It is shown that when the density of flow of electrons is the same, the generation of bisolitons is energetically much more favourable than solitons.

17.2.3 Solitons and the Molecular Mechanism of the Effect of Radiation of Cells

Devyatkov and his coworkers[22] found resonant effects of millimeter electromagnetic waves with millimeter on the growth rate of some bacteria which have also been observed by others. These responses of living organisms to electromagnetic radiation at sufficiently low intensities that produce no appreciable heating of tissues, are referred to as athermal effects. One of the essential features of athermal effects is the sharp resonance frequently observed. As Devyatkov emphasizes[22]: "The explanation of the mechanism of resonance radiation effect as well as other related peculiarities are of great importance for science. A rigorous scientific explanation of the effect of millimeter-electromagnetic waves has not yet been given".

A new hypothesis on the mechanism of such a resonance effect was suggested by Eremko and myself[23–25]. This hypothesis is based on concept of solitons which play an active role in all vital activities. As have already been pointed out, solitons are ideal carriers of the energy of ATP hydrolysis along proteins. It was shown[23,25] that under the influence of electromagnetic radiation with a definite wavelength, solitons can transform to metastable excitons. The latter are rapidly deccelarated, dissipating their energy to the environment. Therefore, the transformation of solitons into excitons destroys the efficiency of the transmission of energy and information.

The energy of transition from the soliton state with a local chain deformation (when $s^2 \approx 0$) to the bottom of the exciton band without deformation is given by the quantity,

$$\Delta E = \frac{ma^2\mathcal{L}^4}{24\kappa^2\hbar^2} = \mathcal{L}^4/48\kappa J \,. \tag{17.30}$$

where $\mathcal{L}$ is the coupling parameter of intramolecular excitations with displacements.

The optical transition from the soliton to the exciton state occurs so rapidly that the local deformation has no time to annihilate. For this reason, together with ΔE, the energy of optical transition should involve the deformation energy which is equal to $2\Delta E$. Hence, the total energy of transition from the soliton to the exciton state with a wave number k is given by the formula,

$$\hbar\omega_k = 2\Delta E + \hbar k^2/2m \ . \qquad (17.31)$$

In previous publications[23,25], it was demonstrated that for slow solitons ($s^2 \ll 1$), the probability of absorption of electromagnetic radiation with friequency ω, wave number k, and electric field amplitude $\vec{\mathcal{E}}$,is determined by the expression,

$$\mathcal{P}(\omega) = \sum_k \mid \vec{\mathcal{E}}\, \vec{\mathcal{D}}_k \mid^2 \ G_k(\omega) \ , \qquad (17.32)$$

where

$$\vec{\mathcal{D}}_k \approx \vec{\mathcal{D}}\, \mathrm{sech}(\pi k/2Q) \ , \ Q = \chi^2/4a\kappa J \ . \qquad (17.33)$$

$\mathcal{D}_k$ is the dipole electric moment of quantum transition from the soliton state to the exciton state with wave number k. The parameter $1/Q$ in Eq. (17.33) characterizes the spatial width of the soliton. The function $G_k(\omega)$ depends on the dimensionless parameter, $\mu = 2\chi^2 M^{1/2}/\hbar\sqrt{\mathcal{L}}$.

In Eq. (17.32), a summation is carried out for all of wave numbers belonging to the exciton band. Due to the factor of Eq. (17.33) the creation of the exciton with wavenumber $k \approx 0$ becomes more probable if $\vec{\mathcal{E}} \parallel \vec{\mathcal{D}}$. Therefore, the resonance transition frequency takes the value,

$$\hbar\omega_0 = 3\Delta E \ . \qquad (17.34)$$

The square of maximum displacement of the equilibrium position of peptide groups is determined by $\beta^2_{max} = \mathcal{L}^2/\kappa^2$ under a soliton excitation, and the amplitude of elastic zero vibrations of particle of mass M. In the chain with elasticity coefficient κ is equal to $< \chi^2 >= \hbar/2\sqrt{\kappa M}$. Therefore, the dimensionless parameter μ is given by the ratio,

$$\mu = \beta^2_{max}/ < \chi^2 > \ . \qquad (17.35)$$

If $\mu \gg 1$, the function $G_k(\omega)$ at $k \approx 0$ takes a Gaussian form,

$$G_0(\omega) = \frac{\sqrt{2\pi B}}{\hbar} \exp(-\frac{(\omega - \omega_0)^2}{2B^2}) \ , \qquad (17.36)$$

where $B = 3.6\chi^2 Q^2 V_0/\kappa a\hbar\pi^3$.

In the second case, when the square of the maximum displacement of the equilibrium position of the molecules is considerably less than the square of the amplitude of the zeroth vibration ($\mu \ll 1$), the function $G_0(\omega)$ takes the Lorentz form,

$$G_0(\omega) = \frac{2\gamma_0}{pi}[(\omega - \omega_0)^2 + \gamma_0^2]^{-1} \quad \gamma_0 = \frac{\mathcal{L}^6\sqrt{M}}{48\hbar^2\kappa^{3/2}} \, . \tag{17.37}$$

Using the values of the parameters for α-helix proteins,

$$J = 19.5 \text{ Nm}^{-1}, \quad \mathcal{L} = 8 \cdot 10^{-11} N, \quad M \approx 70\mu_p,$$

we obtain $\mu \approx 16 \cdot 10^{-2}$. A rather small probability for the transformation of soliton into an exciton can take place under Raman scattering of light. Eremko *et al.*[26] have demonstrated that the frequency of scattering of light by solitons shifts into the infra-red region. The frequency shift, $\Delta\omega$, depends on the value of parameter μ. For this process, a photon expends its energy on the transformation of soliton into exciton with wave number $k \approx 0$. If $\mu \gg 1$, the frequency shift is

$$\Delta\omega_1 = 3\Delta E/\hbar.$$

In this case due to the Frank-Condon principle, there occurs a dissociation of the soliton which preserves the local chain deformation. For this reason, the energy of the final state is given by the sum of energy of soliton formation, ΔE, and deformation energy $2\Delta E$.

In the opposite case, $\mu \ll 1$, the frequency shift is rather weak,

$$\Delta\omega_2 = \Delta E/\hbar \, ,$$

and the energy of the final state corresponds only to that of soliton deformation, without the energy of local deformation. One can suggest that the second case, with,

$$\Delta\omega \approx (0.4 \; - \; 0.7) cm^{-1},$$

takes place for vibrational solitons of alpha-helical protein molecules which correspond to the intrapeptide vibration of Amide-1 type.

17.2.4 Possible Mechanism of General Anaesthesia

A class of molecules, used as intravenous anaesthetics, can form hydrogen bonds with protein molecules situated inside and outside the cell membrane. The binding of anaesthetic molecules to proteins inhibits the normal function of a cell. One of the possible mechanisms of such an inhibition by the binding of barbiturates to protein was proposed by Layne[27] in 1984 on the basis of a soliton model.

Barbiturates contain the HNCO group of atoms which are very similar to the peptide groups in proteins. When a barbiturate molecule attaches to a protein molecule, the hydrogen bonds between neighbouring peptide groups in the protein are broken, and new hydrogen bonds are formed with the oxygen and hydrogen atoms of the barbiturate molecule. At the site of dissociating hydrogen bonds in the protein molecule, the interpeptide separation increases, and this impedes the free motion of solitons, and hence, interferes with the information process and energy transduction along the protein molecule.

Layne suggested that a decrease in efficiency of the transduction of energy and information by solitons due to the binding of barbiturates to proteins, affects in particular the following cellular structures: a) the helical protein of inner mitochondrial membranes which participates in ATP synthesis; b) the helical proteins of the cell membrane which are responsible for the chemical reception and transduction of signals into the cell.

17.2.5 Intracellular Dynamics and Solitons

The alpha-helical structure is often found at those domains of protein molecules where energy is transduced from one end to the other, and which are often involved in coupling several processes to one another. For instance, the haemoglobin molecule of red blood corpuscules which transports oxygen, possesses 32 helical domains. The bacteriorhodopsin of *Halobacteria* which live in salty lakes and basins, traverses the cell membrane seven times in the form of parallel alpha-helical segments. The same alpha-helical protein segments are present in cytochromes and ot her proteins spanning the inner mitochondrial membrane.

Recent investigations of cellular structure reveal that the entire interior of the cell is criss-crossed by a network of protein microfilaments and microtubes which anchors all intercellular organelles, determine the shape of the cell and any changes thereof, and is responsible for all intracellular transport of materials. Such a network of microfilaments and micrcotubes is referred to as the *cytoskeleton.*

Protein molecules incorporated into the cytoskeleton release transduction energy along the network, which is responsible for all intracellular coupling. These cytoskeletal processes depend on the energy released in ATP hydrolysis, which is catalyzed by enzymes and controlled by calcium ions, in a manner similar to muscle fibres. The existence of contractile proteins, such as actin, myosin and troponin, decnin and tubulin, in nonmuscle cells confirm the hypothesis that a universal mechanism exists for converting chemical energy of ATP hydrolysis into mechanical motion. Nevertheless, the mechanism in the cytoskeleton differs in some detail from that which applies to muscle fibers. In muscle fibers, thin and thick filaments are located in definite positions relative to each other within the sarcomere. In nonmuscle cells, however, myosin filaments have no fixed positions, and seem to float freely within the cytoplasm. Helical segments make up a considerable part of the structure of the cytoskeleton. Besides being cent ral to motility, they also transfer energy and information from one place to another in a cell by means of the propagation of soliton excitations[28].

Transmembrane glycoproteins play an important role in the life of a cell. Glycoproteins are formed by covalent bonding of a protein with a carbohydrate residue, or a polysaccharide. The longitudinal protein fractions have a helical structure which spans the whole thickness of the cellular membrane. A polysaccharide component is hydrophilic and is located outside the cell. The interior glycoproteins are strongly coupled with the microfilaments and microtubules of the cytoskeleton. Thus glycoproteins provide a coupling of the cell interior and the cell exterior. Based on the nonlinear dynamics of solitons, one can undoubtedly proceed in understanding the mechanism of transduction of information from the exterior to the interior of a cell. Glycoproteins determine cellular individuality, their adhesion, and intercellular interaction. They transmit signals into the cell from its environment through the binding of hormones, neurotransmitters, immunoglobulins and other molecules.

One can conclude that the nonlinear dynamics of solitons provides a key to understanding the mechanism of the transduction of information from the exterior to the interior of a cell as well as within the cell.

17.3 Appendix

17.3.1 The Generation of Solitons and Bisolitons in Electron Transfer

As the solitons transferring electrons are topologically stable, they can be generated only at the ends of molecular chains. The mechanism of soliton and bisoliton generation at the chain end under the influence of an electron flow incident upon it was investigated by Davydov and Ermakov[21] using a simple model of a system consisting of two semi-finite chains: a soft and a rigid one joined at the point $x = 0$.

In a soft chain the electron interacting with a local deformation caused by it can form a soliton, lowering its energy from the value ζ_0, which corresponds to the energy of the bottom of the conduction band of a free quasi- particle. In a rigid chain the electron with effective mass m^* produces no local deformation and its states are described by plane waves.

We now investigate a flow of electrons coming from a rigid chain (or a metalic electrode) to a soft chain. In a stationary state this flow partially penetrates in a soft chain and partially reflects. At the energy $E = \hbar^2 k^2/2m^* < \zeta_0$, the state of an electron flow with density D^2 in a rigid chain ($x \leq 0$) is described by the function,

$$\psi_1(x,t) = (De^{ikx} + Re^{-ikx}) \exp(-iEt/\hbar) . \tag{17.38}$$

In a soft chain, when we have the resonance interaction between neighbouring molecules which are at a distance a apart, and the dimensionless parameter g for deformation interaction, the state of an electron with the same energy E can be described by,

$$\psi_2(x,t) = a^{-1/2}\Phi_2(\xi) \exp(i\varphi - iEt/\hbar) . \tag{17.39}$$

Using the dimensionless variables,

$$\xi = x/a; \quad \gamma = \sqrt{(\varepsilon_0 - E)/J}; \quad \varepsilon_0 = \zeta_0 + W; \tag{17.40}$$

the function $\Phi_2(\xi)$, decreasing rapidly as $v \to \infty$, has the form,

$$\Phi_2(\xi) = \gamma g^{-1/2} \mathrm{sech}[\gamma(\xi - \xi_0)] . \tag{17.41}$$

The value W is the energy of a local deformation of a soft chain:

$$W = \sigma^2(8a\kappa)^{-1} \int_0^\infty \Phi_2^4(\xi y)\, d\xi . \tag{17.42}$$

The probability of finding an electron in a soft chain is defined by the integral,

$$I = \int_0^\infty \Phi_2^2(\xi)\, d\xi . \tag{17.43}$$

The position ξ_0 of the maximum of the function (17.42) for every energy E is expressed through g and I using the equation,

$$\xi_0 = (2\gamma)^{-1} \ln\, [gI/2\gamma - gI)] . \tag{17.44}$$

Using the standard method of combining the functions ψ_1, ψ_2 and their derivatives at the point $\xi = 0$, we can express the phase φ and the integral I through the density D^2 of an incident electron flow:

$$\varphi = \arctan\left(\frac{\gamma - gI}{ka}\right); \quad (2\gamma - gI)I = 4D^2 . \tag{17.45}$$

If for the fixed energy E of a flow of electrons the equality,

$$D_s^2 \equiv \gamma^2/4g = (\varepsilon_0 - E)/4gI \tag{17.46}$$

is valid, the maximum of the function (17.42) lies at the boundary when $\xi_0 = 0$. At the flow densities $D^2 < D_s^2$, the maximum is shifted toward the negative values, $\xi_0 < 0$. Total reflection of the electron is observed in these cases.

At electron densities exceeding D_s^2, there are no stationary states. To investigate non-stationary states, it is necessary to take into account the time dependence of the functions $D(t)$, $R(t)$, $\varphi(t)$ and $\Phi(x,t)$ involved in Eqs. (17.38) and (17.39). From the conditions that the derivatives and the functions are combined at the boundary $\xi = 0$, we obtain a set of equations for an auxiliary function $Z(t)$, which is used to express the position of the maximum $\xi_0(t)$:

$$Z^3(t) - Z(t) + 4kaD\sin\varphi(t)/g^{3/2}I^2(t) = 0 \tag{17.47}$$

and the derivatives $d\varphi/dt$ and dI/dt. Analysis of these equations shows that at $D^2 > D_s^2$, the function $Z^2(t)$ is a period function of the time. At $t = t_0$, when the phase $\varphi(t_0) = 0$, Eq. (17.47) gives $Z^2(t) = 1$. In this case the maximum of the function (17.24) lies at the value $\xi_0 = 0$. The function $Z^2(t)$ decreases with increasing time. When the function reaches the value 1/3, the position of the maximum is shifted to the point $\xi_0 = 3\ln 3/4gI$. By this time the soliton has been fully produced in a soft chain and it breaks off with a velocity,

$$V = \frac{\hbar k(E - E_{s,cr})}{2m(\zeta_0 - E_{s,cr})} \tag{17.48}$$

from the boundary $\xi = 0$. At the same moment the function $Z^2(t)$ increases rapidly to values exceeding unity, then decreases again, and when it passes over the range from 1 to 1/3, a new soliton begins to be generated.

At a fixed energy of incident electrons that determines the value D_s^2 according to Eq. (17.46), the solitons are created over the time intervals,

$$T_s = 3am\log 3\frac{[J(\xi_0 - E_{s,cr})]^{1/2}}{4\hbar k(E - E_{s,cr})}. \tag{17.49}$$

The condition $D^2 > D_s^2$ under which the solitons are generated in a soft chain is equivalent to the energy inequality,

$$E > E_{s,cr} \equiv \zeta_0 - g^2 J/12 \,. \tag{17.50}$$

As mentioned in section 2.2, the soliton pairing that generates bisolitons is advantageous in a soft chain. According to Eqs. (17.16) and (17.29), the wave functions describing the movement of one quasi-particle in a bisoliton can be obtained from the function (17.16) by formally replacing the parameter g by $2g$. Therefore, when a flow of pairs and electrons of density $2D^2$ is incident on the boundary of a soft chain, bisolitons are generated in it, beginning with the critical energy,

$$E_s s, cr = \zeta_0 - 2g^2 J/3 . \tag{17.51}$$

Therefore, when the density of a flow of incident electrons, D^2, is the same, the generation of bisolitons requires less energy.

When the density of incident electrons with energy less than $E_{s,cr}$ but greater than $E_{ss,cr}$ is fixed, the bisoliton generation seems to proceed in two stages. A soft chain is first hit by one electron described by the tail of the function (17.33) with its maximum at $\xi_0 < 0$. This electron stays bound to the chain edge. When another electron with the same energy and an opposite spin penetrates the chain, the local deformation is restructured and a bisoliton is produced. At an energy E exceeding $E_{ss,cr}$, it breaks off and moves along a soft chain.

Figuratively, the electron that has hit a soft chain waits for another electron with an opposite spin to come and jointly create a bisoliton which, breaking off from the chain edge, carries a double electric charge. The energy condition for bisoliton generation at the chain edge are therefore much more advantageous than those for the creation of solitary electrosolitons.

Bibliography

[1] A.S.Davydov and N.C.Kislukha, *Phys. Stat.Sol.(b)* **75**(1976) 735.

[2] A.S.Davydov, *J. Theor. Biol.* **66**(1977) 379.

[3] A.S.Davydov, *Intern. J. Quant. Chem.* **26** (1979) 5.

[4] A.S.Davydov, *Phys. Scripta* **20**(1979) 20, 387.

[5] A.S.Davydov and A.A.Eremko, *Ukr. Fiz. Zh.***22**(1977) 881.

[6] Yu.N.Chirgadze and N.A.Nevskaya, *Dokl. ANSSSR* **208**(1973) 447.

[7] J.M.Hyman, D.W.McLanghlin and A.C.Scott, *Physica D* (1981) 23.

[8] J.P.Gottingham and J.V.Schweitzer, *Phys. Rev. Lett* **62**(1985) 1752.

[9] A.S.Davydov, *Journ Biol. Phys.* **18**(1991) 111.

[10] J.C.Eilbec, *Davydov Soliton* (A 16 mm mute film. I Wood Road Winbleton, London S.W. 200 Hm, England,1979).

[11] A.S.Davydov, *Ukr. Fiz. Zh.* **20** (1975) 179.

[12] A.S.Davydov, in *Biophysica and Biochemistry of Contracting Muscles*, (M.Nauka, 1976) pp. 254-257.

[13] A.S.Davydov, *Studia Biophysica* **47** (1974) 221.

[14] I.E.Turner, V.Anderson and E.Fox, *Phys.Rev.* **174**(1968) 81.

[15] A.S.Davydov, in *Biology and Quantum Mechanics* (Pergamon Press, Oxford, 1982).

[16] A.S.Davydov, in *Solitons in Molecular Systems*, ed D.Reidel (Dordrecht, 1985).

[17] A.S.Davydov, in *Solitons in Bioenergetic*, ed. D. Reidel (Dordrecht, 1985) p. 318.

[18] L.S.Brizhik and A.S.Davydov, *Fiz. Nzk.Temp.* **10** (1984) 358 and 748.

[19] L.S.Brizhik and A.S.Davydov, *Prepr. Inst. Theor. Phys. Acad. Nauk Ukr. SSR* **ITP 87-53E** (1987).

[20] P.C.Hinke and R.E.McCarty, *Sci. Am.* **238**(1978) 104.

[21] A.S.Davydov and V.N.Ermakov, *Ukr. Fiz. Zh.* **13**(1988) 1633.

[22] N.D.Devyatkov and M.B.Golant, *Letters in ZhTF* **8** (1982) 39.

[23] A.A.Eremko, *Dokl. Acad. Sci. Ucr. SSR* **A3** (1984) 51.

[24] A.A.Eremko, in *Nonlinear and Turbulent Processes in Physica* **2** (Harward) pp. 757-761.

[25] A.S.Davydov and A.A.Eremko, *Ukr. Fiz. Jour.* **22** (1977) 881.

[26] A.A.Eremko, Yu.B.Gaididei and A.A.Vakhnenko, *Phys. Stat. Solidi(b)* **127** (1965) 703.

[27] S.P.Layne, *Los Alamos Sci N 10* (1984) 27.

[28] A.S.Davydov, in *Nonlinear Bioenergetic*, ed.W. Cruyter (Berlin - New York, 1968) pp. 69-85.

Chapter 18

Bioelectrodynamics and Biocommunication – An Epilogue

Mae-Wan Ho and Fritz-Albert Popp

18.1 The Relationship between Bioelectrodynamics and Biocommunication

'Bioelectrodynamics' refers to processes endogenous to the living system, which are responsible for the organisms' sensitivity to weak, external electromagnetic fields. Bioelectrodynamic processes are also the basis of biological organization, or 'biocommunication'. In the history of how the present subject unfolds (Chapter 1), the two aspects of the organism's own electrodynamical processes and its sensitivity to external electromagnetic fields were closely intertwined: 'animal electricity' and 'animal magnetism' serving both as the endogenous vital force, and as the explanation of the organism's sensitivity to external electric and magnetic fields.

The contemporary explanation of the physical basis of bioelectrodynamical processes is given in Chapter 2 under the rubric of 'bioelectromagnetics', which is to be distinguished from 'electro-magneto-biology'. The latter covers the effects of external electric and magnetic fields. A more inclusive class of electrodynamical processes is mentioned in Chapter 7 in connection with the 'morphogenetic field'.

The term, 'biocommunication' is significant in several respects. First, it denotes the processes of signal generation, transmission and transduction covering the entire electromagnetic spectrum and in both the 'phonon' and 'photon' modes, that are involved in the coordination of biological functions. Thus, it has the character of the 'language' (Chapter 5) whereby cells, organs and organisms communicate within and among themselves. However, it is more than just a symbolic language, which has to be interpreted by something else, for the signals and signal transduction processes are themselves responsible for the interpretation. In other words, they are also the motive, organizational forces which give rise to specific biological effects. Bioelectrodynamical processes are the means whereby both communication and the effects of communication are achieved. It becomes arbitrary to separate the signalling process and call it 'information', for the signalling, the reception and the transduction form an indivisible unit which gives content to the so-called information. Finally, biocommunication is the key to biological organization. An organic whole, in contrast to a mechanical whole, does not consist of a hierarchy of parts which exert control over other parts. Instead, it is a maximally responsive and transparent system in which changes and adjustments propagate simultaneously 'upwards', 'downwards' and 'sideways' in the maintenance of the whole. So, instead of 'control', it is much more appropriate to think in terms of 'communication'.

In this volume, we have covered the whole range of biocommunication from the ELF (Chapters 3, 5, 6, 7, 9 and 13) to the microwave (Chapters 3, 4, 16 and 17), and the visible range (Chapters 8, 11 and 12). These involve interactions with external fields in membrane macromolecules and cells (Chapters 4-6, 9-11 and 16), whole organisms (Chapters 3 and 7) and interactions among individuals in populations of cells and organisms (Chapters 11 and 12).

18.2 Direct Measurements of Electrodynamical Activities - Noninvasive Technologies

Direct measurements of electrodynamical activities in cells and organisms have been carried out using noninvasive techniques of great sensitivity. Iaonnides (Chapter 13) describes the most significant recent development in this regard - the extremely sensitive SQUID magnetometer - which can detect the *femto*-Tesla level magnetic fields generated by coherent, electrical activities within the human brain, which turns out to have applications in the diagnosis of neurological disorders. Similarly, the detection of 'biophotons', or ultraweak light emission from living organisms with a super-sensitive cathode-photomultiplier (range of detection, 200 to 900nm), is providing a unique insight into the biophysics of coherence in living organization (Chapters 11 and 12). This technique is currently employed in many aspects of biosensing[1].

A completely novel noninvasive technique which has been developed within the past two years is the imaging of live organisms by interference colours (Chapter 8). It is based on detecting dynamically ordered, coherent liquid-crystalline mesophases of the macromolecules making up living tissues. As such, it can be used to detect directly the state of dynamic, or energetic order within organisms, complementing the other techniques mentioned above.

These measurements demonstrate that organisms are active dielectric, optical systems, and cannot be adequately described in terms of passive properties (see Chapter 3).

18.3 Bioelectrodynamical Interactions with External Fields

Evidence for biological effects of external electromagnetic fields are presented in many chapters. Smith (Chapter 3) and also Warnke (Chapter 15) present a general review of the literature on the biological effects of electromagnetic fields, including the epidemiological studies linking environmental electromagnetic fields to leukemia and other diseases, and the relevant biochemical, cellular aspects and clinical findings. Edmonds (Chapter 6) does not regard the epidemiological evidence convincing, but cites evidence of sensitivity of animals to weak electric and magnetic fields, offering some plausible mechanisms to account for the sensitivity. The electromagnetic sensitivity of both animals and humans is extensively reviewed by Warnke (Chapter 15), who also comments on electromagnetic therapy.

Tsong and Gross (Chapter 5) describe their studies on the effect of ELF electric fields on membrane ion-transport and synthesis of ATP, and give an interesting model of 'electroconformational' coupling of membrane proteins to external electric fields which enables the protein to use the energy in the external fields for ion-transport and ATP synthesis. Liburdy (Chapter 6) also concentrates on the interaction of external microwave electromagnetic fields with cell membranes at phase transition temperatures, which are of a more general nature. The effects include protein-shedding and influx of external ions into cells, and leakage of drugs enclosed in artifical liposomes. He suggests that the effects are due to membrane instability during phase-transition. Chang *et al* (Chapter 10) investigate the dynamics of membrane changes subsequent to single short pulses of strong electric fields, and propose 'electropore' formation as being responsible for the permeation of normally impermeant molecules. Both microwaves and single pulse electric fields may have applications in targetting drugs into cells and in *in vitro* cell fusion.

A particularly interesting example of bioelectrodynamical interactions with external ELF fields is described by Pethig (Chapter 9). The electrophoretic mobility of a cell subjected to a DC electric field depends on the permanent electric charges on its surface. A different kind of mobility occurs when the cell is exposed to an AC electric field, which then depends on the nature of induced charges both at the surface and within the cell. This is responsible for the phenomena of electrorotation and dielectrophoresis, both of which give specific information concerning the dielectric and electrodynamical properties of the cell, and can hence be used to separate mixtures of cells as well as to detect specific subpopulations.

The effect of weak external ELF magnetic fields on pattern formation is described by

Ho *et al* (Chapter 8). They show that specific, global perturbations of body-pattern result from brief exposures to external static magnetic fields during early embryogenesis. The results suggest the involvement of coherent electrodynamical activities in pattern determination; more specifically, they suggest that the morphogenetic field is a globally coherent electrodynamical domain. If so, it may serve as a biological detector of the Aharanov-Bohm effect (see also Chapter 3).

18.4 Coherence and Biological Organization

Central to biocommunication is the coherent, collective modes of activity which can arise in single protein molecules, as excitons, solitons and bisolitons (Davydov, Chapter 17, also Wu, Chapter 16), in macromolecular aggregates and membranes (Chapters 5, 16), in macroscopic organs such as the human brain (Chapter 13), and in whole organisms (Chapters 7, 8 and 16). Ultimately, the organism is a domain of coherent, electrodynamical activities extending simultaneously from the molecular, or microscopic level to the macroscopic domain. A striking demonstration of this multiplex, multilevel coherence is the image of live organisms produced by visualizing the dynamically ordered liquid-crystalline mesophases of diverse molecular arrays making up all the tissues in the body, even as the organism is engaged in a macroscopically observable activity (Chapter 8).

This coherence can extend to whole populations of cells and organisms. Popp *et al* (Chapter 12) present convincing evidence of biocommunication between previously separated populations which are brought into contact, and also among individuals of a population maintained at different densities. They point out the distinctive properties of coherence for biocommunication: extreme sensitivity (high signal/noise ratio), specificity and rapidity, which are just the distinguishing characteristics of living systems.

A frequently raised objection to the hypothesis of coherence in biological systems is that the collective, or coherent state is readily destroyed by thermal fluctuations. Davydov (Chapter 17) and Wu (Chapter 16) have specifically addressed this issue. Davydov shows that solitons in proteins, which arise from the combination of resonant dipole-dipole interactions and mechanical deformations of the peptide chain, are stable collective modes, which can transfer energy without loss. Wu, similarly, demonstrates that the 'Fröhlich state' of coherent excitations arises spontaneously as the result of energy pumping beyond a certain threshold in a two-dimensional system capable of exchanging and storing energy (as phonons). It turns out that such collective, coherent states not only possess stability (see also Ho[2]), but are global attractors (see Duffield[3]). In other words, given the most general conditions of energy pumping in a system capable of exchanging and storing energy, the system will 'condense', or undergo phase-transition into coherent modes which maximizes biocommunication. As demonstrated by Zhang and Popp in Chapter 14, the coherent state which maximizes communication gives a log-normal distribution for its physiological measurements, relfecting the maximization of all possible relationships, or potential degrees of freedom, which allows the system to most readily access the single degree of freedom that is required for coherent action[4].

The idea of electrodynamical coherence as the basis of biological organization has many empirical and theoretical, as well as practical implications. We have only scratched the surface of this new and exciting area of research.

Bibliography

[1] F.A. Popp, K.H. Li and Q. Gu (eds.), *Recent Advances in Biophoton Research and its Applications*, World Scientific, Singapore, 1992.

[2] M.W. Ho, *The Rainbow and the Worm - The Physics of Organisms*, World Scientific, Singapore, 1993.

[3] N.G. Duffield, *J. Phy. A: Math. Gen.* **21**, 625, 1988.

[4] M.W. Ho, submitted for publication.